Grundlehren der mathematischen Wissenschaften 278

A Series of Comprehensive Studies in Mathematics

Grundlehren der mathematischen Wissenschaften

A Series of Comprehensive Studies in Mathematics

A Selection

190. Faith: Algebra: Rings, Modules, and Categories I
191. Faith: Algebra II, Ring Theory
192. Mallcev: Algebraic Systems
193. Pólya/Szegö: Problems and Theorems in Analysis I
194. Igusa: Theta Functions
195. Berberian: Baer*-Rings
196. Athreya/Ney: Branching Processes
197. Benz: Vorlesungen über Geometric der Algebren
198. Gaal: Linear Analysis and Representation Theory
199. Nitsche: Vorlesungen über Minimalflächen
200. Dold: Lectures on Algebraic Topology
201. Beck: Continuous Flows in the Plane
202. Schmetterer: Introduction to Mathematical Statistics
203. Schoeneberg: Elliptic Modular Functions
204. Popov: Hyperstability of Control Systems
205. Nikollskii: Approximation of Functions of Several Variables and Imbedding Theorems
206. André: Homologie des Algébres Commutatives
207. Donoghue: Monotone Matrix Functions and Analytic Continuation
208. Lacey: The Isometric Theory of Classical Banach Spaces
209. Ringel: Map Color Theorem
210. Gihman/Skorohod: The Theory of Stochastic Processes I
211. Comfort/Negrepontis: The Theory of Ultrafilters
212. Switzer: Algebraic Topology—Homotopy and Homology
213. Shafarevich: Basic Algebraic Geometry
214. van der Waerden: Group Theory and Quantum Mechanics
215. Schaefer: Banach Lattices and Positive Operators
216. Pólya/Szegö: Problems and Theorems in Analysis II
217. Stenström: Rings of Quotients
218. Gihman/Skorohod: The Theory of Stochastic Process II
219. Duvant/Lions: Inequalities in Mechanics and Physics
220. Kirillov: Elements of the Theory of Representations
221. Mumford: Algebraic Geometry I: Complex Projective Varieties
222. Lang: Introduction to Modular Forms
223. Bergh/Löfström: Interpolation Spaces. An Introduction
224. Gilbarg/Trudinger: Elliptic Partial Differential Equations of Second Order
225. Schütte: Proof Theory
226. Karoubi: K-Theory, An Introduction
227. Grauert/Remmert: Theorie der Steinschen Räume
228. Segal/Kunze: Integrals and Operators
229. Hasse: Number Theory
230. Klingenberg: Lectures on Closed Geodesics
231. Lang: Elliptic Curves: Diophantine Analysis
232. Gihman/Skorohod: The Theory of Stochastic Processes III
233. Stroock/Varadhan: Multi-dimensional Diffusion Processes
234. Aigner: Combinatorial Theory

Continued after Index

Michael Barr
Charles Wells

Toposes, Triples and Theories

Springer-Verlag
New York Berlin Heidelberg Tokyo

Michael Barr
McGill University
Department of Mathematics
and Statistics
Montreal, Quebec
Canada H3A 2K6

Charles Wells
Case Western Reserve University
Department of Mathematics
and Statistics
Cleveland, Ohio 44106
U.S.A.

AMS Classification: 18-02, 18B25, 18C15, 18C10, 18F10, 18F20

Library of Congress Cataloging in Publication Data
Barr, Michael.
Toposes, triples and theories.
(Grundlehren der mathematischen Wissenschaften; 278)
Biblography: p.
Includes index.
1. Categories (Mathematics) 2. Toposes. 3. Triples,
Theory of. I. Wells, Charles. II. Title. III. Series.
QA169.B346 1984 512.'55 84-20233

Printed and bound by R.R. Donnelley & Sons, Harrisonburg, Virginia.
Printed in the United States of America.

9 8 7 6 5 4 3 2 1

ISBN 0-387-96115-1 Springer-Verlag New York Berlin Heidelberg Toyko
ISBN 3-540-96115-1 Springer-Verlag Berlin Heidelberg New York Toyko

To Marcia and Jane

Table of Contents

Preface

As its title suggests, this book is an introduction to three ideas and the connections between them. Before describing the content of the book in detail, we describe each concept briefly. More extensive introductory descriptions of each concept are in the introductions and notes to Chapters 2, 3 and 4.

A topos is a special kind of category defined by axioms saying roughly that certain constructions one can make with sets can be done in the category. In that sense, a topos is a generalized set theory. However, it originated with Grothendieck and Giraud as an abstraction of the properties of the category of sheaves of sets on a topological space. Later, Lawvere and Tierney introduced a more general idea which they called "elementary topos" (because their axioms did not quantify over sets), and they and other mathematicians developed the idea that a theory in the sense of mathematical logic can be regarded as a topos, perhaps after a process of completion.

The concept of triple originated (under the name "standard constructions") in Godement's book on sheaf theory for the purpose of computing sheaf cohomology. Then Peter Huber discovered that triples capture much of the information of adjoint pairs. Later Linton discovered that triples gave an equivalent approach to Lawvere's theory of equational theories (or rather the infinite generalizations of that theory). Finally, triples have turned out to be a very important tool for deriving various properties of toposes.

Theories, which could be called categorical theories, have been around in one incarnation or another at least since Lawvere's Ph.D. thesis. Lawvere's original insight was that a mathematical theory—corresponding roughly to the definition of a class of mathematical objects—could be usefully regarded as a category with structure of a certain kind, and a model of that theory—one of those objects—as a set-valued functor from that category which preserves the structure. The structures involved are more or less elaborate, depending on the kind of objects involved. The most elaborate of these use categories which have all the structure of a topos.

Chapter 1 is an introduction to category theory which develops the basic constructions in categories needed for the rest of the book. All the category theory the reader needs to understand the book is in it, but the reader should be warned that if he has had no prior exposure to categorical reasoning the book might be tough going.

Chapters 2, 3 and 4 introduce each of the three topics of the title and develop them independently up to a certain point. Each of them can be read immediately after Chapter 1. Chapter 5 develops the theory of toposes further, making heavy use of the theory of triples from Chapter 3. Chapter 6 covers various fundamental constructions which give toposes, with emphasis on the idea of "topology", a concept due to Grothendieck which enables us through Giraud's theorem to partially recapture the original idea that toposes are abstract sheaf categories. Chapter 7 provides the basic representation theorems for toposes. Theories are then carried further in Chapter 8, making use of the representation theorems and the concepts of topology and sheaf. Chapter 9 develops further topics in triple theory, and may be read immediately after Chapter 3. Thus in a sense the book, except for for Chapter 9, converges on the exposition of theories in Chapters 4 and 8. We hope that the way the ideas are applied to each other will give a coherence to the many topics discussed which will make them easier to grasp.

We should say a word about the selection of topics. We have developed the introductory material to each of the three main subjects, along with selected topics for each. The connections between theories as developed here and mathematical logic have not been elaborated; in fact, the point of categorical theories is that it provides a way of making the intuitive concept of theory precise without using concepts from logic and the theory of formal systems. The connection between topos theory and logic via the concept of the language of a topos has also not been described here. Categorical logic is the subject of forthcoming book by J. Lambek and P. Scott which is nicely complementary to our book.

Another omission, more from lack of knowledge on our part than from any philosophical position, is the intimate connection between toposes and algebraic geometry. In order to prevent the book from growing even more, we have also omitted the connection between triples and cohomology, an omission we particularly regret. This, unlike many advanced topics in the theory of triples, has been well covered in the literature.

Chapters 2, 3, 5, 6 and 7 thus form a fairly thorough introduction to the theory of toposes, covering topologies and the representation theorems but omitting the connections with algebraic geometry and logic. Adding chapters 4 and 8 provides an introduction to the concept of categorical theory, again without the connection to logic. On the other hand, Chapters 3 and 9 provide an introduction to the basic ideas of triple theory, not including the connections with cohomology.

It is clear that among the three topics, topos theory is "more equal" than the others in this book. That reflects the current state of development and, we believe, importance of topos theory as compared to the other two.

Foundational questions

It seems that no book on category theory is considered complete without some remarks on its set-theoretic foundations. The well-known set theorist Andreas Blass gave a talk (published in Gray [1984]) on the interaction between category theory and set theory in which, among other things, he offered three set-theoretic foundations for category theory. One was the universes of Grothendieck (of which he said that one advantage was that it made measurable cardinals respectable in France) and another was systematic use of the reflection principle, which probably does provide a complete solution to the problem; but his first suggestion, and one that he clearly thought at least reasonable, was: None. This is the point of view we shall adopt.

For example, we regard a topos as being defined by its elementary axioms, saying nothing about the set theory in which its models live. One reason for our attitude is that many people regard topos theory as a possible new foundation for mathematics. When we refer to "the category of sets" the reader may choose between thinking of a standard model of set theory like ZFC and a topos satisfying certain additional requirements, including at least two-valuedness and choice.

We will occasionally use procedures which are set-theoretically doubtful, such as the formation of functor categories with large exponent. However, our conclusions can always be justified by replacing the large exponent by a suitable small subcategory.

Terminology and notation

With a few exceptions, we usually use established terminology and standard notation; deviations from customary usage add greatly to the difficulties of the reader, particularly the reader already somewhat familiar with the subject, and should be made only when the gain in clarity and efficiency are great enough to overcome the very real inconvenience they cause. In particular, in spite of our recognition that there are considerable advantages to writing functions on the right of the argument instead of the left and composing left to right, we have conformed reluctantly to tradition in this respect: in this book, functions are written on the left and composition is read right to left.

We often say "arrow" or "map" for "morphism", "source" for "domain" and "target" for "codomain". We generally write "αX" instead of "α_X" for the component at X of the natural transformation α, which avoids double subscripts and is generally easier to read. It also suppresses the distinction between the component of a natural transformation at a functor and a functor applied to a natural transformation. Although these two notions are semantically distinct, they are syntactically identical; much progress in mathematics comes about from muddying such distinctions.

Our most significant departures from standard terminology are the adoption of Freyd's use of "exact" to denote a category which has all finite limits and colimits or for a functor which preserves them and the use of "sketch" in a sense different from that of Ehresmann. Our sketches convey the same information while being conceptually closer to naive theories.

There are two different categories of toposes: one in which the geometric aspect is in the ascendent and the other in which the logic is predominant. The distinction is analogous to the one between the categories of complete Heyting algebras and that of locales. Thinking of toposes as models of a theory emphasizes the second aspect and that is the point of view we adopt. In particular, we use the term "subtopos" for a subcategory of a topos which is a topos, which is different from the geometric usage in which the *right* adjoint is supposed an embedding.

Historical notes

At the end of many of the chapters we have added historical notes. It should be understood that these are not History as that term is understood by the historian. They are at best the raw material of history.

At the end of the first draft we made some not very systematic attempts to verify the accuracy of the historical notes. We discovered that our notes were divided into two classes: those describing events that one of us had directly participated in and those that were wrong! The latter were what one might conjecture on the basis of the written record, and we discovered that the written record is invariably misleading. Our notes now make only statements we could verify from the participants. Thus they are incomplete, but we have some confidence that those that remain have some relation to the actual events.

What is expected from the reader

We assume that the reader is familiar with concepts typically developed in first-year graduate courses, such as group, ring, topological space, and so on. The elementary facts about sheaves which are needed are developed in the book. The reader who is familiar with the elements of category theory including adjoint functors can skip nearly all of Chapter 1; he may need to learn the element notation introduced in Section 1.4 and the square bracket notation defined in Sections 1.6 and 1.7.

Most of the exercises provide examples or develop the theory further. Those marked with a diamond $^\diamond$ contain results that are used in the text. We have mostly avoided including exercises asking for routine verifications or giving trivial examples. On the other hand, most routine verifications are omitted from the text; usually, in a proof, the basic construction is given and the verification that it works is left to the reader (but the first time a verification of a given type is used it is given in more detail). This means that if you want to gain a thorough understanding of the material, you should be prepared to stop every few sentences (or even every sentence) and verify the claims made there in detail. You should be warned that a statement such as, "It is easy to see..." does not mean it is necessarily easy to see without pencil and paper!

Acknowledgments

We are grateful to Barry Jay, Peter Johnstone, Anders Linnér, John A. Power and Philip Scott for reading portions of the manuscript and making many corrections and suggestions for changes; we are particularly grateful to Barry Jay, who up to two weeks before the final printout was still finding many obscurities and typoes and some genuine mathematical errors. We have benefited from stimulating and informative discussions with many people including, but not limited to Marta Bunge, Radu Diaconescu, John W. Duskin, Michael Fourman, Peter Freyd, John Gray, Barry Jay, Peter Johnstone, André Joyal, Joachim Lambek, F. William Lawvere, Colin McLarty, Michael Makkai and Myles Tierney. We would like to give especial thanks to Roberto Minio who expended enormous effort in turning a string of several million zeroes and ones into the text you see before you; John Aronis also helped in this endeavor, which took place at Carnegie-Mellon University with the encouragement and cooperation of Dana Scott.

We are also grateful to Beno Eckmann, who brought us together at the Forschungsinstitut für Mathematik, ETH Zürich. If Eilenberg and Mac Lane were the fathers of categorical algebra, Eckmann was in a very real sense the godfather. Many of the most important developments in categorical algebra and categorical logic took place in the offices of the Forschungsinstitut on Zehnderweg.

Portions of this book were written while both authors were on sabbatical leave from their respective institutions. The first author was supported during the writing by grants from the National Science and Engineering Research Council, by a team grant from the Ministère de l'Éducation du Québec and by a grant to the Groupe Interuniversitaire en Études

Catégories, also from the Ministère de l'Éducation du Québec. The second author was partially supported by DOE contract DE-AC01-80RA5256. In addition we received considerable free computing time from the McGill University Computing Centre.

Chapter dependency chart

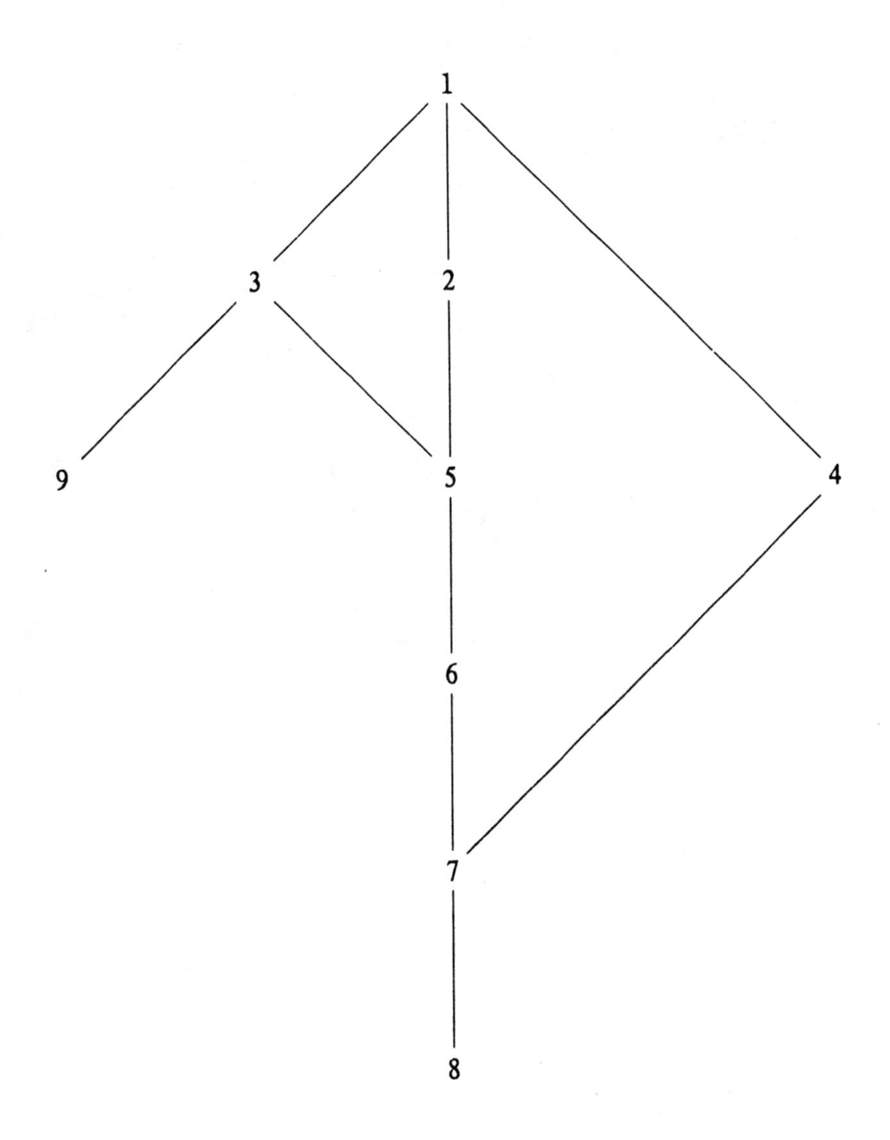

Chapter 1.

Categories

1.1. Definition of Category

A **category** $\mathcal{C}$ consists of two collections, $\mathcal{O}b(\mathcal{C})$, whose elements are the **objects** of $\mathcal{C}$, and $\mathcal{A}r(\mathcal{C})$, the **arrows** (or **morphisms** or **maps**) of $\mathcal{C}$. To each arrow is assigned a pair of objects, called the **source** (or **domain**) and the **target** (or **codomain**) of the arrow. The notation $f : A \to B$ means that f as an arrow with source A and target B. If $f : A \to B$ and $g : B \to C$ are two arrows, there is an arrow $g \circ f : A \to C$ called the **composite** of g and f. The composite is not defined otherwise. We often write gf instead of $g \circ f$ when there is no danger of confusion. For each object A there is an arrow id_A (often written 1_A or just 1, depending on the context), called the **identity** of A, whose source and target are both A. These data are subject to the following axioms:

(i) For $f : A \to B$,

$$f \circ \mathrm{id}_A = \mathrm{id}_B \circ f = f;$$

(ii) For $f : A \to B$, $g : B \to C$, $h : C \to D$,

$$h \circ (g \circ f) = (h \circ g) \circ f.$$

The set of arrows with source A and target B is denoted $\mathrm{Hom}_{\mathcal{C}}(A, B)$. We will omit the subscript denoting the category whenever we can get away with it. A set of the form $\mathrm{Hom}(A, B)$ is called a **homset**.

Many familiar examples of categories will occur immediately to the reader, such as the category $\mathcal{S}et$ of sets and set functions, the category $\mathcal{G}rp$ of groups and homomorphisms, and the category $\mathcal{T}op$ of topological spaces and continuous maps. In each of these cases, the composition operation on arrows is the usual composition of functions.

A more interesting example is the category whose objects are topological spaces and whose arrows are homotopy classes of continuous maps.

Because homotopy is compatible with composition, homotopy classes of continuous functions behave like functions (they have sources and targets, they compose, etc.) but are not functions. This category is usually known as the category of homotopy types.

All but the last example are of categories whose objects are sets with mathematical structure and the morphisms are functions which preserve the structure. Many mathematical structures are themselves categories. For example, one can consider any group G as a category with exactly one object; its arrows are the elements of G regarded as having the single object as both source and target. Composition is the group multiplication, and the group identity is the identity arrow. This construction works for monoids as well. In fact, a monoid can be defined as a category with exactly one object.

A poset (partially ordered set) can also be regarded as a category: its objects are its elements, and there is exactly one arrow from an element x to an element y if and only if $x \leq y$; otherwise there are no arrows from x to y. Composition is forced by transitivity and identity arrows by reflexivity. Thus a category can be thought of as a generalized poset. This perception is important, since many of the fundamental concepts of category theory specialize to nontrivial and often well-known concepts for posets (the reader is urged to fill in the details in each case).

In the above examples, we have described categories by specifying both their objects and their arrows. Informally, it is very common to name the objects only; the reader is supposed to supply the arrows based on his general knowledge. If there is any doubt, it is, of course, necessary to describe the arrows as well. For example, in an exercise on cyclic groups in Section 8.4, we have specified that the morphisms are not assumed to preserve the generator. Sometimes there are two or more categories in general use with the same objects but different arrows. For example, the following five categories all have the same objects: complete sup-semilattices, complete inf-semilattices, complete lattices and the first two with the preservation of the top (resp. bottom). In such cases we even distinguish the objects by name even though they are the same class.

Some constructions for categories

A **subcategory** $\mathcal{D}$ of a category $\mathcal{C}$ is a pair of subsets D_O and D_A of the objects and arrows of $\mathcal{C}$ respectively, with the properties

(i) If $f \in D_A$ then the source and target of f are in D_O.
(ii) If $C \in D_O$, then $\mathrm{id}_C \in D_A$.
(iii) If f, g are in D_A then so is $g \circ f$.

The subcategory is **full** if for any $C, D \in \mathcal{D}_O$, if $f : C \to D$ in $\mathcal{C}$, then $f \in \mathcal{D}_A$. For example, the category of Abelian groups is a full subcategory of the category of groups (every homomorphism of groups between Abelian groups is a homomorphism of Abelian groups), whereas the category of monoids (semigroups with identity element) is a subcategory, but not a full subcategory, of the category of semigroups (a semigroup homomorphism need not preserve 1).

One also constructs the **product** $\mathcal{C} \times \mathcal{D}$ of two categories $\mathcal{C}$ and $\mathcal{D}$ in the obvious way: the objects of $\mathcal{C} \times \mathcal{D}$ are pairs (A, B) with A an object of $\mathcal{C}$ and B an object of $\mathcal{D}$. An arrow

$$(f, g) : (A, B) \to (A', B')$$

has $f : A \to A'$ in $\mathcal{C}$ and $g : B \to B'$ in $\mathcal{D}$. Composition is coordinatewise.

To define the next concept, we need the idea of commutative diagram. A diagram is said to commute if any two paths between the same nodes compose to give the same morphism.

If A is any object of a category $\mathcal{C}$, the **slice category** $\mathcal{C}/A$ of objects of $\mathcal{C}$ **over** A has as objects all arrows of $\mathcal{C}$ with target A. An arrow of $\mathcal{C}/A$ from $f : B \to A$ to $g : C \to A$ is an arrow $h : B \to C$ making the following diagram commute.

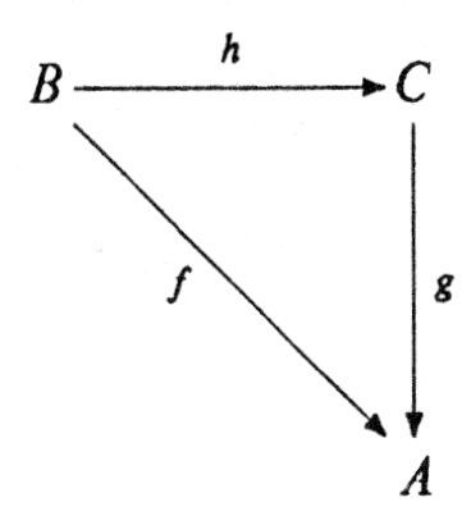

In this case, one sometimes writes $h : f \to g$ **over** A.

It is useful to think of an object of $\mathcal{S}et/A$ as an A-indexed family of disjoint sets (the inverse images of the elements of A). The commutativity of the above diagram means that the function h is consistent with the decomposition of B and C into disjoint sets.

Definitions without using elements

The introduction of categories as a part of the language of mathematics has made possible a fundamental, intrinsically categorical technique: the element-free definition of mathematical properties by means of commutative diagrams, limits and adjoints. (Limits and adjoints are defined later in this chapter.) By the use of this technique, category theory has made mathematically precise the unity of a variety of concepts in different branches of mathematics, such as the many product constructions which occur all over mathematics (described in Section 1.7) or the ubiquitous concept of isomorphism, discussed below. Besides explicating the unity of concepts, categorical techniques for defining concepts without mentioning elements have enabled mathematicians to provide a useful axiomatic basis for algebraic topology, homological algebra and other theories.

Despite the possibility of giving element-free definitions of these constructions, it remains intuitively helpful to think of them as being defined with elements. Fortunately, this can be done: In Section 1.4, we introduce a more general notion of element of an object in a category (more general even when the category is *Set*) which in many circumstances makes categorical definitions resemble familiar definitions involving elements of sets, and which also provides an explication of the old notion of variable quantity.

Isomorphisms and terminal objects

The notion of isomorphism can be given an element-free definition for any category: An arrow $f : A \to B$ in a category is an **isomorphism** if it has an **inverse**, namely an arrow $g : B \to A$ for which $f \circ g = \mathrm{id}_B$ and $g \circ f = \mathrm{id}_A$. In other words, the following diagrams must commute:

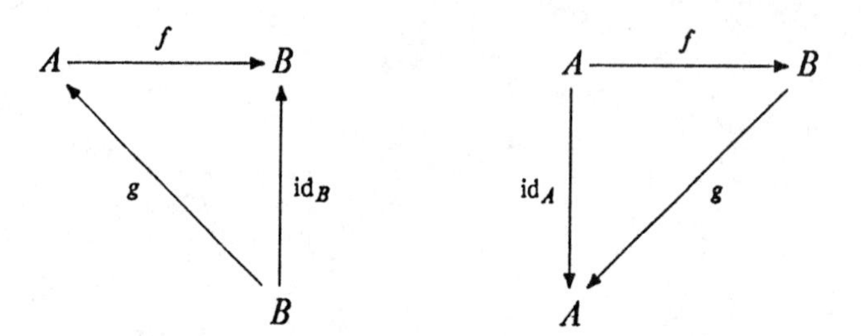

In a group regarded as a category, every arrow is invertible, whereas in a poset regarded as a category, the only invertible arrows are the identity arrows (which are invertible in any category).

It is easy to check that an isomorphism in *Grp* is what is usually called an isomorphism (commonly defined as a bijective homomorphism, but some newer texts give the definition above). An isomorphism in *Set* is a bijective function, and an isomorphism in *Top* is a homeomorphism.

Singleton sets in *Set* can be characterized without mentioning elements, too. A **terminal object** in a category C is an object T with the property that for every object A of C there is exactly one arrow from A to T. It is easy to see that terminal objects in *Set* , *Top*, and *Grp* are all one element sets with the only possible structure in the case of the last two categories.

Duality

If $\mathcal{C}$ is a category, then we define $\mathcal{C}^{op}$ to be the category with the same objects and arrows as $\mathcal{C}$, but an arrow $f : A \to B$ in $\mathcal{C}$ is regarded as an arrow from B to A in $\mathcal{C}^{op}$. In other words, for all objects A and B of $\mathcal{C}$,

$$\mathrm{Hom}_{\mathcal{C}}(A, B) = \mathrm{Hom}_{\mathcal{C}^{op}}(B, A).$$

If $f : A \to B$ and $g : B \to C$ in $\mathcal{C}$, then the composite $f \circ g$ in $\mathcal{C}^{op}$ is by definition the composite $g \circ f$ in $\mathcal{C}$. $\mathcal{C}^{op}$ is called the **opposite category** of $\mathcal{C}$.

If P is a property that objects or arrows in a category may have, then the **dual** of P is the property of having P in the opposite category. As an example, consider the property of being a terminal object. If an object A of a category $\mathcal{C}$ is a terminal object in $\mathcal{C}^{op}$, then $\mathrm{Hom}_{\mathcal{C}}(B, A)$ has exactly one arrow for every object B of $\mathcal{C}$. Thus the dual property of being a terminal object is the property: $\mathrm{Hom}(A, B)$ has exactly one arrow for each object B. An object A with this property is called an **initial object**. In $\mathcal{Set}$ and $\mathcal{Top}$, the empty set is the initial object (see "Fine points" below). In $\mathcal{Grp}$, on the other hand, the one-element group is both an initial and a terminal object.

Clearly if property P is dual to property Q then property Q is dual to property P. Thus being an initial object and being a terminal object are dual properties. Observe that being an isomorphism is a self-dual property.

Constructions may also have duals. For example, the dual to the category of objects over A is the category of objects **under** A. An object is an arrow *from* A and an arrow from the object $f : A \to B$ to the object $g : A \to C$ is an arrow h from B to C for which $h \circ f = g$.

Often a property and its dual each have their own names; when they don't (and sometimes when they do) the dual property is named by prefixing "co-". For example, one could, and some sources do, call an initial object "coterminal", or a terminal object "coinitial".

Definition of category by commutative diagrams

The notion of category itself can be defined in an element-free way. We describe the idea behind this alternate definition here, but some of the sets we construct are defined in terms of elements. In Section 1.6, we show how to define these sets without mentioning elements (by pullback diagrams).

A category consists of two sets A and O and four functions $d^0, d^1 : A \to O$, $u : O \to A$ and $m : P \to A$, where P is the set

$$\{(f, g) \mid d^0(f) = d^1(g)\}$$

of composable pairs of arrows for which the following diagrams (i) through (iv) commute. For example, the left diagram of (ii) below says that $d^0 \circ p_1 = d^0 \circ m$. We will treat diagrams more formally in Section 1.7.

In these diagrams we use the following notational devices, which will recur throughout the book.

(a) If X and Y are sets, $p_1 : X \times Y \to X$ and $p_2 : X \times Y \to Y$ are the coordinate projections.

(b) If X, Y and Z are sets and $f : X \to Y$, $g : X \to Z$ are functions,

$$(f, g) : X \to Y \times Z$$

is the function whose value at $a \in X$ is $(f(a), g(a))$.

(c) If X, Y, Z, and W are sets and $f : X \to Z$, $g : Y \to W$ are functions, then

$$f \times g : X \times Y \to Z \times W$$

is the function whose value at (a, b) is $(f(a), g(b))$. This notation is also used for maps defined on subsets of product sets (as in (iv) below).

In (iv), Q is the set of composable triples of arrows:

$$Q = \{(f, g, h) \mid d^1(h) = d^0(g) \text{ and } d^1(g) = d^0(f)\}.$$

Here are the diagrams:

(i)

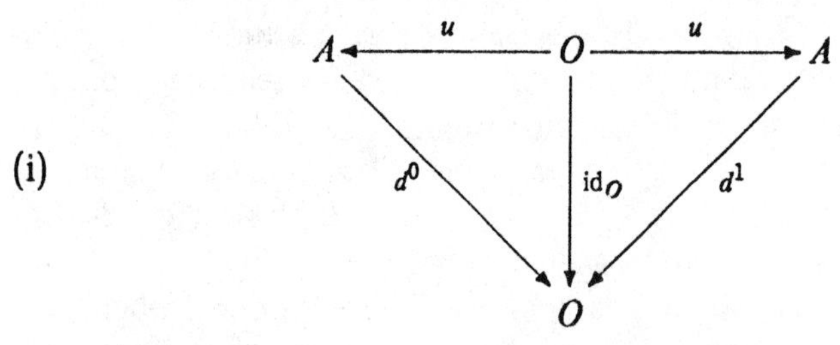

This says that the source and target of id_X is X.

(ii)

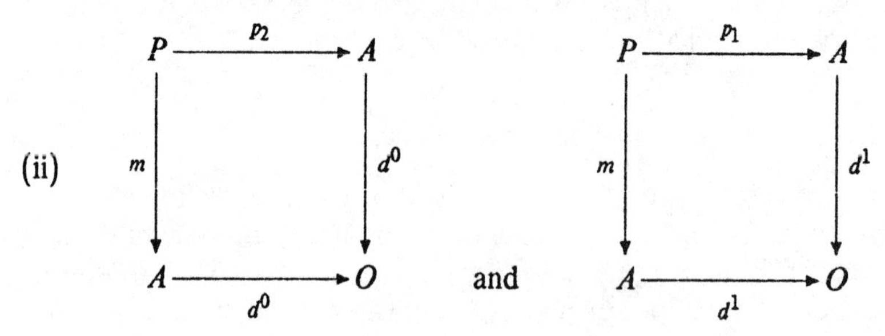

This says that the source of $f \circ g$ is that of g and its target is that of f.

(iii)

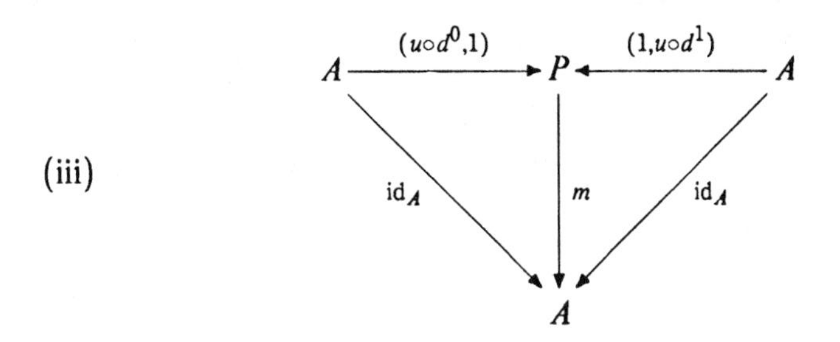

This characterizes the left and right identity laws.

(iv)

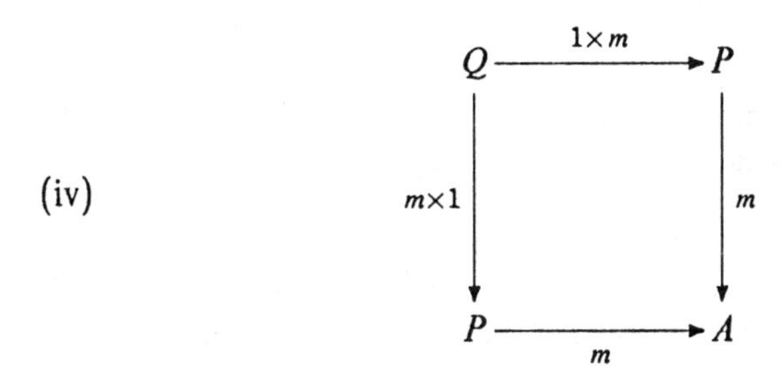

This is associativity of composition.

It is straightforward to check that this definition is equivalent to the first one.

The diagrams just given actually describe geometric objects, namely the classifying space of the category. Indeed, the functions between O, A, P and Q generated by u, d^0, d^1, m and the coordinate maps form a simplicial set truncated in dimension three. The reader does not need to know simplicial theory to read this book, however.

A third way of defining category is given in Exercise (SGPOID).

Fine points

Note that a category may be empty, that is have no objects and (of course) no arrows. Observe that a subcategory of a monoid regarded as a category may be empty; if it is not empty, then it is a submonoid. This should cause no more difficulty than the fact that a submonoid of a group may not be a subgroup.

It is important to observe that in categories such as *Set*, *Grp* and *Top* in which the arrows are actually functions, the definition of category requires that the function have a uniquely specified domain and codomain, so that for example in *Top* the continuous function from the set R of real numbers to the set R^+ of nonnegative real numbers which takes a number to its square is different from the function from R to R which does the same thing, and both of these are different from the squaring function from R^+ to R^+.

A definition of "function" in *Set* which fits this requirement is this: A **function** is an ordered triple (A, G, B) where A and B are sets and G is a subset of $A \times B$ with the property that for each $x \in A$ there is exactly one $y \in B$ such that $(x, y) \in G$. This is equivalent to saying that the composite

$$G \to A \times B \to A$$

is an isomorphism (the second function is projection on the first coordinate). Then the domain of the function is the set A and the codomain is B. As a consequence of this definition, A is empty if and only if G is empty, but B may or may not be empty. Thus there is exactly one function, namely $(\emptyset, \emptyset, B)$, from the empty set to each set B, so that the empty set is the initial object in *Set* , as claimed previously. (Note also that if (A, G, B) is a function then G uniquely determines A but not B. This asymmetry is reversed in the next paragraph.)

An equivalent definition of function is a triple (A, G^*, B) where G^* is the quotient of the disjoint union $A + B$ by an equivalence relation for which each element of B is contained in exactly one equivalence class. In other words, the composite

$$B \to A + B \to G^*$$

is an isomorphism. This actually corresponds to the intuitive picture of function frequently drawn for elementary calculus students which illustrates the squaring function from $\{-2, -1, 0, 1, 2\}$ to $\{0, 1, 2, 3, 4\}$ this way:

-2 2	4
-1 1	1
0	0
	2
	3

We will see in Section 1.8 that the two notions of function are dual to each other.

Exercises 1.1

(SGRPOID)°. Show that the following definition of category which is sometimes used is equivalent to the definition given in this section: A **category** is a set with a partially defined binary operation denoted "$\circ$" with the following properties:

(i) The following statements are equivalent:

(a) $f \circ g$ and $g \circ h$ are both defined.
(b) $f \circ (g \circ h)$ is defined.
(c) $(f \circ g) \circ h$ is defined.

(ii) If $(f \circ g) \circ h$ is defined, then $(f \circ g) \circ h = f \circ (g \circ h)$.
(iii) For any f, there are elements e and e' for which $e \circ f$ is defined and equal to f and $f \circ e'$ is defined and equal to f.

(CCON)°. Verify that the following constructions produce categories.

(a) For any category $\mathcal{C}$, the **arrow category** $Ar(\mathcal{C})$ of arrows of $\mathcal{C}$ has as objects the arrows of $\mathcal{C}$, and an arrow from $f : A \to B$ to $g : A' \to B'$ is a pair of arrows $h : A \to A'$ and $k : B \to B'$ making the following diagram commute:

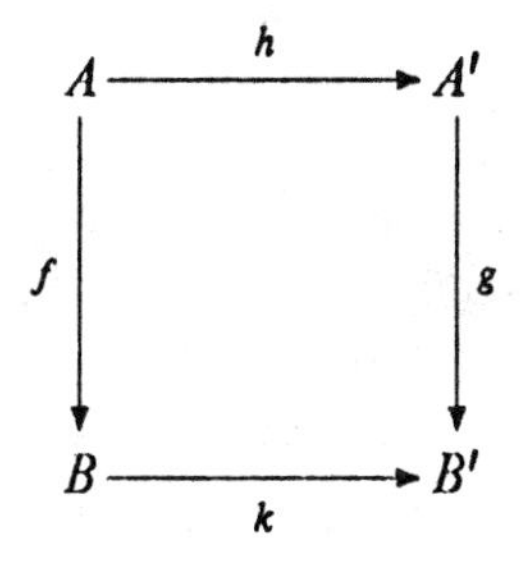

(b) The **twisted arrow category of** $\mathcal{C}$ is defined the same way as the arrow category except that the direction of k is reversed.

(ISO)°. (a) Show that $h : f \to g$ is an isomorphism in the category of objects of $\mathcal{C}$ over A if and only if h is an isomorphism of $\mathcal{C}$ and $g \circ h = f$.
(b) Give an example of objects A, B and C in a category $\mathcal{C}$ and arrows $f : B \to A$ and $g : C \to A$ such that B and C are isomorphic in $\mathcal{C}$ but f and g are not isomorphic in $\mathcal{C}/A$.

(IIT). Describe the isomorphisms, initial objects, and terminal objects (if they exist) in each of the categories in Exercise (CCON).

(IPOS). Describe the initial and terminal objects, if they exist, in a poset regarded as a category.

(TISO)°. Show that any two terminal objects in a category are isomorphic by a unique isomorphism.

(SKEL). (a) Prove that for any category $\mathcal{C}$ and any arrows f and g of $\mathcal{C}$ such that the target of g is isomorphic to the source of f, there is an arrow f' which (i) is isomorphic to f in $Ar(\mathcal{C})$ and (ii) has source the same as the target of g. ($Ar(\mathcal{C})$ is defined in Exercise (CCON) above.)

(b) Use the fact given in (a) to describe a suitable definition of domain, codomain and composition for a category with one object chosen for each isomorphism class of objects of $\mathcal{C}$ and one arrow from each isomorphism class of objects of $Ar(\mathcal{C})$. Such a category is called a **skeleton** of $\mathcal{C}$.

(COMP). A category is **connected** if it is possible to go from any object to any other object of the category along a path of "composable" *forward or backward* arrows. Make this definition precise and prove that every category is a union of disjoint connected subcategories in a unique way.

(PREOR). A **preorder** is a set with a reflexive, transitive relation defined on it. Explain how to regard a preorder as a category with at most one arrow from any object A to any object B.

(OPP). (a) Describe the opposite of a group regarded as a category. Show that it is isomorphic to, but not necessarily the same as, the original group.

(b) Do the same for a monoid (set with associative binary operation and identity = semigroup with identity), but show that the opposite need not be isomorphic to the original monoid.

(c) Do the same as (b) for posets.

(QUOT)°. A **congruence** on a category $\mathcal{C}$ is an equivalence relation E on the arrows for which

(i) fEf' implies that f and f' have the same domain and codomain.
(ii) If fEf' and gEg' and $f \circ g$ is defined, then $(f \circ g)E(f' \circ g')$.

(a) Show that any relation R on the arrows of $\mathcal{C}$ generates a unique congruence on $\mathcal{C}$.

(b) Given a congruence E on $\mathcal{C}$, define the **quotient category** $\mathcal{C}/E$ in the obvious way (same objects as $\mathcal{C}$) and show that it is a category. This notation conflicts with the slice notation, but context should make it clear. In any case, quotient categories are not formed very often. (Thus any set of diagrams in $\mathcal{C}$ generate a congruence E on $\mathcal{C}$ with the property that $\mathcal{C}/E$ is the largest quotient in which the diagrams commute.)

(PID)°. Show that in a category with an initial object 0 and a terminal object 1, $0 \cong 1$ if and only if there is a map $1 \to 0$.

1.2. Functors

Like every other kind of mathematical structured object, categories come equipped with a notion of morphism. It is natural to define a morphism of categories to be a map which takes objects to objects, arrows to arrows, and preserves source, target, identities and composition.

If $\mathcal{C}$ and $\mathcal{D}$ are categories, a **functor** $F : \mathcal{C} \to \mathcal{D}$ is a map for which

(i) If $f : A \to B$ is an arrow of $\mathcal{C}$, then $Ff : FA \to FB$ is an arrow of $\mathcal{D}$;
(ii) $F(\mathrm{id}_A) = \mathrm{id}_{FA}$; and
(iii) If $g : B \to C$, then $F(g \circ f) = Fg \circ Ff$.

If $F : \mathcal{C} \to \mathcal{D}$ is a functor, then $F^{\mathrm{op}} : \mathcal{C}^{\mathrm{op}} \to \mathcal{D}^{\mathrm{op}}$ is the functor which does the same thing as F to objects and arrows.

A functor $F : \mathcal{C}^{\mathrm{op}} \to \mathcal{D}$ is called a **contravariant functor** from $\mathcal{C}$ to $\mathcal{D}$. In this case, F^{op} goes from $\mathcal{C}$ to $\mathcal{D}^{\mathrm{op}}$. For emphasis, a functor from $\mathcal{C}$ to $\mathcal{D}$ is occasionally called a **covariant functor.**

$F : \mathcal{C} \to \mathcal{D}$ is **faithful** if it is injective when restricted to each homset, and it is **full** if it is surjective on each homset, i.e., if for every pair of objects A and B, every arrow in $\mathrm{Hom}(FA, FB)$ is F of some arrow in $\mathrm{Hom}(A, B)$. Some sources use the phrase "fully faithful" to describe a functor which is full and faithful.

F **preserves** a property P that an arrow may have if $F(f)$ has property P whenever f has. It **reflects** property P if f has the property whenever $F(f)$ has. For example, any functor must preserve isomorphisms (Exercise (PISO)), but a functor need not reflect them.

Here are some examples of functors:

(i) For any category $\mathcal{C}$, there is an identity functor $\mathrm{id}_{\mathcal{C}} : \mathcal{C} \to \mathcal{C}$.
(ii) The categories $\mathcal{G}rp$ and $\mathcal{T}op$ are typical of many categories considered in mathematics in that their objects are sets with some sort of structure on them and their arrows are functions which preserve that structure. For any such category $\mathcal{C}$, there is an **underlying set functor** $U : \mathcal{C} \to \mathcal{S}et$ which assigns to each object its set of elements and to each arrow the function associated to it. Such a functor is also called a **forgetful functor,** the idea being that it forgets the structure on the set. Such functors are always faithful and rarely full.

(iii) Many other mathematical constructions, such as the double dual functor on vector spaces, the commutator subgroup of a group or the fundamental group of a path connected space, are the object maps of functors (in the latter case the domain is the category of pointed topological spaces and base-point-preserving maps). There are, on the other hand, some canonical constructions which do not extend to maps. Examples include the center of a group or ring, and groups of automorphisms quite generally. See Exercises (CTR) and (AUT).

(iv) For any set A, let FA denote the free group generated by A. The defining property of free groups allows you to conclude that if $f : A \to B$ is any function, there is a unique homomorphism $Ff : FA \to FB$ with the property that $Ff \circ i = j \circ f$, where $i : A \to FA$ and $j : B \to FB$ are the inclusions. It is an easy exercise to see that this makes F a functor from $\mathcal{Set}$ to $\mathcal{Grp}$. Analogous functors can be defined for the category of monoids, the category of Abelian groups, and the category of R-modules for any ring R.

(v) For a category $\mathcal{C}$, $\text{Hom}_{\mathcal{C}} = \text{Hom}$ is a functor in each variable separately, as follows: For fixed object A, $\text{Hom}(A,f) : \text{Hom}(A,B) \to \text{Hom}(A,C)$ is defined for each arrow $f : B \to C$ by requiring that $\text{Hom}(A,f)(g) = f \circ g$ for $g \in \text{Hom}(A,B)$; this makes $\text{Hom}(A,-) : \mathcal{C} \to \mathcal{Set}$ a functor. Similarly, for a fixed object B, $\text{Hom}(-,B)$ is a functor from $\mathcal{C}^{\text{op}}$ to $\mathcal{Set}$; $\text{Hom}(h,B)$ is composition with h on the right instead of on the left. $\text{Hom}(A,-)$ and $\text{Hom}(-,B)$ are the **covariant** and **contravariant hom functors**, respectively. $\text{Hom}(-,-)$ is also a $\mathcal{Set}$-valued functor, with domain $\mathcal{C}^{\text{op}} \times \mathcal{C}$. A familiar example of a contravariant homfunctor is the functor which takes a vector space to the underlying set of its dual.

(vi) The powerset (set of subsets) of a set is the object map of an important contravariant functor $\mathbf{P}$ from $\mathcal{Set}$ to $\mathcal{Set}$ which plays a central role in this book. The map from $\mathbf{P}B$ to $\mathbf{P}A$ induced by a function $f : A \to B$ is the inverse image map; precisely, if $B_0 \in \mathbf{P}B$, i.e. $B_0 \subseteq B$, then

$$\mathbf{P}f(B_0) = \{x \in A \mid f(x) \in B_0\}.$$

$\mathbf{P}$ can also be made into a covariant functor, in at least two different ways (Exercise (Pow)).

(vii) If G and H are groups considered as categories with a single object, then a functor from G to H is exactly a group homomorphism.

(viii) If P and Q are posets, a functor from P to Q is exactly a nondecreasing map. A contravariant functor is a nonincreasing map.

Isomorphism and equivalence of categories

The composite of functors is a functor, so the collection of categories and functors is itself a category, denoted $\mathcal{C}at$. If $\mathcal{C}$ and $\mathcal{D}$ are categories and $F : \mathcal{C} \to \mathcal{D}$ is a functor which has an inverse $G : \mathcal{D} \to \mathcal{C}$, so that it is an isomorphism in the category of categories, then naturally $\mathcal{C}$ and $\mathcal{D}$ are said to be **isomorphic.**

However, the notion of isomorphism does not capture the most useful sense in which two categories can be said to be essentially the same; that is the notion of equivalence. A functor $F : \mathcal{C} \to \mathcal{D}$ is said to be an **equivalence** if it is full and faithful and has the property that for any object B of $\mathcal{D}$ there is an object A of $\mathcal{C}$ for which $F(A)$ is isomorphic to B. The definition appears asymmetrical but in fact given the axiom of choice if there is an equivalence from $\mathcal{C}$ to $\mathcal{D}$ then there is an equivalence from $\mathcal{D}$ to $\mathcal{C}$ (Exercise (Equ)).

The notion of equivalence captures the perception that, for example, for most purposes you are not changing group theory if you want to work in a category of groups which contains only a countable number (or finite, or whatever) of copies of each isomorphism type of groups and all the homomorphisms between them.

Statements in Section 1.1 like, "A group may be regarded as a category with one object in which all arrows are isomorphisms" can be made precise using the notion of equivalence: The category of groups and homomorphisms is equivalent to the category of categories with exactly one object in which each arrow is an isomorphism, and all functors between them. Any isomorphism between these categories would seem to require an axiom of choice for proper classes.

Comma categories

Let $\mathcal{A}$, $\mathcal{C}$ and $\mathcal{D}$ be categories and $F : \mathcal{C} \to \mathcal{A}$, $G : \mathcal{D} \to \mathcal{A}$ be functors. From these ingredients we construct the **comma category** (F, G) which is a generalization of the slice $\mathcal{A}/A$ of a category over an object discussed in Section 1.1. The objects of (F, G) are triples (C, f, D) with $f : FC \to GD$ an arrow of $\mathcal{A}$ and C, D objects of $\mathcal{C}$ and $\mathcal{D}$ respectively. An arrow $(h, k) : (C, f, D) \to (C', f', D')$ consists of $h : C \to C'$ and $k : D \to D'$ making

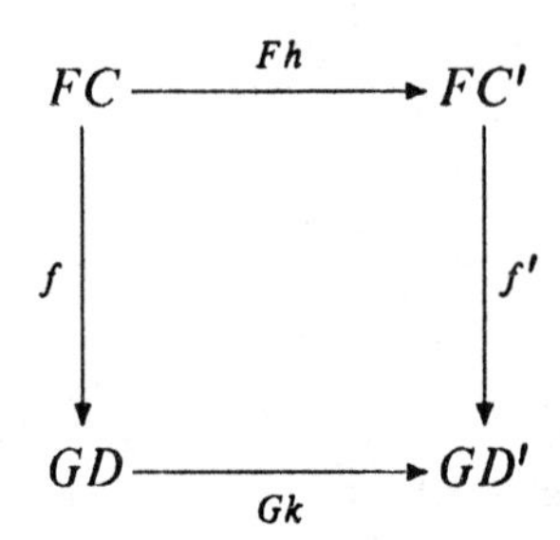

commute. It is easy to verify that coordinatewise composition makes (F, G) a category.

When A is an object of $\mathcal{A}$ and G is taken to be the constant functor embedding A and its identity arrow as a trivial subcategory of $\mathcal{A}$, then $(\mathrm{id}_{\mathcal{A}}, G)$ is the slice $\mathcal{A}/A$ defined in Section 1.1. The category of arrows under an object is similarly a comma category.

Each comma category (F, G) is equipped with two **projections** $p_1 : (F, G) \to \mathcal{C}$ projecting objects and arrows onto their first coordinates, and $p_2 : (F, G) \to \mathcal{D}$ projecting objects onto their third coordinates and arrows onto their second.

Exercises 1.2

(PISO). Show that functors preserve isomorphisms, but do not necessarily reflect them.

(AC). Use the concept of arrow category to describe a functor which takes a group homomorphism to its kernel.

(EAAM). Show that the following define functors:

(a) The projection map from a product $\mathcal{C} \times \mathcal{D}$ of categories to one of them.

(b) For $\mathcal{C}$ a category and A an object of $\mathcal{C}$, the constant map from a category $\mathcal{B}$ to $\mathcal{C}$ which takes every object to A and every arrow to id_A.

(c) The "forgetful" functor from the category $\mathcal{C}/A$ of objects over A to $\mathcal{C}$ which takes an object $B \to A$ to B and an arrow $h : B \to C$ over A to itself.

(PPP). (a) Show that the functor **P** of example (vi) is faithful but not full.

(b) Show that **P** reflects isomorphisms.

(FTI). Give examples showing that functors need not preserve or reflect initial or terminal objects.

(POW). Show that the map which takes a set to its powerset is the object map of at least two covariant functors from $\mathcal{S}et$ to $\mathcal{S}et$: If $f : A \to B$, one functor takes a subset A_0 of A to its image $f_!(A_0) = f(A_0)$, and the other takes A_0 to the set $f_*(A_0) =$

$$\{y \in B \mid \text{if } f(x) = y \text{ then } x \in A_0\} = \{y \in B \mid f^{-1}(y) \subseteq A_0\}$$

Show that $f^{-1}(B) \subseteq A$ if and only if $B \subseteq f_*(A)$ and that $A \subseteq f^{-1}(B)$ if and only if $f_!(A) \subseteq B$.

(FRG). Show that the definition given in Example (iv) makes the free group construction F a functor.

(CTR). Show that there is no functor from $\mathcal{G}rp$ to $\mathcal{G}rp$ which takes each group to its center. (Hint: Consider the group G consisting of all pairs (a, b) where a is any integer and b is 0 or 1, with multiplication

$$(a, b)(c, d) = (a + (-1)^b c, b + d),$$

the addition in the second coordinate being $(\bmod\ 2)$.)

(AUT). Show that there is no functor from $\mathcal{G}rp$ to $\mathcal{G}rp$ which takes each group to its automorphism group.

(SKEL2). Show that every category is equivalent to its skeleton (see Exercise (SKEL) of Section 1.1).

(EQU). Show that equivalence is an equivalence relation on any set of categories. (This exercise is easier to do after you do Exercise (EQUII) of Section 1.3).

(POSEQ). (a) Make the statement "a preordered set can be regarded as a category in which there is no more than one arrow between any two objects" precise by defining a subcategory of the category of categories and functors that the category of preordered sets and order-preserving maps is equivalent to (see Exercise (PREO) of Section 1.1).

(b) Show that, when regarded as a category, every preordered set is equivalent to a poset.

(BOOL). An **atom** in a Boolean algebra is an element greater than 0 but with no elements between it and 0. A Boolean algebra is **atomic** if every element x of the algebra is the join of all the atoms smaller than x. A Boolean algebra is **complete** if every subset has an infimum and a supremum. A **CABA** is a complete atomic Boolean algebra. A CABA homomorphism is a Boolean algebra homomorphism between CABA's which preserves all infs and sups (not just finite ones, which any Boolean algebra homomorphism would do). Show that the opposite of the category of sets is equivalent to the category of CABA's and CABA homomorphisms.

(Usl). An **upper semilattice** is a partially ordered set in which each finite subset (including the empty set) of elements has a least upper bound. Show that the category of upper semilattices and functions which preserve the least upper bound of any finite subset (and hence preserve the ordering) is equivalent to the category of commutative monoids in which every element is idempotent and monoid homomorphisms.

(Coma). Show that the arrow and twisted arrow categories of Exercise (Ccon) of Section 1.1 are comma categories.

(Neq). (a) Show that the category $\mathcal{Set}$ of sets and functions is not equivalent to $\mathcal{Set}^{\text{op}}$. (Hint: Find a property of the category for which the dual property is not satisfied.)
(b) Do the same for the category $\mathcal{Ab}$ of abelian groups.

1.3. Natural Transformations

In topology, a homotopy from $f : A \to B$ to $g : A \to B$ is given by a path in B from fx to gx for each element $x \in A$ such that the paths fit together continuously. A natural transformation is analogously a deformation of one *functor* to another.

If $F : \mathcal{C} \to \mathcal{D}$ and $G : \mathcal{C} \to \mathcal{D}$ are two functors, $\lambda : F \to G$ is a **natural transformation** from F to G if λ is a collection of arrows $\lambda C : FC \to GC$, one for each object C of $\mathcal{C}$, such that for each arrow $g : C \to C'$ of $\mathcal{C}$ the following diagram commutes:

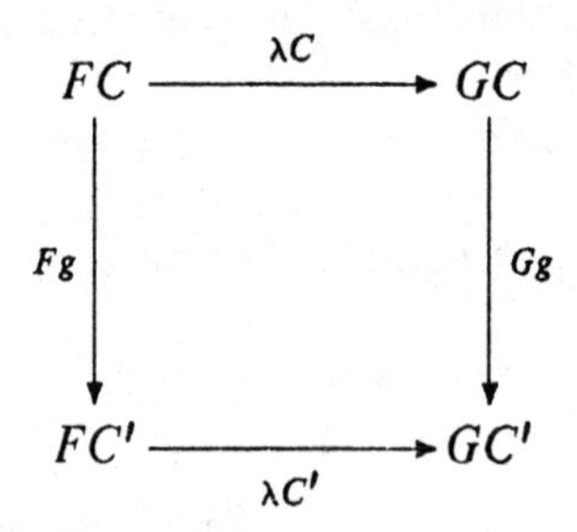

The arrows λC are the **components** of λ.

The natural transformation λ is a **natural equivalence** if each component of λ is an isomorphism in $\mathcal{D}$.

The natural map of a vector space to its double dual is a natural transformation from the identity functor on the category of vector spaces and linear maps to the double dual functor. When restricted to finite dimensional vector spaces, it is a natural equivalence. As another example, let $n > 1$ be a positive integer and let GL_n denote the functor from the category of commutative rings with unity to the category of groups which takes a ring to the group of invertible $n \times n$ matrices with entries from the ring, and let Un denote the group of units functor (which is actually GL_1). Then the determinant map is a natural transformation from GL_n to Un. The Hurewicz transformation from the fundamental group of a topological space to its first homology group is also a natural transformation of functors.

Natural transformations occur throughout this book. If you ever have trouble verifying a statement involving a natural transformation, try using the commutative diagram given in the definition.

Functor categories

If $\mathcal{C}$ and $\mathcal{D}$ are categories, the set $\mathrm{Func}(\mathcal{C}, \mathcal{D})$ of functors from $\mathcal{C}$ to $\mathcal{D}$ is actually a category with natural transformations as arrows. If $\lambda : F \to G$ and $\mu : G \to H$ are natural transformations, their composite $\mu \circ \lambda$ is defined by requiring that its component at C to be $\mu C \circ \lambda C$. Of course, $\mathrm{Func}(\mathcal{C}, \mathcal{D})$ is just $\mathrm{Hom}_{\mathcal{C}at}(\mathcal{C}, \mathcal{D})$, and so is already a functor in each variable to $\mathcal{S}et$. It is easy to check that for any $F : \mathcal{D} \to \mathcal{E}$,

$$\mathrm{Func}(\mathcal{C}, F) : \mathrm{Func}(\mathcal{C}, \mathcal{D}) \to \mathrm{Func}(\mathcal{C}, \mathcal{E})$$

is actually a functor and not only a $\mathcal{S}et$-function, and similarly for $\mathrm{Func}(F, \mathcal{C})$, so that in each variable Func is actually a $\mathcal{C}at$-valued functor.

Since $\mathrm{Func}(\mathcal{C}, \mathcal{D})$ is a category, it has a homfunctor which we denote $\mathrm{Nat}(F, G)$ for functors $F, G : \mathcal{C} \to \mathcal{D}$.

A category of the form $\mathrm{Func}(\mathcal{C}, \mathcal{D})$ is called a **functor category** and is frequently denoted $\mathcal{D}^{\mathcal{C}}$ especially in the later chapters on sheaves.

Notation for natural transformations

Suppose there are categories and functors as shown in this diagram:

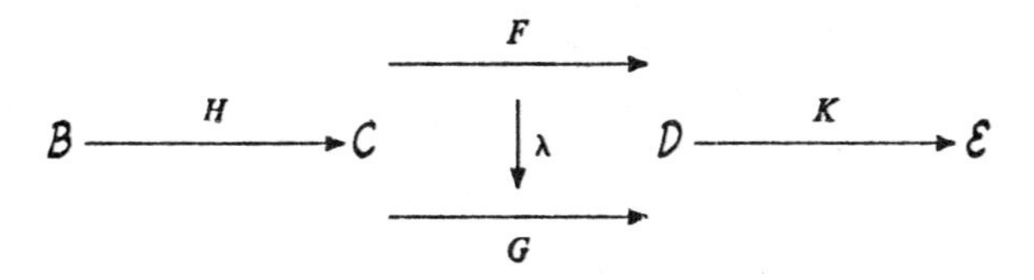

Suppose $\lambda : F \to G$ is a natural transformation. Then λ induces two natural transformations $K\lambda : KF \to KG$ and $\lambda H : FH \to GH$. The component of $K\lambda$ at an object C of $\mathcal{C}$ is

$$K(\lambda C) : KFC \to KGC.$$

$K\lambda$ is a natural transformation simply because K, like any functor, takes commutative diagrams to commutative diagrams. The component of λH at an object B of $\mathcal{B}$ is the component of λ at HB. λH is a natural transformation because H is defined on morphisms.

We should point out that although the notations $K\lambda$ and λH look formally dual, they are quite different in meaning. The first is the result of applying a functor to a value of a natural transformation (which is a morphism in the codomain category) while the second is the result of taking the component of a natural transformation at a value of a functor. Nonetheless, the formal properties of the two quite different operations are the same. This is why we use the parallel notation when many other writers use distinct notation. (Compare the use of $\langle f, v\rangle$ for $f(v)$ by many analysts.) Thus advances mathematics.

Exercise (GOD) below states a number of identities which hold for natural transformations. Some of them are used later in the book, particularly in triple theory.

Exercises 1.3

(NATISO). Show that a natural equivalence from $F : \mathcal{C} \to \mathcal{D}$ to $G : \mathcal{C} \to \mathcal{D}$ is an isomorphism in the functor category $\text{Func}(\mathcal{C}, \mathcal{D})$.

(NTF). Show how to describe a natural transformation as a functor from an arrow category to a functor category.

(NTG). What is a natural transformation from one group homomorphism to another?

(HMNAT)°. Let $R : \mathcal{C} \to \mathcal{D}$ be a functor. Show that $f \to Rf$ is a natural transformation $\text{Hom}_{\mathcal{C}}(C, -) \to \text{Hom}_{\mathcal{D}}(RC, R(-))$ for any object C of $\mathcal{C}$.

(UCU). (a) Show that the inclusion of a set A into the free group FA generated by A determines a natural transformation from the identity functor on $\mathcal{S}et$ to the functor UF where U is the underlying set functor.

(b) Find a natural transformation from $FU : \mathcal{G}rp \to \mathcal{G}rp$ to the identity functor on $\mathcal{G}rp$ which takes a one letter word of FUG to itself. Show that there is only one such.

(SING). In Section 1.2, we mentioned three ways of defining the powerset as a functor. (See Exercise (POW).) For which of these definitions do the maps which take each element x of a set A to the set $\{x\}$ (the "singleton" maps) form a natural transformation from the identity functor to the powerset functor?

(GOD)°. Let categories and functors be given as in the following diagram.

$$\mathcal{B} \underset{G}{\overset{F}{\rightrightarrows}} \mathcal{C} \underset{K}{\overset{H}{\rightrightarrows}} \mathcal{D}$$

Suppose $\kappa : F \to G$ and $\mu : H \to K$ are natural transformations.

(a) Show that this diagram commutes:

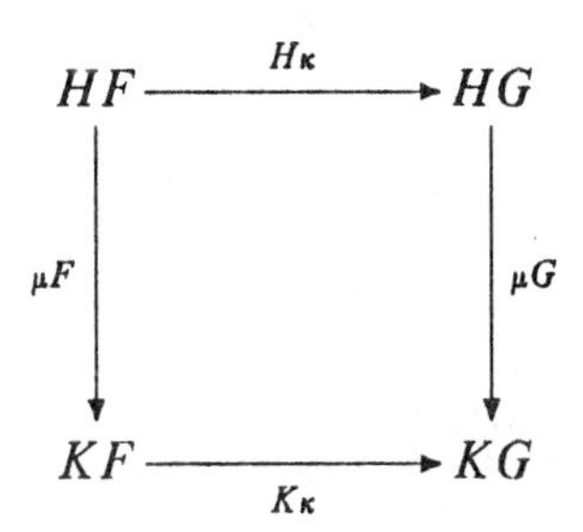

(b) Define $\mu\kappa$ by requiring that its component at B be $\mu GB \circ H\kappa B$, which by (a) is $K\kappa B \circ \mu FB$. Show that $\mu\kappa$ is a natural transformation from $H \circ F$ to $K \circ G$. This defines a composition operation, called **application**, on natural transformations. Although it has the syntax of a composition law, as we will see below, semantically it is the result of *applying* μ to κ. In many, especially older works, it is denoted $\mu * \kappa$, and these books often use juxtaposition to denote composition.

(c) Show that $H\kappa$ and μG have the same interpretation whether thought of as instances of application of a functor to a natural transformation, resp. evaluation of a natural transformation at a functor, or as examples of an application operation where the name of a functor is used to stand for the identity natural transformation. (This exercise may well take longer to understand than to do.)

(d) Show that application as defined above is associative in the sense that if $(\mu\kappa)\beta$ is defined, then so is $\mu(\kappa\beta)$ and they are equal.

(e) Show that the following rules hold, where $\circ$ denotes the composition of natural transformations defined earlier in this chapter. These are called **Godement's rules**. In each case, the meaning of the rule is that if one side is defined, then so is the other and they are equal. They all refer to this diagram, and the name of a functor is used to denote the identity natural transformation from that functor to itself. The other natural transformations are

$$\kappa : F_1 \to F_2,\ \lambda : F_2 \to F_3,\ \mu : G_1 \to G_2, \text{ and } \nu : G_2 \to G_3.$$

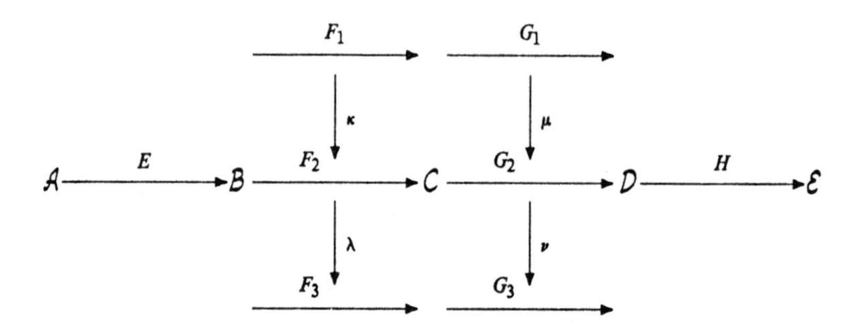

(i) (The interchange law)

$$(\nu \circ \mu)(\lambda \circ \kappa) = (\nu\lambda) \circ (\mu\kappa).$$

(ii) $(H \circ G_1)\kappa = H(G_1\kappa)$.
(iii) $\mu(F_1 \circ E) = (\mu F_1)E$.
(iv) $G_1(\lambda \circ \kappa)E = (G_1\lambda E) \circ (G_1\kappa E)$.
(v) $(\mu F_2) \circ (G_1\kappa) = (G_2\kappa) \circ (\mu F_1)$.

(EQUII)$^\diamond$. Show that two categories $\mathcal{C}$ and $\mathcal{D}$ are equivalent if and only if there are functors $F : \mathcal{C} \to \mathcal{D}$ and $G : \mathcal{D} \to \mathcal{C}$ such that $G \circ F$ is naturally isomorphic to $\mathrm{id}_{\mathcal{C}}$ and $F \circ G$ is naturally isomorphic to $\mathrm{id}_{\mathcal{D}}$.

1.4. Elements and Subobjects

One of the important perceptions of category theory is that an arrow $x : T \to A$ in a category can be regarded as an **element of A defined over T**. The idea is that x is a *variable* element of A, meaning about the same thing as the word "quantity" in such sentences as, "The quantity x^2 is nonnegative," found in older calculus books.

One must not get carried away by this idea and introduce elements everywhere. One of the main benefits of category theory is you don't have to do things in terms of elements unless it is advantageous to. In example (iv) of Section 3.1 is a construction that is almost impossible to understand in terms of elements, but is very easy with the correct conceptual framework. On the other hand, we will see many examples later in which the use of elements leads to a substantial simplification. The point is not to allow a tool to become a straitjacket.

When $x : T \to A$ is thought of as an element of A defined on T, we say that T is the **domain of variation** of the element x. It is often useful to think of x as an element of A defined in terms of a parameter in T.

A related point of view is that x is a set of elements of A indexed by T. By the way, this is distinct from the idea that x is a family of disjoint subsets of T indexed by A, as mentioned in Section 1.1. The latter idea also plays an important role in topos theory, and in fact it is from one point of view the dual of the notion of element.

The notation "$x \in^T A$" is a useful quick way of saying that x is an element of A defined on T. This notation will be extended when we consider subobjects later in this section.

If $x \in^T A$ and $f : A \to B$, then $f \circ x \in^T B$; thus morphisms can be regarded as functions taking elements to elements. The Yoneda Lemma, Theorem 2 of the next section, says (among other things) that any function which takes elements to elements in a coherent way in a sense that will be defined precisely "is" a morphism of the category. Because of this, we will write $f(x)$ for $f \circ x$ when it is helpful to think of x as a generalized element.

Note that every object A has at least one element id_A, its **generic element.**

If A is an object of a category $\mathcal{C}$ and $F : \mathcal{C} \to \mathcal{D}$ is a functor, then F takes any element of A to an element of FA in such a way that (i) generic elements are taken to generic elements, and (ii) the action of F on elements commutes with change of the domain of variation of the element. (If you spell those two conditions out, they are essentially the definition of functor.)

Isomorphisms can be described in terms of elements, too: An arrow $f : A \to B$ is an isomorphism if and only if f (thought of as a function) is a bijection between the elements of A defined on T and the elements of B defined on T for all objects T of $\mathcal{C}$. (To get the inverse, apply this fact to the element $\mathrm{id}_A : A \to A$.) And a terminal object is a singleton in a very strong sense — for any domain of variation it has exactly one element.

In the rest of this section we will develop the idea of element further and use it to define subobjects, which correspond to subsets of a set.

Monomorphisms and epimorphisms

An arrow $f : A \to B$ is a **monomorphism** (or just a "mono", adjective "monic"), if f (i.e., $\mathrm{Hom}(T,f)$) is injective (one to one) on elements defined on each object T — in other words, for every pair x,y of elements of A defined on T, $f(x) = f(y)$ implies $x = y$.

In terms of composition, this says that f is left cancellable, i.e, if $f \circ x = f \circ y$, then $x = y$. This has a dual concept: The arrow f is an **epimorphism** ("epi", "epic") if it is *right* cancellable. This is true if and only if the contravariant functor $\mathrm{Hom}(f, T)$ is injective (not surjective!)

for every object T. Note that surjectivity is not readily described in terms of generalized elements.

In $\mathcal{Set}$, every monic is injective and every epic is surjective (onto). The same is true of $\mathcal{Grp}$, but the fact that epis are surjective in $\mathcal{Grp}$ is moderately hard to prove (Exercise (SIG)). On the other hand, any dense map, surjective or not, is epi in the category of Hausdorff spaces and continuous maps.

An arrow $f : A \to B$ which is "surjective on elements", in other words for which $\mathrm{Hom}(T,f)$ is surjective for every object T, is necessarily an epimorphism and is called a **split epimorphism**. An equivalent definition (Exercise (SPL)) is that there is an arrow $g : B \to A$ which is a right inverse to f, so that $f \circ g = \mathrm{id}_B$. The Axiom of Choice is equivalent to the statement that every epi in $\mathcal{Set}$ is split. In general, in categories of sets with structure and structure preserving functions, split epis are surjective and (as already pointed out) surjective maps are epic (see Exercise (UND), but the converses often do not hold. We have already mentioned Hausdorff spaces as a category in which there are nonsurjective epimorphisms; another example is the embedding of the ring of integers in the field of rational numbers in the category of rings and ring homomorphisms. As for the other converse, in the category of groups the (unique) surjective homomorphism from the cyclic group of order 4 to the cyclic group of order 2 is an epimorphism which is not split.

An arrow with a *left* inverse is necessarily a monomorphism and is called a **split monomorphism**. Split monos in $\mathcal{Top}$ are called retractions; in fact the word "retraction" is sometimes used to denote a split mono in any category.

The property of being a split mono or split epi is necessarily preserved by any functor. The property of being monic or epic is certainly not in general preserved by any functor. Indeed, if Ff is epi for every functor F, then f is necessarily a split epi. (Exercise (GLEP).)

Notation: In diagrams, we usually draw an arrow with an arrowhead at its tail:

$$\rightarrowtail,$$

to indicate that it is a monomorphism. The usual dual notation for an epimorphism is

$$\twoheadrightarrow .$$

However in this book we reserve that latter notation for *regular* epimorphisms to be defined in Exercise (REGMON) of Section 1.7.

Subobjects

We now define the notion of subobject of an object in a category; this idea partly captures and partly generalizes the concept of "subset", "subspace", and so on familiar in many branches of mathematics.

If $i : A_0 \to A$ is a monomorphism and $a : T \to A$, we say a **factors through** i (or factors through A_0 if it is clear which monomorphism i is meant) if there is an arrow j for which

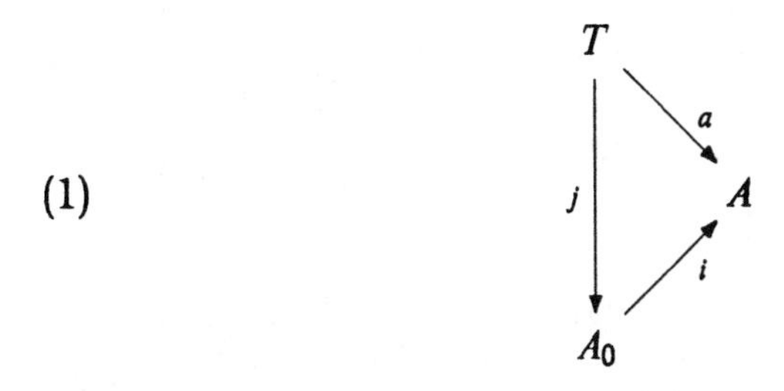

commutes. In this situation we extend the element point of view and say that the element a of A is an element of A_0 (or of i if necessary). This is written "$a \in_A^T A_0$". The subscript A is often omitted if the context makes it clear.

Lemma 1. *Let $i : A_0 \to A$ and $i' : A_0' \to A$ be monomorphisms in a category $\mathcal{C}$. Then A_0 and A_0' have the same elements of A if and only if they are isomorphic in the category $\mathcal{C}/A$ of objects over A, in other words if and only if there is an isomorphism $j : A \to A'$ for which*

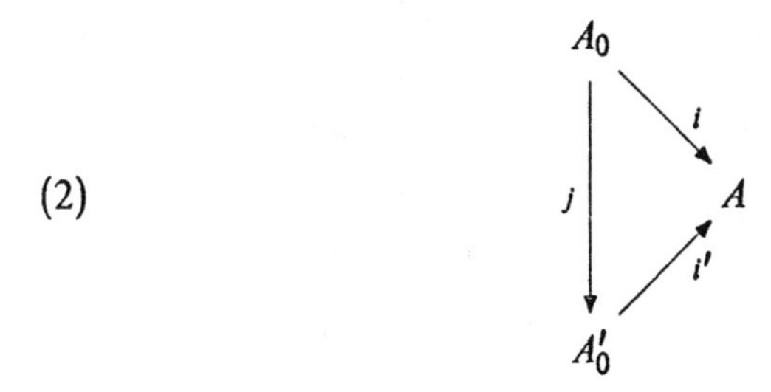

commutes.

Proof. Suppose A_0 and A_0' have the same elements of A. Since $i \in_A^{A_0} A_0$, it factors through A_0', so there is an arrow $j : A_0 \to A_0'$ such that (2) commutes. Interchanging A_0 and A_0' we get $k : A_0' \to A_0$ such that $i \circ k = i'$. Using the fact that i and i' are monic, it is easy to see that j and k must be inverses to each other, so they are isomorphisms.

Conversely, if j is an isomorphism making (2) commute and $a \in_A^T A_0$, so that $a = i \circ u$ for some $u : T \to A_0$, then $a = i' \circ j \circ u$ so that $a \in_A^T A_0'$. A similar argument interchanging A_0 and A_0' shows that A_0 and A_0' have the same elements of A.

Two monomorphisms which have the same elements are said to be **equivalent**. A **subobject** of A is an equivalence class of monomorphisms into A. We will frequently refer to a subobject by naming one of its members, as in "Let $A_0 \rightarrowtail A$ be a subobject of A".

In *Set* , each subobject of a set A contains exactly one inclusion of a subset into A, and the subobject consists of those injective maps into A which has that subset as image. Thus "subobject" captures the notion of "subset" in *Set* exactly.

Any map from a terminal object in a category is a monomorphism and so determines a subobject of its target. Because any two terminal objects are isomorphic *by a unique isomorphism* (Exercise (TISO) of Section 1.1), that subobject contains exactly one map based on each terminal object. We will henceforth assume that in any category we deal with, we have picked a particular terminal object (if it has one) as the canonical one and call it "*the* terminal object".

Global elements

In the category of sets, an element in the ordinary sense of a set B is essentially the same thing as an arrow from the terminal object of *Set* to B. In general, an arrow in some category from the terminal object to some object is called a **global element** of that object, for reasons which will become apparent in the next paragraph. In most categories which arise in practice, *except Set* , an object is not determined by its global elements. For example, in *Grp* , each group has exactly one global element.

A more interesting example arises in connection with continuous functions. This example is worth studying in detail because it illustrates and motivates much of sheaf theory. Let A be a topological space and let $\mathbb{R}$ denote the set of real numbers. Let $\mathcal{O}(A)$ denote the category whose objects are the open sets of A and whose arrows are the inclusion maps of one open set into another. Let $C : \mathcal{O}(A)^{\mathrm{op}} \to Set$ denote the contravariant functor which takes each open set U to the set of real-valued continuous functions defined on U, and to each inclusion of an open set U of A into an open set V associates the map from $C(V)$ to $C(U)$ which restricts a continuous function defined on V to U. An important point about these restriction maps is that they are not in general surjective — that is, there are in general functions defined on an open set which cannot be extended to a bigger open set. Think of $f(x) = 1/x$, for example.

This functor C is an object in the category $\mathcal{F} = \mathrm{Func}(\mathcal{O}(A)^{\mathrm{op}}, Set)$. The terminal object of $\mathcal{F}$ is the functor which associates a singleton set to each open set of A and the only possible map to each arrow (inclusion map) of $\mathcal{O}(A)$. It is a nice exercise to prove that a global element of C

is precisely a continuous real-valued function defined on all of A. This example is explored further in Exercise (RSHF).

Exercises 1.4

(IEL). Describe initial objects using the terminology of elements, and using the terminology of indexed families of subsets.

(ISS). (a) Show that in *Set*, a function if injective if and only if it is a monomorphism and surjective if and only if it is an epimorphism.

(b) Show that every epimorphism in *Set* is split. (This is the Axiom of Choice).

(ISA). (a) Show that in the category of Abelian groups and group homomorphisms, a homomorphism is injective if and only if it is a monomorphism and surjective if and only if it is an epimorphism.

(b) Show that neither monos nor epis are necessarily split in the category of Abelian groups.

(SIG). Show that in *Grp*, every homomorphism is injective if and only if it is a monomorphism and surjective if and only if it is an epimorphism. (If you get stuck trying to show that an epimorphism in *Grp* is surjective, see the hint on p.21 of Mac Lane [1971].)

(SURTOP). Show that all epimorphisms are surjective in *Top*, but not in the category of all Hausdorff spaces and continuous maps.

(SURRING). Show that the embedding of an integral domain (assumed commutative with unity) into its field of quotients is an epimorphism in the category of commutative rings and ring homomorphisms. When is it a split epimorphism?

(SPL). (a) Show that the following two statements about an arrow $f : A \to B$ in a category $\mathcal{C}$ are equivalent:

(i) $\mathrm{Hom}(T,f)$ is surjective for every object T of $\mathcal{C}$.

(ii) There is an arrow $g : B \to A$ such that $f \circ g = \mathrm{id}_B$.

(b) Show that any arrow satisfying the conditions of (a) is an epimorphism.

(GLEP)°. Show that if Ff is epi for every functor F, then f is a split epi.

(UND). Let $U : \mathcal{C} \to \mathit{Set}$ be a faithful functor and f an arrow of $\mathcal{C}$. (Note that the functors we have called "forgetful" — we have not defined that word formally — are obviously faithful.) Prove:

(a) If Uf is surjective then f is an epimorphism.

(b) If f is a split epimorphism then Uf is surjective.

(c) If Uf is injective then f is a monomorphism.
(d) If f is a split monomorphism, then Uf is injective.

(SUBF)°. A **subfunctor** of a functor $F : \mathcal{C} \to \mathbf{Set}$ is a functor G with the properties

(i) $GA \subseteq FA$ for every object A of $\mathcal{C}$.
(ii) If $f : A \to B$, then $Gf(GA) \subseteq GB$.

Show that the subfunctors of a functor are the "same" as subobjects of the functor in the category $\text{Func}(\mathcal{C}, \mathbf{Set})$.

1.5. The Yoneda Lemma

A functor $F : \mathcal{C} \to \mathbf{Set}$ is an object in the functor category $\text{Func}(\mathcal{C}, \mathbf{Set})$: an "element" of F is therefore a natural transformation into F. The Yoneda Lemma, Lemma 1 below, says in effect that the elements of a $\mathbf{Set}$-valued functor F defined (in the sense of Section 1.4) on the hom-functor $\text{Hom}(A, -)$ for some object A of $\mathcal{C}$ are essentially the same as the (ordinary) elements of the set FA. To state this properly requires a bit of machinery.

If $f : A \to B$ in $\mathcal{C}$, then f induces a natural transformation from $\text{Hom}(B, -)$ to $\text{Hom}(A, -)$ by composition: the component of this natural transformation at an object C of $\mathcal{C}$ takes an arrow $h : B \to C$ to $h \circ f : A \to C$. This construction defines a contravariant functor from $\mathcal{C}$ to $\text{Func}(\mathcal{C}, \mathbf{Set})$ called the **Yoneda map**. It is straightfoward and very much worthwhile to check that this construction really does give a natural transformation for each arrow f and that the resulting Yoneda map really is a functor.

Because $\text{Nat}(-, -)$ is contravariant in the first variable (it is a special case of Hom), the map which takes an object B of $\mathcal{C}$ and a functor $F : \mathcal{C} \to \mathbf{Set}$ to $\text{Nat}(\text{Hom}(B, -), F)$ is a functor from $\mathcal{C} \times \text{Func}(\mathcal{C}, \mathbf{Set})$ to $\mathbf{Set}$. Another such functor is the evaluation functor which takes (B, F) to FB, and (g, λ), where $g : B \to A \in \mathcal{C}$ and $\lambda : F \to G$ is a natural transformation, to $Gg \circ \lambda B$. Remarkably, these two functors are naturally isomorphic; it is in this sense that the elements of F defined on $\text{Hom}(B, -)$ are the ordinary elements of FB.

Lemma 1 (Yoneda). *The map* $\varphi : \text{Nat}(\text{Hom}(B, -), F) \to FB$ *defined by* $\varphi(\lambda) = \lambda B(\text{id}_B)$ *is a natural isomorphism of the functors defined in the preceding paragraph.*

Proof. The inverse of φ takes an element u of FB to the natural transformation λ defined by requiring that $\lambda A(g) = Fg(u)$ for $g \in$

$\mathrm{Hom}(B, A)$. The rest of proof is a routine verification of the commutativity of various diagrams required by the definitions.

The first of several important consequences of this lemma is the following embedding theorem, which in one way or another is used in practically every mathematical argument in this book. This theorem is obtained by taking F in the Lemma to be $\mathrm{Hom}(A, -)$, where A is an object of $\mathcal{C}$; this results in the statement that there is a natural bijection between arrows $g : A \to B$ and natural transformations from $\mathrm{Hom}(B, -)$ to $\mathrm{Hom}(A, -)$.

Theorem 2 (Yoneda Embeddings). (i) *The map which takes $f : A \to B$ to the induced natural transformation*

$$\mathrm{Hom}(B, -) \to \mathrm{Hom}(A, -)$$

is a full and faithful contravariant functor from $\mathcal{C}$ to $\mathrm{Func}(\mathcal{C}, \mathcal{S}et)$.

(ii) *The map taking f to the natural transformation*

$$\mathrm{Hom}(-, A) \to \mathrm{Hom}(-, B)$$

is a full and faithful functor from $\mathcal{C}$ to $\mathrm{Func}(\mathcal{C}^{\mathrm{op}}, \mathcal{S}et)$.

Proof. It is easy to verify that the maps defined in the Theorem are functors. The fact that the first one is full and faithful follows from the Yoneda Lemma with $\mathrm{Hom}(A, -)$ in place of F. The other proof is dual.

The induced maps in the Theorem deserve to be spelled out. If $f : S \to T$, the natural transformation corresponding to f given by (i) has component $\mathrm{Hom}(f, A) : \mathrm{Hom}(T, A) \to \mathrm{Hom}(S, A)$ at an object A of $\mathcal{C}$ — this is composing by f on the right. If $x \in^T A$, the action of $\mathrm{Hom}(f, A)$ "changes the parameter" in A along f.

The other natural transformation corresponding to f is $\mathrm{Hom}(T, f) : \mathrm{Hom}(T, A) \to \mathrm{Hom}(T, B)$; since the Yoneda embedding is faithful, we can say that f is essentially the same as $\mathrm{Hom}(-, f)$. If x is an element of A based on T, then $\mathrm{Hom}(T, f)(x) = f \circ x$. Since "$f$ is essentially the same as $\mathrm{Hom}(-, f)$", this justifies the notation $f(x)$ for $f \circ x$ introduced in Section 1.4.

The fact that the Yoneda embedding is full means that *any natural transformation* $\mathrm{Hom}(-, A) \to \mathrm{Hom}(-, B)$ determines a morphism $f : A \to B$, namely the image of id_A under the component of the transformation at A. Spelled out, this says that if f is any function which assigns to every element $x : T \to A$ an element $f(x) : T \to B$ with the property that for all $T : S \to T$, $f(x \circ t) = f(x) \circ t$ (this is the "coherence condition" mentioned in Section 1.4) then f "is" (via the Yoneda embedding) a morphism, also called f to conform to our conventions, from A to B. One says such an arrow exists "by Yoneda".

In the same vein, if $g : 1 \to A$ is a morphism of $\mathcal{C}$, then for any object T, g determines an element $g(\ )$ of A defined on T by composition with the unique element from T to 1, which we denote $(\ )$. This notation captures the perception that a global element depends on no arguments. We will extend the functional notation to more than one variable in Section 1.7.

Universal elements

Another special case of the Yoneda Lemma occurs when one of the elements of F defined on $\mathrm{Hom}(A, -)$ is a natural isomorphism. If $\beta : \mathrm{Hom}(A, -) \to F$ is such a natural isomorphism, the (ordinary) element $u \in FA$ corresponding to it is called a **universal element** for F, and F is called a **representable functor**, represented by A. It is not hard to see that if F is also represented by A', then A and A' are isomorphic objects of $\mathcal{C}$. (See Exercise (UNIQ), which actually says more than that.)

The following lemma gives a characterization of universal elements which in many books is given as the definition.

Lemma 3. *Let $F : \mathcal{C} \to \mathcal{S}et$ be a functor. Then $u \in FA$ is a universal element for F if and only if for every object B of $\mathcal{C}$ and every element $t \in FB$ there is exactly one arrow $g : A \to B$ such that $Fg(u) = t$.*

Proof. If u is such a universal element corresponding to a natural isomorphism $\beta : \mathrm{Hom}(A, -) \to F$, and $t \in FB$, then the required arrow g is the element $(\beta^{-1}B)(t)$ in $\mathrm{Hom}(A, B)$. Conversely, if $u \in FA$ satisfies the conclusion of the Lemma, then it corresponds to some natural transformation $\beta : \mathrm{Hom}(A, -) \to F$ by the Yoneda Lemma. It is routine to verify that the map which takes $t \in FB$ to the arrow $g \in \mathrm{Hom}(A, B)$ given by the assumption constitutes an inverse in $\mathrm{Func}(\mathcal{C}, \mathcal{S}et)$ to βB.

In this book, the phrase "$u \in FA$ is a universal element for F" carries with it the implication that u and A have the property of the lemma. (It is possible that u is also an element of FB for some object B but not a universal element in FB).

As an example, let G be a free group on one generator g. Then g is the "universal group element" in the sense that it is a universal element for the underlying set functor $U : \mathcal{G}rp \to \mathcal{S}et$ (more precisely, it is a universal element in UG). This translates into the statement that for any element x in any group H there is a unique group homomorphism

$F : G \to H$ taking g to x, which is exactly the definition of "free group on one generator g".

Another example which will play an important role in this book concerns the contravariant powerset functor $\mathbf{P} : \mathcal{S}et \to \mathcal{S}et$ defined in Section 1.2. It is straightforward to verify that a universal element for $\mathbf{P}$ is the subset $\{1\}$ of the set $\{0,1\}$; the function required by the Lemma for a subset B_0 of a set B is the characteristic function of B_0. (A universal element for a contravariant functor, as here — meaning a universal element for $\mathbf{P} : \mathcal{S}et^{\mathrm{op}} \to \mathcal{S}et$ —is often called a "couniversal element").

Exercises 1.5

(UNIV). Find a universal element for the functor

$$\mathrm{Hom}(-, A) \times \mathrm{Hom}(-, B) : \mathcal{S}et^{\mathrm{op}} \to \mathcal{S}et,$$

for any two sets A and B. (If $h : U \to V$, this functor takes a pair (f, g) to $(h \circ f, h \circ g)$.)

(GPA). (a) Show that an action of a group G on a set A is essentially the same thing as a functor from G regarded as a category to $\mathcal{S}et$.

(b) Show that such an action has a universal element if and only if for any pair x and y of elements of A there is exactly one element g of G for which $gx = y$.

(UPOW). Are either of the covariant powerset functors defined in Exercise (POW) of Section 1.2 representable?

(UNIQ)°. Let $F : \mathcal{C} \to \mathcal{S}et$ be a functor and $u \in FA$, $u' \in FA'$ be universal elements for F. Show that there is a unique isomorphism $\varphi : A \to A'$ such that $F\varphi(u) = u'$.

(FRGP). Let $U : \mathcal{G}rp \to \mathcal{S}et$ be the underlying set functor, and $F : \mathcal{S}et \to \mathcal{G}rp$ the functor which takes a set A to the free group on A. Show that for any set A, the covariant functor $\mathrm{Hom}_{\mathcal{S}et}(A, U(-))$ is represented by FA, and for any group G, the contravariant functor $\mathrm{Hom}_{\mathcal{G}rp}(F(-), G)$ is represented by UG.

1.6. Pullbacks

The set P of composable pairs of arrows used in Section 1.1 in the alternate definition of category is an example of a "fibered product" or "pullback". A pullback is a special case of "limit", which we treat in Section 1.7. In this section, we discuss pullbacks in detail and use them to define the subobject functor, which will be central to our definition of topos.

Let us consider the following diagram D in a category $\mathcal{C}$.

(1)

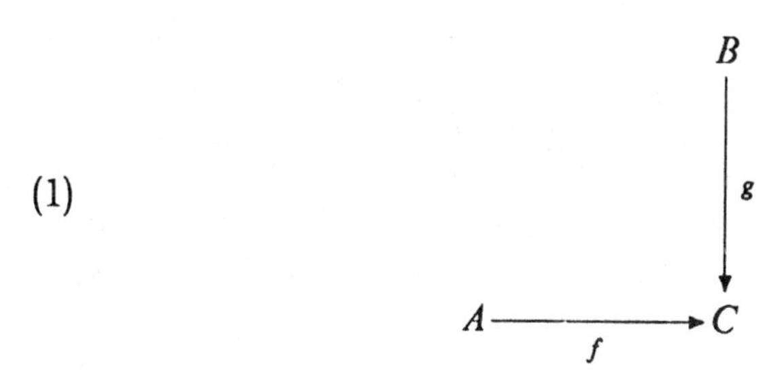

We would like to *objectify* the set $\{(x, y) \mid f(x) = g(y)\}$ in $\mathcal{C}$; that is, find an object of $\mathcal{C}$ whose elements are those pairs (x, y) with $f(x) = g(y)$. Observe that for a pair (x, y) to be in this set, x and y must be elements of A and B respectively defined over the same object T.

The set of composable pairs of arrows in a category (see Section 1.1) are a special case in $\mathcal{S}et$ of this, with $A = B$ being the set of arrows and $f = d^0$, $g = d^1$.

Thus we must consider commutative diagrams like

(2)

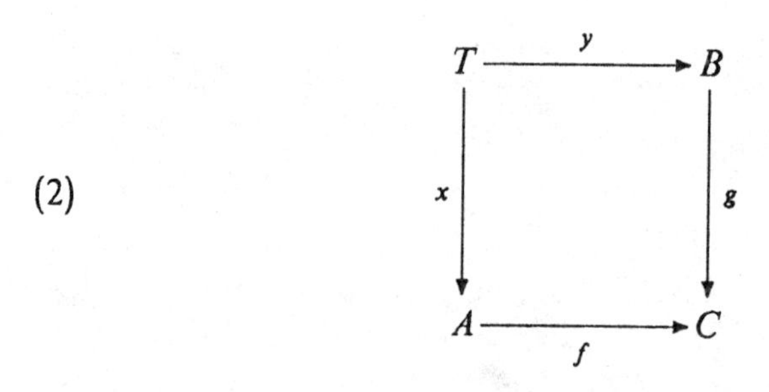

In this situation, (T, x, y) is called a **commutative cone over D based on T**, and the set of commutative cones over D based on T is denoted $\text{Cone}(T, D)$. A commutative cone based on T over D may usefully be regarded as an element of D defined on T. In Section 1.7, we will see that a commutative cone is actually an arrow in a certain category, so that this idea fits with our usage of the word "element".

Our strategy will be to turn $\text{Cone}(-, D)$ into a functor; then we will say that an object represents (in an informal sense) elements of D, in other words pairs (x, y) for which $f(x) = g(y)$, if that object represents (in the precise technical sense) the functor $\text{Cone}(-, D)$.

We make will make $\text{Cone}(-, D)$ into a contravariant functor to $\mathcal{S}et$: If $h : W \to T$ is an arrow of $\mathcal{C}$ and (T, x, y) is a commutative cone over (1), then

$$\text{Cone}(h, D)(T, x, y) = (W, x \circ h, y \circ h),$$

which it is easy to see is a commutative cone over D based on W.

An element (P, p_1, p_2) of D which is a universal element for Cone$(-, D)$ (so that Cone$(-, D)$ is representable) is called the **pullback** or the **fiber product** of the diagram D. The object P is often called the pullback, with p_1 and p_2 understood. As the reader can verify, this says that (P, p_1, p_2) is a pullback if

(3)

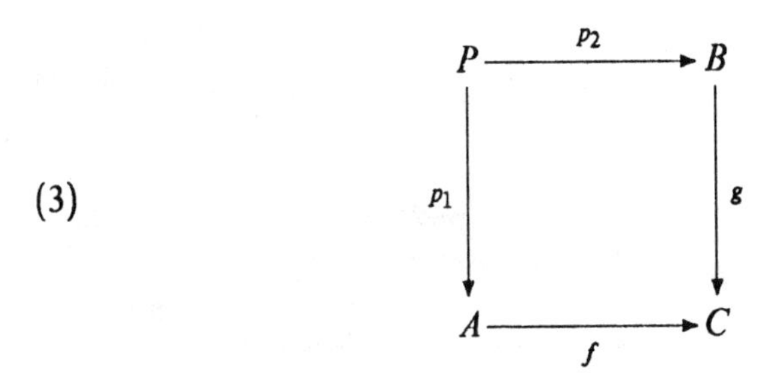

commutes and for any element of D based on T, there is a unique element of P based on T which makes

(4)

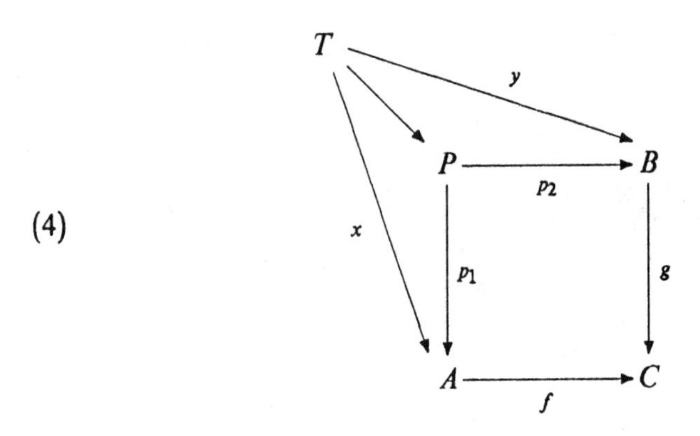

commute. Thus there is a bijection between the elements of the diagram D defined on T and the elements of the fiber product P defined on T. When a diagram like (3) has this property it is called a **pullback diagram.**

The Cone functor exists for any category, but a particular diagram of the form (1) need not have a pullback.

Proposition 1. *If diagram* (3) *is a pullback diagram, then the cone in diagram* (2) *is also a pullback of diagram* (1) *if and only if the unique arrow from T to P making everything in diagram* (4) *commute is an isomorphism.*

Proof. (This theorem actually follows from Exercise (Uniq) of Section 1.5, but we believe a direct proof is instructive.) Assume that (2) and (3) are both pullback diagrams. Let $u : T \to P$ be the unique arrow given because (3) is a pullback diagram, and let $v : P \to T$ be the unique arrow given because (2) is a pullback diagram. Then both for $g = u \circ v : P \to P$ and $g = \mathrm{id}_P$ it is true that $p_1 \circ g = p_1$ and $p_2 \circ g = p_2$. Therefore by the uniqueness part of the definition of universal element, $u \circ v = \mathrm{id}_P$. Similarly, $v \circ u = \mathrm{id}_T$, so that u is an isomorphism between T and P making everything commute. The converse is easy.

The preceding argument is typical of many arguments making use of the uniqueness part of the definition of universal element. We will usually leave arguments like this to the reader.

A consequence of Proposition 1 is that a pullback of a diagram in a category is not determined uniquely but only up to a "unique isomorphism which makes everything commute". This is an instance of a general fact about constructions defined as universal elements which is made precise in Proposition 1 of Section 1.7.

Notation for pullbacks

We have defined the pullback P of diagram (1) so that it objectifies the set $\{(x, y) \mid f(x) = g(y)\}$. This fits nicely with the situation in $\mathcal{Set}$, where one pullback of (1) *is* the set $\{(x, y) \mid f(x) = g(y)\}$ together with the projection maps to A and B, and any other pullback is in one to one correspondence with this one by a bijection which commutes with the projections. This suggest the introduction of a setlike notation for pullbacks: We let $[(x, y) \mid f(x) = g(y)]$ denote a pullback of (1). In this notation, $f(x)$ denotes $f \circ x$ and $g(y)$ denotes $g \circ y$ as in Section 1.4, and (x, y) denotes the unique element of P defined on T which exists by definition of pullback. It follows that $p_1(x, y) = x$ and $p_2(x, y) = y$, where we write $p_1(x, y)$ (not $p_1((x, y))$) for $p_1 \circ (x, y)$.

The idea is that square brackets around a set definition denotes an object of the category which represents the set of arrows listed in curly brackets—"represents" in the technical sense, so that the set in curly brackets has to be turned into the object map of a set-valued functor. The square bracket notation is ambiguous. Proposition 1 spells out the ambiguity precisely.

We could have defined a commutative cone over (1) in terms of *three* arrows, namely a cone (T, x, y, z) based on T would have $x : T \to A$, $y : T \to B$ and $z : T \to C$ such that $f \circ x = g \circ y = z$. Of course, z is redundant and in consequence the Cone functor defined this way would be naturally isomorphic to the Cone functor defined above, and so would have the same universal elements. (The component of the natural isomorphism at T takes (T, x, y) to $(T, x, y, f \circ x)$). Thus the pullback of (1) also represents the set $\{(x, y, z) \mid f(x) = g(y) = z\}$, and so could be denoted $[(x, y, z) \mid f(x) = g(y) = z]$. Although this observation is inconsequential here, it will become more significant when we discuss more general constructions (limits) defined by cones.

There is another way to construct a pullback in *Set* when the map g is monic. In general, when g is monic, $\{(x, y) \mid f(x) = g(y)\} \cong \{x \mid f(x) \in g(B)\}$, which in *Set* is often denoted $f^{-1}(B)$. In general, a pullback along a subobject can be interpreted as an inverse image which as we will see is again a subobject.

The pullback diagram (3) is often regarded as a sort of generalized inverse image construction even when g is not monic. In this case, it is called the "pullback of g along f". Thus when P is regarded as the fiber product, the notion of pullback is symmetrical in A and B, but when it is regarded as the generalized inverse image of B then the diagram is thought of as asymmetrical.

A common notation for the pullback of (1) reflecting the perception of a pullback as fiber product is "$A \times_C B$".

The subobject functor

In this section, we will turn the subobject construction into a contravariant functor, by using the inverse image construction described above. To do this, we need to know first that the inverse image of a monomorphism is a monomorphism:

Lemma 2. *In any category* $\mathcal{C}$, *in a pullback diagram* (3), *if* g *is monic then so is* p_1.

Proof. Consider the diagram below, in which the square is a pullback.

(5)

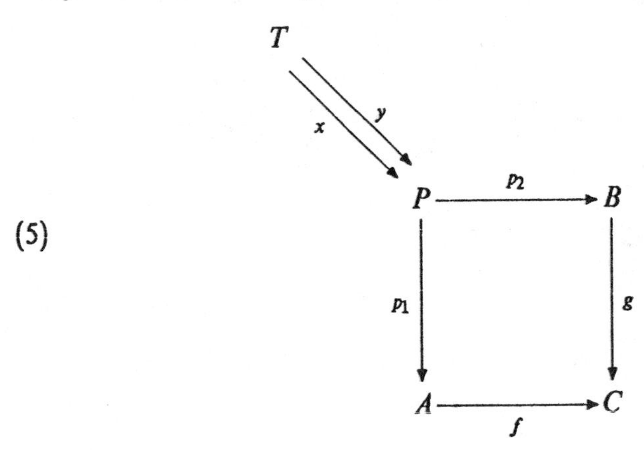

Then $P = [(x, y) \mid fx = gy]$. To show that p_1 is monic is the same as showing that if $(x, y) \in^T P$ and $(x, y') \in^T P$ then $y = y'$. But if (x, y)

and (x, y') are in P, then $g(y) = f(x) = g(y')$, so $y = y'$ since g is monic.

To turn the subobject construction into a functor, we need more than that the pullback of monics is monic. We must know that the pullback of a subobject is a well-defined subobject. In more detail, for A in $\mathcal{C}$, Sub A will be the set of subobjects of A. If $f : B \to A$, then for a subobject represented by a monic $g : U \to A$, $\mathrm{Sub}(f)(g)$ will be the pullback of g along f. To check that $\mathrm{Sub}(f)$ is well-defined, we need:

Theorem 3. *If $g : U \rightarrowtail A$ and $h : V \rightarrowtail A$ determine the same subobject, then the pullbacks of g and h along $f : B \to A$ represent the same subobjects of B.*

Proof. This follows because the pullback of g is $[y \mid f(y) \in_A^P U]$ and the pullback of h is $[y \mid f(y) \in_A^P V]$, which has to be the same since by definition a subobject is entirely determined by its elements.

The verification that $\mathrm{Sub}(f)$ is a functor is straightforward and is omitted.

Exercises 1.6

(Gp). Show how to describe the kernel of a group homomorphism $f : G \to H$ as the pullback of f along the map which takes the trivial group to the identity of H.

(Ep). Give an example of a pullback of an epimorphism which is not an epimorphism.

(Pbm)°. Prove that an arrow $f : A \to B$ is monic if and only if the diagram

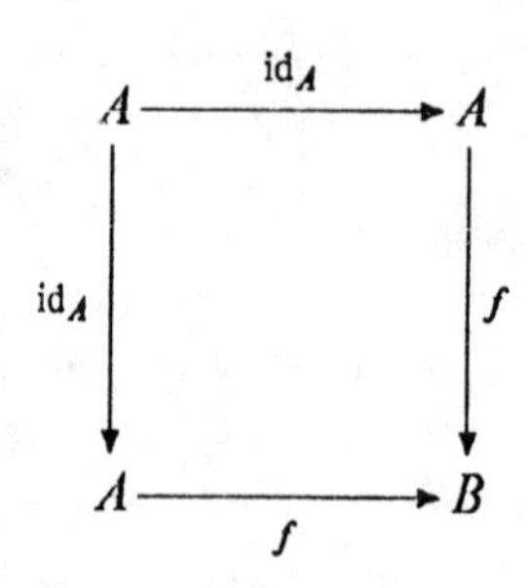

is a pullback.

(PBS). (a) Suppose that

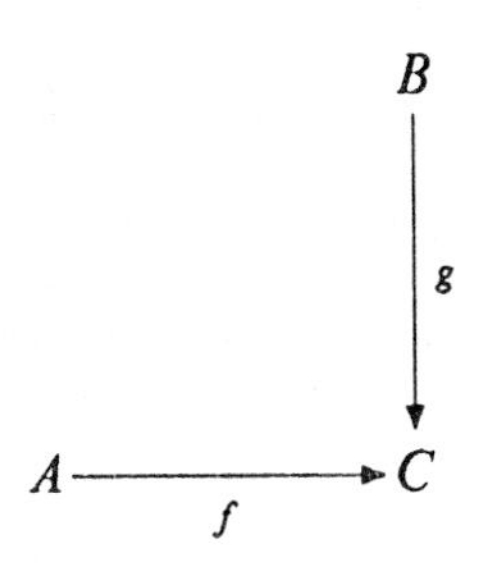

is a diagram in *Set* with g an inclusion. Construct a pullback of the diagram as a fiber product and as an inverse image of A along f, and describe the canonical isomorphism between them.

(b) Suppose that g is injective, but not necessarily an inclusion. Find two ways of constructing the pullback in this case, and find the isomorphism between them.

(c) Suppose f and g are both injective. Construct the pullback of diagram (2) in four different ways: (i) fiber product, (ii) inverse image of the image of g along f, (iii) inverse image of the image of f along g, (iv) and the intersection of the images of f and g. Find all the canonical isomorphisms.

(d) Investigate which of the constructions in (c) coincide when one or both of f and g are inclusions.

(INVIM). When g is monic in diagram (1), redefine "Cone" so that

(a) $\mathrm{Cone}(T, D) = \{(x, z) \mid z \in B \text{ and } f(x) = z\}$, or equivalently
(b) $\mathrm{Cone}(T, D) = \{x \mid f(x) \in B\}$.

Show that each definition gives a functor naturally isomorphic to the Cone functor originally defined.

(PPOS). Identify pullbacks in a poset regarded as a category. Apply this to the powerset of a set, ordered by inclusion.

(LAT)°. For two subobjects $g : U \to A$ and $h : V \to A$, say that $U \leq V$ (or $g \leq h$) if g factors through h. Show that this makes the set of subobjects of A a partially ordered set with a maximum element.

1.7. Limits

Graphs

A limit is the categorical way of defining an object by means of equations between elements of given objects. The concept of pullback as described in Section 1.6 is a special case of limit, but sufficiently complicated to be characteristic of the general idea. To give the general definition, we need a special notion of "graph". What we call a graph here is what a graph theorist would probably call a "directed multigraph with loops".

Formally, a **graph** $\mathcal{G}$ consists of two sets, a set O of **objects** and a set A of **arrows**, and two functions $d^0, d^1 : A \to O$. Thus a graph is a "category without composition" and we will use some of the same terminology as for categories: O is the set of **objects** (or sometimes **nodes**) and A is the set of **arrows** of the graph; if f is an arrow, $d^0(f)$ is the **source** of f and $d^1(f)$ is the **target** of f.

A **homomorphism** $F : \mathcal{G} \to \mathcal{H}$ from a graph $\mathcal{G}$ to a graph $\mathcal{H}$ is a function taking objects to objects and arrows to arrows and preserving source and target; in other words, if $f : A \to B$ in $\mathcal{G}$, then $F(f) : F(A) \to F(B)$ in $\mathcal{H}$.

It is clear that every category $\mathcal{C}$ has an **underlying graph** which we denote $|\mathcal{C}|$; the objects, arrows, source and target maps of $|\mathcal{C}|$ are just those of $\mathcal{C}$. Moreover, any functor $F : \mathcal{C} \to \mathcal{D}$ induces a graph homomorphism $|F| : |\mathcal{C}| \to |\mathcal{D}|$. It is easy to see that this gives an **underlying graph functor** from the category of categories and functors to the category of graphs and homomorphisms.

A **diagram** in a category $\mathcal{C}$ (or in a graph $\mathcal{G}$—the definition is the same) is a graph homomorphism $D : \mathcal{I} \to |\mathcal{C}|$ for some graph $\mathcal{I}$. $\mathcal{I}$ is the **index graph** of the diagram. Such a diagram is called a **diagram of type** $\mathcal{I}$. For example, a diagram of the form of (1) of Section 1.6 (which we used to define pullbacks) is a diagram of type $\mathcal{I}$ where $\mathcal{I}$ is the graph

$$1 \to 2 \leftarrow 3.$$

D is called a **finite diagram** if the index category has only a finite number of nodes and arrows.

We will write $D : \mathcal{I} \to \mathcal{C}$ instead of $D : \mathcal{I} \to |\mathcal{C}|$; this conforms to standard notation.

Observe that any object A of $\mathcal{C}$ is the image of a constant graph homomorphism $K : \mathcal{I} \to \mathcal{C}$ and so can be regarded as a degenerate diagram of type I.

If D and E are two diagrams of type $\mathcal{I}$ in a category $\mathcal{C}$, a **natural transformation** $\lambda : D \to E$ is defined in exactly the same way as a natural transformation of functors (which does not involve the composition of arrows in the domain category anyway); namely, λ is a family of arrows

$$\lambda i : D(i) \to E(i)$$

of $\mathcal{C}$, one for each object i of $\mathcal{I}$, for which

(1)

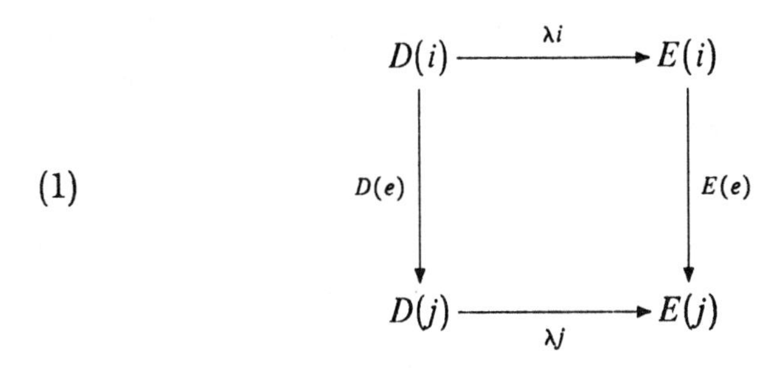

commutes for each arrow $e : i \to j$ of $\mathcal{I}$.

A **commutative cone** with vertex W over a diagram $D : \mathcal{I} \to \mathcal{C}$ is a natural transformation α from the constant functor with value W on $\mathcal{I}$ to D. We will refer to it as the "cone $\alpha : W \to D$". This amounts to giving a compatible family $\{\alpha i\}$ of elements of the vertices $D(i)$ based on W. This commutative cone α is an element (in the category of diagrams of type $\mathcal{I}$) of the diagram D based on the constant diagram W. The individual elements αi (elements in $\mathcal{C}$) are called the **components** of the element α.

Thus to specify a commutative cone with vertex W, one must give for each object i of $\mathcal{I}$ an element αi of $D(i)$ based on W (that is what makes it a cone) in such a way that if $e : i \to j$ is an arrow of $\mathcal{I}$, then $D(e)(\alpha i) = \alpha j$ (that makes it commutative). This says that the following diagram must commute for all $e : i \to j$.

(2)

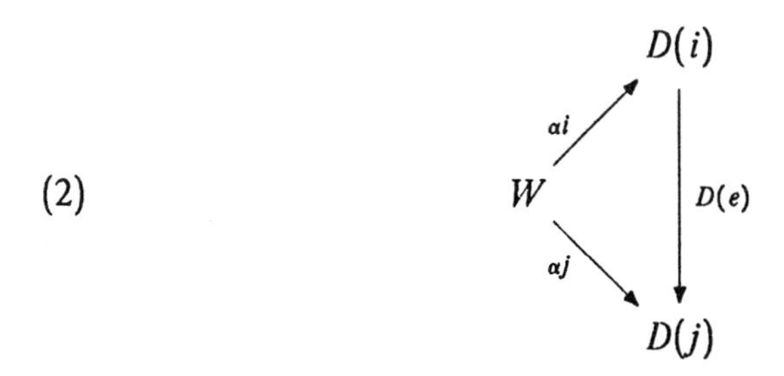

Note that in Section 1.6 our definition of commutative cone for pullbacks does not fit our present definition, since we give no arrow to C in diagram (2). Of course, this is only a technicality, since there is an *implied* arrow to C which makes it a commutative cone. This is why we gave an alternative, but equivalent construction in terms of three arrows in Section 1.6.

Just as in the case of pullbacks, an arrow $W' \to W$ defines a commutative cone over D with vertex W' by composition, thus making $\text{Cone}(-, D) : \mathcal{C} \to \mathcal{S}et$ a contravariant functor. ($\text{Cone}(W, D)$ is the set of commutative cones with vertex W.) Then a **limit** of D, denoted $\lim D$, is a universal element for $\mathrm{Cone}(-,D)$.

Any two limits for D are isomorphic via a unique isomorphism which makes everything commute. This is stated precisely by the following proposition, whose proof is left as an exercise.

Proposition 1. *Suppose $D : \mathcal{I} \to \mathcal{C}$ is a diagram in a category $\mathcal{C}$ and $\alpha : W \to D$ and $\beta : V \to D$ are both limits of D. Then there is a unique isomorphism $u : V \to W$ such that for every object i of $\mathcal{I}$, $\alpha i \circ u = \beta i$.*

The limit of a diagram D objectifies the set

$$\{x \mid x(i) \in D(i) \text{ and for all } e : i \to j, D(e)(x(i)) = x(j)\},$$

and so will be denoted

$$[x \mid x(i) \in D(i) \text{ and for all } e : i \to j, D(e)(x(i)) = x(j)].$$

As in the case of pullbacks, implied arrows will often be omitted from the description. In particular, when $y \in^T B$ and $g : A \to B$ is a monomorphism we will often write "$y \in A$" or if necessary $\exists x(g(x) = y)$ when it is necessary to specify g.

By taking limits of different types of diagrams one obtains many well known constructions in various categories. We can recover subobjects, for example, by noting that the limit of the diagram $g : A \to B$ is the commutative cone with vertex A and edges id_A and g. Thus the description of this limit when g is monic is $[(x,y) \mid gx = y] = [y \mid y \in A]$, which is essentially the same as the subobject determined by g since a subobject is determined entirely by its elements. In other words, the monomorphisms which could be this limit are precisely those equivalent to (in the same subobject as) g in the sense of Section 1.6.

A category $\mathcal{C}$ is **complete** if every diagram in the category has a limit. It is **finitely complete** if every finite diagram has a limit. $\mathcal{S}et$, $\mathcal{G}rp$ and $\mathcal{T}op$ are all complete (Exercise (SETC) and Exercise (SGTC) of this section).

Products

A **discrete** graph is a graph with no arrows. If the set $\{1,2\}$ is regarded as a discrete graph $\mathcal{I}$, then a diagram of type $\mathcal{I}$ in a category $\mathcal{C}$ is simply an ordered pair of objects of $\mathcal{C}$. A commutative cone over the diagram (A,B) based on T is simply a pair (x,y) of elements of A and B. Commutativity in this case is a vacuous condition.

Thus a limit of this diagram represents the set $\{(x,y) \mid x \in A, y \in B\}$ and is called the **product** of A and B. It is denoted $A \times B = [(x,y) \mid$

$x \in A, y \in B]$. The object $B \times A = [(y,x) \mid y \in B, x \in A]$ is differently defined, but it is straightforward to prove that it must be isomorphic to $A \times B$.

It follows from the definition that $A \times B$ is an object P together with two arrows $p_1 : P \to A$ and $p_2 : P \to B$ with the property that for any elements x of A and y of B based on T there is a unique element (x,y) of $A \times B$ based on T such that $p_1(x,y) = x$ and $p_2(x,y) = y$. These arrows are conventionally called the **projections** , even though they need not be epimorphisms. Conversely, any element h of $A \times B$ based on T must be of the form (x,y) for some elements of A and B respectively based on T: namely, $x = p_1(h)$ and $y = p_2(h)$. In other words, there is a canonical bijection between $\mathrm{Hom}(T, A \times B)$ and $\mathrm{Hom}(T,A) \times \mathrm{Hom}(T,B)$ (this is merely a rewording of the statement that $A \times B$ represents $\{(x,y) : x \in A, y \in B\}$).

Note that (x,x') and (x',x) are distinct elements of $A \times A$ if x and x' are distinct, because $p_1(x,x') = x$, whereas $p_1(x',x) = x'$. In fact, $(x,x') = (p_2,p_1) \circ (x,x')$.

If $f : A \to C$ and $g : B \to D$, then we define

$$f \times g = (f \circ p_1, g \circ p_2) : A \times B \to C \times D.$$

Thus for elements x of A and y of B defined on the same object, $(f \times g)(x,y) = (f(x), g(y))$.

It should be noted that the notation $A \times B$ carries with it the information about the arrows p_1 and p_2. Nevertheless, one often uses the notation $A \times B$ to denote the object P; the assumption then is that there is a well-understood pair of arrows which make it the genuine product. We point out that in general there may be no canonical choice of which object to take be $X \times Y$, or which arrows as projections. There is apparently such a canonical choice in $\mathcal{S}et$ but that requires one to choose a canonical way of defining ordered pairs.

In a poset regarded as a category, the product of two elements is their infimum, if it exists. In a group regarded as a category, products don't exist unless the group has only one element. The direct product of two groups is the product in $\mathcal{G}rp$ and the product of two topological spaces with the product topology is the product in $\mathcal{T}op$. There are similar constructions in a great many categories of sets with structure.

The product of any indexed collection of objects in a category is defined analogously as the limit of the diagram $D : \mathcal{I} \to \mathcal{C}$ where $\mathcal{I}$ is the index set considered as the objects of a graph with no arrows and D is the indexing function. This product is denoted $\prod Di, i \in \mathcal{I}$, although explicit mention of the index set is often omitted. Also, the index is often subscripted as D_i if that is more convenient. There is a general associative law for products which holds up to isomorphism (see Exercise (PROD)).

There is certainly no reason to expect two objects in an arbitrary category to have a product. A category **has products** if any indexed set of objects in the category has a product. It **has finite products** if any finite indexed set of objects has a product, which by an obvious inductive argument is equivalent to requiring that any two objects have a product. Similar terminology is used for other types of limits; in particular, a category $\mathcal{C}$ **has finite limits** if every diagram $D : \mathcal{I} \to \mathcal{C}$ with I having a finite number of objects and arrows has a limit.

Algebraic structures in a category

The concept of product allows us to define certain notions of abstract algebra in a category. Thus a binary operation on an object A of a category is an arrow $m : A \times A \to A$ (so of course the product $A \times A$ must exist). For elements x, y of A defined on T, we write xy for $m(x, y)$ just as in sets. Observe that the expression xy is only defined if x and y are elements of A defined on the same object. We will use infix notation for symbols for binary operations such as $+$.

The operation m is commutative if $xy = yx$ for all elements x and y of A; spelled out, $m(x, y) = m(y, x)$ for all elements x and y of A defined on the same object. The operation is associative if $(xy)z = m(m(x, y), z) = m(x, m(y, z)) = x(yz)$.

Thus a **group** in a category is an object G of the category together with an associative binary operation on G, a function $i : G \to G$, and a global element e of G with the properties that $e(\,)x = xe(\,) = x$ and $xi(x) = i(x)x = e$ for all x in G. (In notation such as "$e(\,)x$", the element $(\,)$ is assumed to have the same domain as x). Abelian groups, rings, R-modules, monoids, and so on can all be defined in this way.

Equalizers

The **equalizer** of two arrows $f, g : A \to B$ (such arrows are said to be **parallel**) is the object $[x \in A \mid f(x) = g(x)]$. As such this does not describe a commutative cone, but the equivalent expression $[(x, y) \mid x \in A, y \in B$ and $f(x) = g(x) = y]$ does describe a commutative cone, so the equalizer of f and g is the limit of the diagram

$$A \underset{g}{\overset{f}{\rightrightarrows}} B$$

We will also call it $Eq(f, g)$. In Set, the equalizer of f and g is of course the set $\{x \in A \mid f(x) = g(x)\}$. In Grp, the kernel of a homomorphism $f : G \to H$ is the equalizer of f and the map which takes everything to the identity.

Equivalence relations and kernel pairs

In *Set*, an equivalence relation E on a set A gives rise to a quotient set A/E (the set of equivalence classes). In this section we will explore two concepts (equivalence relations in an arbitrary category and kernel pairs) which arise from this situation and in exercises here and in Section 1.8 we explore their connection with each other and with the concept of coequalizer which is defined there.

In a category $\mathcal{C}$ which has finite limits, an **equivalence relation** on an object A is a subobject $(u, v) : E \to A \times A$ which is reflexive, symmetric and transitive: for any elements x, y, z of A based on T, the following must be true:

(i) $(x, x) \in^T E$.
(ii) If $(x, y) \in^T E$, then so is (y, x).
(iii) If (x, y) and (y, z) are both in E then so is (x, z).

These definitions can be translated into statements about diagrams (see Exercise (TRAN)).

The two projections

$$E \rightrightarrows A$$

of an equivalence relation $E \rightarrowtail A \times A$ are also called the equivalence relation. Exercise (ER) describes conditions on a parallel pair of arrows which make it an equivalence relation, thus giving a definition which works in categories without products.

Related to this is the concept of kernel pair. If $f : A \to B$ is any arrow of $\mathcal{C}$, a parallel pair of arrows $h : K \to A$, $k : K \to A$ is a **kernel pair** for f if $f \circ h = f \circ k$ and whenever $s, t : L \to A$ is a pair of arrows for which $f \circ s = f \circ t$ then there is a unique arrow $j : L \to K$ for which $s = h \circ j$ and $t = k \circ j$. K is the pullback of f along itself and h and k are the projections (Exercise (KPL)). Thus $K = [(x, x') \mid f \circ x = f \circ x']$. In *Set*, an equivalence relation (u, v) is the kernel pair for its class map.

Existence of limits

The existence of some limits sometimes implies the existence of others. We state a theorem giving the most useful variations on this theme.

Proposition 2. (a) *In any category $\mathcal{C}$, the following are equivalent:*

(i) *$\mathcal{C}$ has all finite limits.*
(ii) *$\mathcal{C}$ has all equalizers of parallel pairs and all finite products.*
(iii) *$\mathcal{C}$ has a terminal object and all pullbacks.*

(b) *A category* $\mathcal{C}$ *has all limits if and only if it has all equalizers of parallel pairs and all products.*

Proof. That (i) implies (iii) is trivial, and that (iii) implies (ii) follows from Exercise (PEPB).

The construction that shows (ii) implies (i) and the construction for the hard half of (b) are essentially the same. We give only the construction, leaving the details to the reader. Let $D : \mathcal{I} \to \mathcal{C}$ be a diagram. Let P be the product of the set $D(i)$ of objects indexed by the objects of $\mathcal{I}$. Let Q be the product of the set $D(d^1(u))$ indexed by all the arrows u of $\mathcal{I}$. (Recall that d^0 takes a map to its source and d^1 takes a map to its target). Let $f : P \to Q$ be the unique map for which $p(u) \circ f = D(u) \circ p(d^0(u))$ for all arrows u of $\mathcal{I}$, and let $g : P \to Q$ be the unique arrow for which $p(u) \circ g = p(d^1(u))$. Then $\lim D$ is the equalizer of f and g.

Preservation of limits

Let $D : \mathcal{I} \to \mathcal{C}$ be a diagram and $F : \mathcal{C} \to \mathcal{B}$ a functor. Let $d : \lim D \to D$ be a universal element of D. We say that F **preserves** $\lim D$ if $Fd : F(\lim D) \to FD$ is a universal element of FD. The following proposition gives an equivalent condition for preserving a limit.

Proposition 3. *F preserves the limit of D if and only if FD has a limit $d' : \lim FD \to FD$ and there is an isomorphism $g : F(\lim D) \to \lim FD$ with the property that for any object T,*

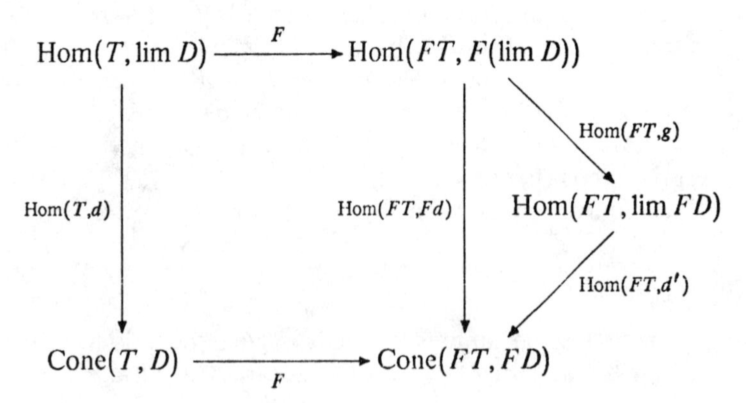

commutes. We use here the obvious property that a functor takes a commutative cone to a commutative cone.

The proof is trivial, but we include this diagram because it is analogous to a later diagram (diagram (3), Section 5.3) which is not so trivial.

Requiring that $\lim FD \cong F(\lim D)$ is not enough for preservation of limits (see Exercise (INFSET)).

Given any arbitrary class of diagrams each of which has a limit in $\mathcal{C}$, the functor F **preserves** that class of limits if it preserves the limit of each diagram in that class. We say, for example, that F preserves all limits (respectively all finite limits) if it preserves the limit of every diagram (respectively every finite diagram). F preserves products (respectively finite products) if F preserves the limit of every discrete diagram (respectively every finite discrete diagram). To preserve finite products it is sufficient to preserve products of two objects.

A functor which preserves finite limits is called **left exact.** This coincides with the concept with the same name when the functor goes from a category of R-modules to $\mathcal{A}b$.

A functor $F : \mathcal{C} \to \mathcal{B}$ **creates** limits of a given type if whenever $D : I \to \mathcal{C}$ is a diagram of that type and $d : \lim FD \to FD$ is a universal element of FD, then there is a *unique* element $u : X \to D$ for which $Fu = d$ and moreover u is a universal element of D. The underlying set functor from $\mathcal{G}rp$ to $\mathcal{S}et$ creates limits. For example, that it creates products is another way of stating the familiar fact that given two groups G and H there is a unique group structure on $G \times H$ (really on $UG \times UH$) making it the product in $\mathcal{G}rp$.

F **reflects** limits of a given type if whenever $D : I \to \mathcal{C}$ is a diagram of that type, $d : \lim FD \to FD$ is a universal element of FD and c is a cone to D for which $Fc = d$, then c is a universal element of D.

Exercises 1.7

(PROD). (a) Let A, B, and C be objects in a category. Show how $(A \times B) \times C$ and $A \times (B \times C)$ can both be regarded as the product of A, B and C (by finding appropriate projection maps) so that they are isomorphic.

(b) Show that if a category has all products of pairs of objects, then it has all finite products.

(c) If you really care, state and prove a general associative law saying that any way of meaningfully parenthesizing a sequence of objects of a category with products gives a product which is isomorphic to any other way of parenthesizing the sequence.

(TERM)$^\diamond$. Show that in a category with a terminal object 1 the product $A \times 1$ exists for any object A and is just A itself equipped with the projections $\mathrm{id} : A \to A$ and $(\,) : A \to 1$.

(PEPB)°. Prove that in a category with finite products, the equalizer of

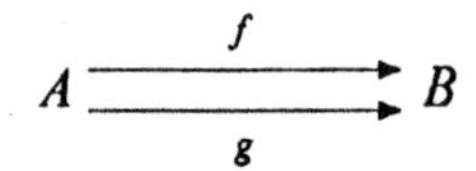

is the pullback of

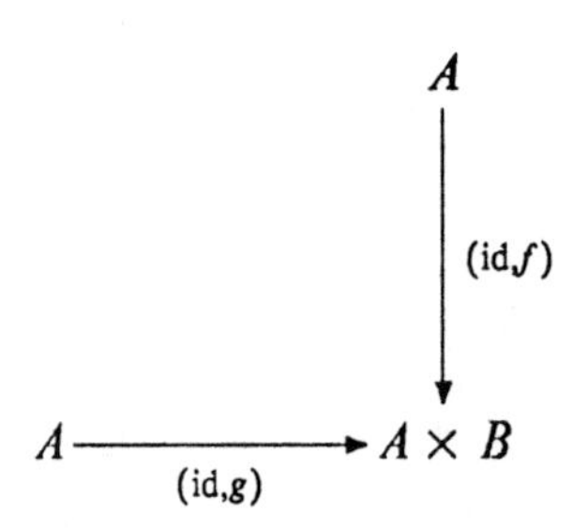

if it exists and in a category with a terminal object, the product of objects A and B is the pullback of

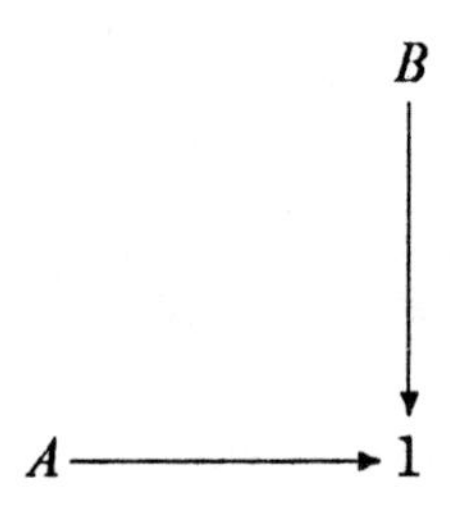

if it exists.

(PIX)°. (a) Let $\mathcal{C}$ be a category and A an object of $\mathcal{C}$. Show that the product of two objects in the category $\mathcal{C}/A$ of objects over A is their pullback over A in $\mathcal{C}$.

(b) Show that the functor $\mathcal{C}/A \to \mathcal{C}$ which takes $B \to A$ to B creates pullbacks.

(CCD). Show that if $\mathcal{D}$ is left exact and $F : \mathcal{D} \to \mathcal{C}$ preserves finite limits, then the comma category $(\mathcal{C}, F)$ is left exact.

(LIMISO)°. Prove Proposition 1.

(TOP)°. Let A be a topological space and let $\mathcal{O}(A)$ denote the set of open sets of A partially ordered by inclusion considered as a category. Show that $\mathcal{O}(A)$ has finite limits. Does $\mathcal{O}(A)$ have all limits?

(REGMON)°. A **monomorphism is regular** if it is the equalizer of two arrows. (The dual notion is called **regular epi**, not "coregular"). Recall from Section 1.4 that a regular epimorphism is denoted in diagrams by a double-headed arrow:

$$\twoheadrightarrow$$

We have no special notation for regular monos nor for ordinary epis. The reason for this asymmetry is basically one of convenience. In most of the situations in this book we are interested in ordinary monos, but only regular epis. Actually, in toposes, where much of our attention will be focused, all epis and all monos will be regular.

(i) Show that any arrow whose domain is the terminal object 1 is a regular mono.
(ii) Show that the pullback of a regular mono is a regular mono.

(SETC). Let $D : \mathcal{I} \to \mathcal{S}et$ be a diagram in $\mathcal{S}et$. Let $*$ be a fixed one-element set. Show that the set of all cones over D with vertex $*$, equipped with the correct projections, can be interpreted as $\lim D$. (This proves that $\mathcal{S}et$ is complete).

(SGTC). Show that $\mathcal{G}rp$ and $\mathcal{T}op$ are complete.

(LFC)°. Let $D : \mathcal{I} \to \mathcal{C}$ be a diagram, and let A be an object of $\mathcal{C}$. Then $D_A = \mathrm{Hom}(A, D(-))$ is a diagram in $\mathcal{S}et$. Let $\mathrm{Cone}(A, D)$ denote the set of cones over D with vertex A. Show that the limit of D_A in $\mathcal{S}et$ is the cone $\alpha : \mathrm{Cone}(A, D) \to D_A$ with αi (for i an object of $\mathcal{I}$) defined by $\alpha i(\beta : A \to D) = \beta i$, for $\beta \in \mathrm{Cone}(A, D)$.

(REPLIM)°. Show that representable functors preserve limits. (Hint: Use Exercises (SETC) and (LFC). A direct proof is also possible).

(HOMLIM). Let $D : \mathcal{I} \to \mathcal{C}$ be a diagram and let $\alpha : W \to D$ be a cone over D. For any object A of $\mathcal{C}$, let $\mathrm{Hom}(A, \alpha) : \mathrm{Hom}(A, W) \to \mathrm{Hom}(A, D(-))$ denote the cone with vertex $\mathrm{Hom}(A, W)$ which is defined by $\mathrm{Hom}(A, \alpha)i = \mathrm{Hom}(A, \alpha i)$. Show that if $\mathrm{Hom}(A, \alpha)$ is a limit of $\mathrm{Hom}(A, D(-))$ for every object A of $\mathcal{C}$, then $\alpha : W \to D$ is a limit of D. (Of course the converse is true by Exercise (REPLIM)).

(INFSET). Let $\mathcal{C}$ be the category of infinite sets and maps between them. Show that the covariant powerset functor $\mathbf{P}$ which takes a map to its image function makes $\mathbf{P}(A \times B)$ isomorphic to $\mathbf{P}A \times \mathbf{P}B$ for any objects A and B but does not preserve products.

(ER). A more general definition of equivalence relation is this: a pair $u : E \to A$, $v : E \to A$ of arrows is jointly monic if for any $f, g : B \to E$, $uf = ug$ and $vf = vg$ imply that $f = g$. Such a pair makes E an equivalence relation on A if for each object B the subset of $\mathrm{Hom}(B, A) \times \mathrm{Hom}(B, A)$ induced by $\mathrm{Hom}(B, E)$ is an equivalence relation (in the usual sense) on $\mathrm{Hom}(B, A)$. Show that this is equivalent to the definition in the text when the product $A \times A$ exists in the category.

(TRAN)°. Show that a relation $(u, v) : R \to A \times A$ in a category with finite limits is transitive if and only if, for the pullback

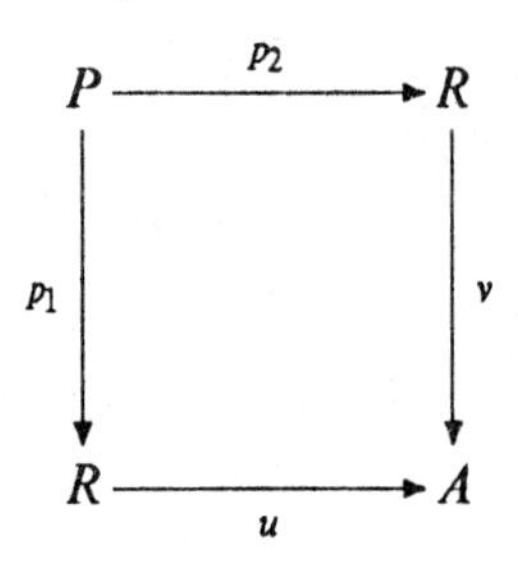

it is true that $(v \circ p_1, u \circ p_2) \in R$.

(KPL)°. Show that $h, k : K \to A$ is a kernel pair of $f : A \to B$ if and only if this diagram is a pullback:

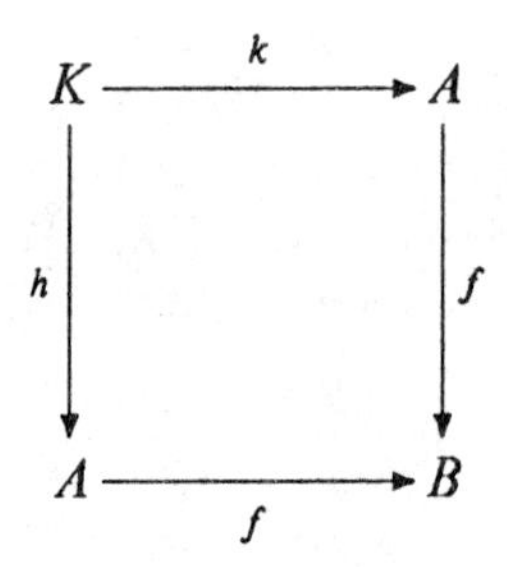

(CREA). (a) Show that the underlying functor from the category of groups creates limits.

(b) Do the same for the category of compact Hausdorff spaces and continuous maps.

(CRRF). Show that if $F : \mathcal{C} \to \mathcal{D}$ is an equivalence of categories, and $U : \mathcal{D} \to \mathcal{A}$ creates limits, then UF reflects limits.

(PER)°. Show that if E is an equivalence relation on A, then $E \times E$ is an equivalence relation on $A \times A$.

(CCL). Let $F : \mathcal{B} \to \mathcal{D}$ and $G : \mathcal{C} \to \mathcal{D}$ be functors. Show that the following diagram is a limit in the category of categories. Here (F, G) is the comma category as defined in Section 1.2.

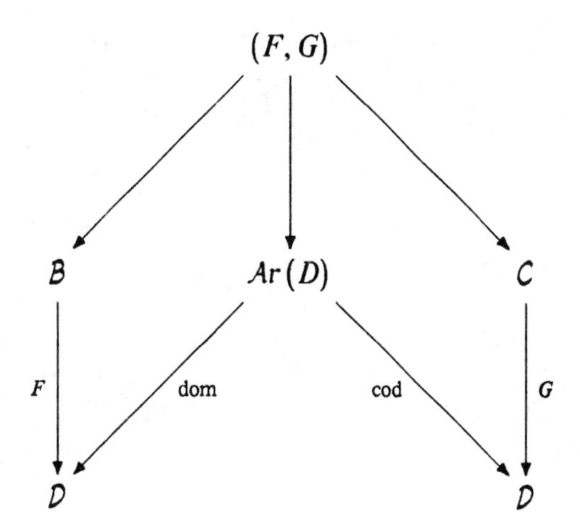

(LIMFUN)°. Show that if $\mathcal{A}$ and $\mathcal{B}$ are categories and $D : \mathcal{I} \to \text{Func}(\mathcal{A}, \mathcal{B})$ is a diagram, and for each object A of $\mathcal{A}$ the diagram $D \circ \text{ev}(A)$ gotten by evaluating at A has a limit, then these limits make up the values of a functor which is the limit of D in the functor category. Conclude that if $\mathcal{B}$ is complete, so is $\text{Func}(\mathcal{A}, \mathcal{B})$.

(PRES)°. Suppose that $\mathcal{A}$ is a category and that $\mathcal{B}$ is a subcategory of $\text{Func}(\mathcal{A}^{\text{op}}, \mathcal{S}et)$ that contains all the representable functors. This means that the Yoneda embedding Y of $\mathcal{A}^{\text{op}}$ into the functor category factors through $\mathcal{B}$ by a functor $y : \mathcal{A}^{\text{op}} \to \mathcal{B}$. Suppose further that a class $\mathcal{C}$ of cones is given in $\mathcal{A}$ with the property that each functor in $\mathcal{B}$ takes every cone in $\mathcal{C}$ to a limit cone. Show that $y^{\text{op}} : \mathcal{A} \to \mathcal{B}^{\text{op}}$ takes every cone in $\mathcal{C}$ to a limit cone.

1.8. Colimits

A **colimit** of a diagram is a limit of the diagram in the opposite category. Spelled out, a **commutative cocone** from a diagram $D : \mathcal{I} \to \mathcal{C}$ with vertex W is a natural transformation from D to the constant diagram with value W. The set of commutative cocones from D to an object A is $\text{Hom}(D, A)$ in the category of diagrams and becomes a covariant functor by composition. A colimit of D is a universal element for $\text{Hom}(D, -)$.

For example, let us consider the dual notion to "product". If A and B are objects in a category, their **sum** (also called **coproduct**) is an object Q together with two arrows $i_1 : A \to Q$ and $i_2 : B \to Q$ for which if $f : A \to C$ and $g : B \to C$ are any arrows of the category, there is a unique arrow $\langle f,g\rangle : Q \to C$ for which $\langle f,g\rangle \circ i_1 = f$ and $\langle f,g\rangle \circ i_2 = g$. The arrows i_1 and i_2 are called the coproduct injections although they need not be monic. Since $\mathrm{Hom}(A+B,C) \cong \mathrm{Hom}(A,B) \times \mathrm{Hom}(A,C)$, $\langle f,g\rangle$ represents an ordered pair of maps, just as the symbol (f,g) we defined when we treated products in Section 1.7.

The sum of two sets in *Set* is their disjoint union, as it is in *Top*. In *Grp* the categorical sum of two groups is their free product; on the other hand the sum of two Abelian groups *in the category of Abelian groups* is their direct sum with the standard inclusion maps of the two groups into their direct sum. The categorical sum in a poset regarded as a category is the supremum. The categorical sum of two posets in the category of posets and nondecreasing maps is their disjoint union, with no element of the one summand related to any element of the other.

The **coequalizer** of two arrows $f,g : A \to B$ is an arrow $h : B \to C$ such that

(i) $h \circ f = h \circ g$, and
(ii) if $k : B \to W$ and $k \circ f = k \circ g$, then there is a unique arrow $u : C \to D$ for which $u \circ h = k$.

The coequalizer of any two functions in *Set* exists but is rather complicated to construct (Exercise (CoEQ)). If K is a normal subgroup of a group G, then the coequalizer of the inclusion of K in G and the map which takes everything in K to the identity of G is the canonical map from G to G/K. See Exercise (CoEQ) for general coequalizers in that category.

The dual concept to "pullback" is "pushout", which we leave to the reader to formulate.

The notion of a functor creating or preserving a colimit, or a class of colimits, is defined analogously to the corresponding notion for limits. A functor which preserves finite colimits is called **right exact**. In general, a categorical concept which is defined in terms of limits and/or colimits is said to be defined by "exactness conditions".

Exercises 1.8

(SUM). Given two arrows $f : A \to C$ and $g : B \to D$, then there is a unique arrow $f+g : A+B \to C+D$ for which $(f+g) \circ i_1 = i_1 \circ f$ and similarly for the second index. Write a formula for $f+g$ in terms of pointed brackets. (Compare the definition of $f \times g$ in Section 1.7.)

(COEQG). Let G be a group with subgroup K (not necessarily normal in G). Describe the coequalizer of the inclusion of K in G and the constant map taking everything in K to the identity.

(COEQ). Show that coequalizers of any two parallel arrows exist in $\mathcal{Set}$ and $\mathcal{Grp}$.

(CBB). (Coequalizers can be big). Let **1** denote that category with one object and one arrow, and **2** the category with two objects and exactly one nonidentity arrow, going from one object to the other. There are exactly two functors from **1** to **2**. Show that their coequalizer in the category of categories and functors is the monoid $(\mathsf{N}, +)$ regarded as a category with one object.

(CAE)°. (a) Show that a coequalizer of two parallel arrows is an epimorphism.

(b) Show that the converse of (a) is true in $\mathcal{Set}$ and $\mathcal{Grp}$, but not in general.

(EQC)°. An equivalence relation is **effective** if it is the kernel pair of some arrow. An epimorphism is **regular** if it is the coequalizer of some pair of arrows. Prove:

(a) If an equivalence relation is effective and has a coequalizer then it is the kernel pair of its coequalizer.

(b) If an epimorphism is regular and has a kernel pair then it is the coequalizer of its kernel pair. (Warning: a parallel pair it coequalizes need not be its kernel pair.)

(FCR)°. Let $\mathcal{C}$ be a small category. Show that any functor $F : \mathcal{C}^{op} \to \mathcal{Set}$ is a colimit of representable functors. (Hint: The required graph $\mathcal{I}$ is constructed as follows. An object of $\mathcal{I}$ is a pair (C, c) where $c \in FC$. Thus the set of objects of $\mathcal{I}$ is the disjoint union of all the individual sets FC over all the objects of $\mathcal{C}$. A morphism in $\mathcal{I}$ from $(C, c) \to (C', c')$ is a morphism $f : C \to C'$ such that $Ff(c') = c$. The diagram $D : \mathcal{I} \to \mathrm{Func}(\mathcal{C}, \mathcal{Set})$ is defined by $D(C, c) = \mathrm{Hom}(-, C)$ and $Df = \mathrm{Hom}(-, f)$. Of course we have cheated a bit in calling the morphisms f as they must also be indexed by the names of their domain and codomain.)

(EAPL)°. Prove the following laws of exponents for the element notation introduced in Section 1.5:

(a) For all objects T_1 and T_2, $\in^{T_1+T_2} = \in^{T_1} \times \in^{T_2}$ (an element of A defined on $T_1 + T_2$ is the same as a pair of elements of A, one defined on T_1 and the other on T_2.

(b) For all objects A_1 and A_2, $\in_{A_1 \times A_2} = \in_{A_1} \times \in_{A_2}$.

1.9. Adjoint Functors

Let A be a set and G a group. We have noted that for any set map from A to G, in other words for any element of $\mathrm{Hom}_{\mathcal{S}et}(A, UG)$ where U is the forgetful functor, there is a unique group homomorphism from the free group FA with basis A to G which extends the given set map. This is actually a bijection

$$\mathrm{Hom}_{\mathcal{G}rp}(FA, G) \to \mathrm{Hom}_{\mathcal{S}et}(A, UG).$$

The inverse simply restricts a group homomorphism from FA to G to the basis A. Essentially the same statement is true for monoids instead of groups (replace FA by the free monoid A^*) and also for the category of Abelian groups, with FA the free Abelian group with basis A.

The bijection just mentioned is a natural isomorphism β of functors of two variables, in other words a natural isomorphism from the functor $\mathrm{Hom}_{\mathcal{G}rp}(F(-), -)$ to $\mathrm{Hom}_{\mathcal{S}et}(-, U(-))$. This means precisely that for all set maps $f : A \to B$ and all group homomorphisms $g : H \to G$,

$$\text{(1)} \qquad \begin{array}{ccc} \mathrm{Hom}_{\mathcal{G}rp}(FA, G) & \xrightarrow{\beta(A,G)} & \mathrm{Hom}_{\mathcal{S}et}(A, UG) \\ \uparrow{\scriptstyle \mathrm{Hom}_{\mathcal{G}rp}(Ff,g)} & & \uparrow{\scriptstyle \mathrm{Hom}_{\mathcal{S}et}(f,Ug)} \\ \mathrm{Hom}_{\mathcal{G}rp}(FB, H) & \xrightarrow[\beta(B,H)]{} & \mathrm{Hom}_{\mathcal{S}et}(B, UH) \end{array}$$

commutes.

The free group functor and the underlying set functor are a typical pair of "adjoint functors". Formally, if $\mathcal{C}$ and $\mathcal{D}$ are categories and $L : \mathcal{C} \to \mathcal{D}$ and $R : \mathcal{D} \to \mathcal{C}$ are functors, then L is **left adjoint** to R and R is **right adjoint** to L if for every object A of $\mathcal{C}$ and B of $\mathcal{D}$ there is an isomorphism

$$\mathrm{Hom}_{\mathcal{C}}(A, RB) \to \mathrm{Hom}_{\mathcal{D}}(LA, B)$$

which is natural in the sense of diagram (1). Informally, elements of RB defined on A are essentially the same as elements of B defined on LA.

In particular, if L is left adjoint to R and A is an object of $\mathcal{C}$, then corresponding to id_{LA} in $\mathrm{Hom}_{\mathcal{C}}(LA, LA)$ there is an arrow $\eta A : A \to RLA$; the arrows ηA form a natural transformation from the identity functor on $\mathcal{C}$ to $R \circ L$. This natural transformation η is the **unit** of the adjunction of L to R. A similar trick also produces a natural transformation $\epsilon : L \circ R \to \mathrm{id}_{\mathcal{D}}$ called the **counit** of the adjunction. The unit and counit essentially determine the adjunction completely (Exercise (Uco)).

Examples

Adjoints are everywhere. Several examples are given in the exercises. We mention three examples in $\mathcal{S}et$ which play an important role in the rest of the book.

(i) If A is a set, let 2^A denote the category of subsets of A ordered by inclusion. If $f : A \to B$ is a function, the direct image functor which assigns to each $A_0 \subseteq A$ its image $f_!(A_0)$ is left adjoint to the inverse image functor $f^{-1} : 2^B \to 2^A$; see Exercise (Pow) of Section 1.2. It follows also from that exercise that the f_* defined there is right adjoint to f^{-1}. Observe that $y \in f_!(A_0)$ (a statement without quantifiers) if and only if there exists an $x \in A_0$ such that $f(x) = y$ (a statement with an existential quantifier). Universal quantifiers may also be introduced using $_*$. In this way, quantifiers may be introduced into the language of a topos. However, we will not be doing that. See Lambek and Scott [1984].

(ii) If A is a fixed set, the functor from $\mathcal{S}et$ to $\mathcal{S}et$ which takes any set B to $B \times A$ is left adjoint to the functor which takes a set T to the set $\mathrm{Hom}_{\mathcal{S}et}(A, T)$ of functions from A to T. In other words,

$$\mathrm{Hom}_{\mathcal{S}et}(B \times A, T) \simeq \mathrm{Hom}_{\mathcal{S}et}(B, \mathrm{Hom}_{\mathcal{S}et}(A, T)).$$

The counit of this adjunction is the map from $\mathrm{Hom}_{\mathcal{S}et}(A, B) \times A$ to B which takes a pair (f, x) to its value $f(x)$ and so is called the **evaluation map**. Note the formal similarity between the evaluation map and the *modus ponens* rule of logic.

(iii) Let $\mathcal{A}$ be an equationally defined category of algebraic structures. (You can skip this example if you don't know about equationally defined theories. They will be treated in detail later.) For example, a group is a set with one nullary operation e (the unit element), one unary operation (which takes an element to its inverse), and one binary operation (the group multiplication), subject to the equations

$$xe(\,) = e(\,)x = x,$$

$$xx^{-1} = x^{-1}x = e(\,),$$

and

$$(xy)z = x(yz)$$

which hold for all x,y,z in the group.

Such a category $\mathcal{A}$ has an underlying set functor $U : \mathcal{A} \to \mathcal{S}et$, and it can be proved that U has a left adjoint $F : \mathcal{S}et \to \mathcal{A}$. It follows from adjointness (Exercise (FRE)) that for any set X and any function $g : X \to UA$ where A is any object of $\mathcal{A}$, there is a unique arrow $g' : FX \to A$ for which

(2)

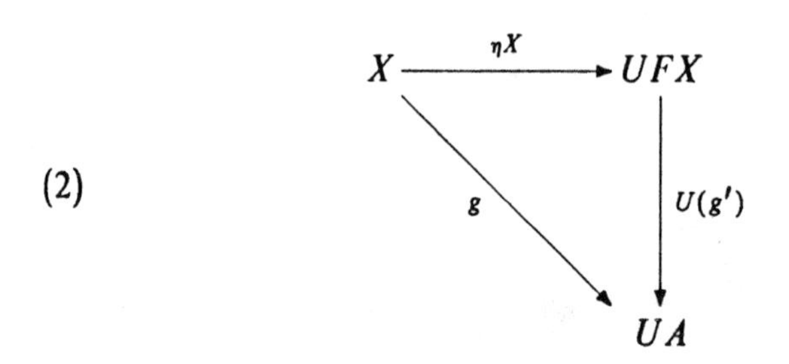

commutes. Thus FX deserves to be called the "free $\mathcal{A}$-structure on X".

Representability and adjointness

The statement that L is left adjoint to R immediately implies that for each object A of $\mathcal{C}$ the functor $\mathrm{Hom}_{\mathcal{C}}(A, R(-)) : \mathcal{D} \to \mathbf{Set}$ is represented by the object LA of $\mathcal{D}$. The universal element for this representation, which must be an element of $\mathrm{Hom}_{\mathcal{C}}(A, RLA)$, is the unit ηA. Dually, the contravariant functor $\mathrm{Hom}_{\mathcal{D}}(L(-), B)$ is represented by RB with universal element εA. The following theorem is a strong converse to these facts.

Theorem 1 ("Pointwise construction of adjoints"). *Let $\mathcal{C}$ and $\mathcal{D}$ be categories .*

(a) *If $R : \mathcal{D} \to \mathcal{C}$ is a functor such that the composite functor $\mathrm{Hom}_{\mathcal{C}}(A, R(-))$ is representable for every object A of $\mathcal{C}$, then R has a left adjoint.*
(b) *If $L : \mathcal{C} \to \mathcal{D}$ is a functor such that $\mathrm{Hom}_{\mathcal{D}}(L(-), B)$ is representable for every object B of $\mathcal{D}$, then L has a right adjoint.*

Proof. We will prove (a); the second statement is dual. We are given for each object A of $\mathcal{C}$ an object LA of $\mathcal{D}$ and a natural isomorphism

$$i(A, -) : \mathrm{Hom}_{\mathcal{D}}(LA, -) \to \mathrm{Hom}_{\mathcal{C}}(A, R(-)).$$

L will be the object function of the left adjoint. Let ηA be the universal element $i(A, LA)(\mathrm{id}_{LA})$; η will be the unit of the adjunction. If $u : LA \to C$ for some object A of $\mathcal{C}$, then the homfunctor

$$\mathrm{Hom}_{\mathcal{D}}(LA, u) : \mathrm{Hom}_{\mathcal{D}}(LA, LA) \to \mathrm{Hom}_{\mathcal{D}}(LA, C)$$

takes id_{LA} to u, so because $i(A, -)$ is natural, $i(A, C)(u)$ has to be given by the formula

$$i(A, C)(u) = Ru \circ \eta A. \tag{3}$$

It also follows from naturality that for $t : LA \to C$ and $g : C \to B$,

$$i(A, B)(g \circ t) = Rg \circ i(A, C)(t). \tag{4}$$

Now for $f : D \to A$ in $\mathcal{C}$, define Lf by the formula

$$i(D, LA)(Lf) = \eta A \circ f. \tag{5}$$

(This defines Lf because $i(D, LA)$ is a bijection). An immediate consequence of (3) and (5) is

$$RLf \circ \eta D = \eta A \circ f. \tag{6}$$

We will use these formulas to prove that L is a functor and that i is a natural transformation of functors of two variables.

To check that L is a functor, let $h : T \to D$ and $f : D \to A$. Then by applying (3) through (6) judiciously, we have

$$\begin{aligned} i(T, A)(L(f \circ h)) &= \eta A \circ f \circ h = RLf \circ \eta D \circ h \\ &= RLf \circ i(T, LA)(Lh) = i(T, A)(Lf \circ Lh). \end{aligned}$$

Naturality requires that for $f : D \to A$ in $\mathcal{C}$ and $g : C \to B$ in $\mathcal{D}$, this diagram must commute:

$$\begin{array}{ccc} \mathrm{Hom}_{\mathcal{D}}(LD, B) & \xrightarrow{i(D,B)} & \mathrm{Hom}_{\mathcal{C}}(D, RB) \\ \Big\downarrow {\scriptstyle \mathrm{Hom}_{\mathcal{D}}(Lf, g)} & & \Big\downarrow {\scriptstyle \mathrm{Hom}_{\mathcal{C}}(f, Rg)} \\ \mathrm{Hom}_{\mathcal{D}}(LA, C) & \xrightarrow[i(A,C)]{} & \mathrm{Hom}_{\mathcal{C}}(A, RC) \end{array} \tag{7}$$

This means we must show that for $u : LA \to C$,

$$i(D, B)(g \circ u \circ Lf) = Rg \circ i(A, C)(u) \circ f.$$

By (3) (applied twice) and (4) this boils down to showing that $RLf \circ \eta D = \eta A \circ f$, which is (6).

Since a covariant representable functor preserves limits and any contravariant one preserves colimits (Exercise (REPLIM) of Section 1.7) we have the following corollary:

Corollary 2. *If a functor has a left adjoint it preserves limits, and if it has a right adjoint it preserves colimits.*

We note that underlying functors in algebra tend to have left adjoints and thereby preserve limits, but rarely have a right adjoint or preserve colimits.

Existence of adjoints

Freyd's Adjoint Functor Theorem (Theorem 3 below) is a partial converse to Corollary 2. To state it, we need a new idea.

Suppose $R : \mathcal{D} \to \mathcal{C}$ is a functor. R satisfies the **solution set condition** if for each object A of $\mathcal{C}$ there is a set $\mathcal{S}$ (the **solution set** for A) of pairs (y, B) with $y : A \to RB$ in $\mathcal{C}$ with the property that for any arrow $d : A \to RD$ there is an element (y, B) of $\mathcal{S}$ and an arrow $f : B \to D$ for which

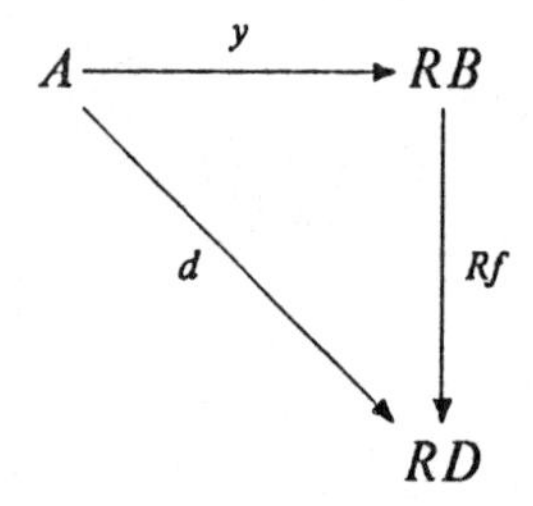

commutes.

If $\mathcal{C}$ is small, then $\mathcal{S}$ can be taken to be the set of all pairs (y, B) with $y : A \to RB$ for all arrows y of $\mathcal{C}$ and objects B of $\mathcal{D}$. (Then f can, for example, be taken to be id_D). On the other hand, if R is already known to have a left adjoint L, then $\mathcal{S}$ can be taken to be the singleton set with $B = LA$ and $y = \eta A$.

If the singleton $\{(y, B)\}$ is a solution set for R, y satisfies the existence but not necessarily the uniqueness property of the definition of universal arrow for $\mathrm{Hom}(A, R(-))$ (Section 1.5). In that case we will say that y is a **weak universal arrow** for R and A.

Theorem 3 (Freyd). *Let $\mathcal{D}$ be a category with all limits. Then a functor $R : \mathcal{D} \to \mathcal{C}$ has a left adjoint if and only if R preserves all limits and satisfies the solution set condition.*

Proof. Let A be an object of $\mathcal{C}$. In order to construct the left adjoint L, it is enough by Theorem 1 and the definition of representable functor (Section 1.5) to construct a universal element for $\mathrm{Hom}(A, R(-))$. We first construct an object WA and a weak universal arrow $\zeta A : A \to RLA$. Then we will use equalizers cleverly to get uniqueness.

The construction of WA is reminiscent of the way one proves that a poset with all infs and a maximum has all sups (to get the sup of a set, take the inf of all the elements bigger than everything in the set). The equalizer construction is not necessary for posets because uniqueness is automatic there.

WA is defined to be the product indexed by all $(y, B) \in S$ of the objects B. WA comes equipped with a projection $WA \to B$ for each pair $(y, B) \in S$, and R preserves the fact that WA is a product with these projections. The arrows Ry collectively induce $\zeta A : A \to RWA$. Given $d : A \to RD$, let $(y, B) \in S$ and $h : B \to D$ be an arrow for which $Rh \circ y = d$, and p the projection of WA onto B indexed by (y, B). Then for $f = h \circ p$, $Rf \circ \zeta A = d$ so that ζA is a weak universal arrow.

To attain the uniqueness condition in the definition of universal arrow, we construct a subobject LA of WA with the property that ζA factors through RLA via a map $\eta A : A \to RLA$. It is easy to set that *any* such ηA is also a weak universal arrow. The idea is to make LA as small as possible so as to obtain the uniqueness property.

The natural thing to do would be to take LA to be the intersection of all the equalizers of maps $f_1, f_2 : WA \to D$ such that $Rf_i \circ \zeta A = d$ for all $d : A \to RD$. The trouble is that these equalizers may not form a set. This is where the clever part of the proof is: Let $U = \{u : WA \to WA \mid Ru \circ \zeta A = \zeta A\}$. U is a set, in fact a submonoid of the endomorphism monoid of WA. (By definition of category, $\mathrm{Hom}(A, B)$ is a set for any objects A and B). Then define $w : LA \to WA$ to be the collective equalizer $[z \in WA \mid u(z) = v(z) \text{ for all } u, v \in U]$, and let $\eta A : A \to RLA$ be the map induced by the facts that Rw must be an equalizer and ζA equalizes the image of U under R. Clearly ηA is a weak universal arrow, and it is easy to see that to get the uniqueness property we had to equalize at *least* the elements of U. We now show that equalizing the elements of U is enough.

Suppose $d : A \to RD$ and for $i = 1, 2$, $g_i : LA \to D$ have the property that $Rg_i \circ \eta A = d$. Let $e : E \to LA$ be the equalizer of g_1 and g_2, so the horizontal part of the following commutative diagram, in which v and

z have yet to be constructed, is an equalizer:

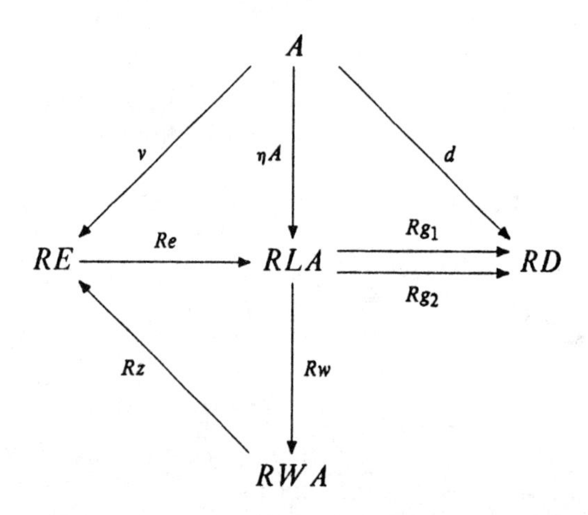

Since Re is an equalizer, there is an arrow v making the upper left triangle commute as shown, and because $\zeta A = Rw \circ \eta A$ is a weak universal arrow, there is an arrow z making $Rz \circ \zeta A = V$. It is easy to see that $wez \in U$, whence $wezw = w$. Because w is monic, this means that zw is a right inverse to e, which implies that $g_1 = g_2$ as required.

The solution set condition is often shown to be satisfied in practice by using a cardinality condition. For example, if U is the underlying functor from $\mathcal{G}rp$ to $\mathcal{S}et$, in constructing a solution set for a particular set A one can clearly restrict one's attention to maps $A \to UG$ (G a group) with the property that the image of A generates G. Since the cardinality of such a group is bounded by some cardinal α, a solution set for A consists of all those pairs (y, B), where B ranges over the distinct (up to isomorphism) groups of cardinality $\leq \alpha$ and y over all functions from A to RB for all such B.

Kan extensions

If $\mathcal{A}$, $\mathcal{C}$ and $\mathcal{D}$ are categories and $F : \mathcal{D} \to \mathcal{C}$ is a functor, then F induces a functor.

$$\mathrm{Func}(F, \mathcal{A}) : \mathrm{Func}(\mathcal{C}, \mathcal{A}) \to \mathrm{Func}(\mathcal{D}, \mathcal{A})$$

which takes a functor $G : \mathcal{C} \to \mathcal{A}$ to $G \circ F$ (like any homfunctor) and a natural transformation $\lambda : G \to H$ to $\lambda F : G \circ F \to H \circ F$. The **left Kan extension** of a functor $T : \mathcal{D} \to \mathcal{A}$ along F is a functor $LF(T) : \mathcal{C} \to \mathcal{A}$ with the property that there is for each $G : \mathcal{C} \to \mathcal{A}$ a bijection

$$\mathrm{Nat}(LF(T), G) \to \mathrm{Nat}(T, G \circ F)$$

which is natural in G.

In the presence of sufficient colimits in $\mathcal{A}$ one can construct the Kan extension of a functor $T : \mathcal{D} \to \mathcal{A}$ provided that $\mathcal{D}$ is a small category. We give the construction and leave the detailed verifications (which are not trivial!) to the reader.

Given $F : \mathcal{D} \to \mathcal{C}$ and $T : \mathcal{D} \to \mathcal{A}$, we must construct $LF(T) : \mathcal{C} \to \mathcal{A}$. For any object C of $\mathcal{C}$, the functor F determines a comma category (F, C) which has a projection p onto $\mathcal{D}$. We define $LF(T)(C)$ to be the colimit of the composite $T \circ p : (F, C) \to \mathcal{A}$. An arrow $f : C \to C'$ in $\mathcal{C}$ determines a functor from (F, C) to (F, C') which by the universal property of colimits determines a map $LF(T)(f) : LF(T)(C) \to LF(T)(C')$. This makes $LF(T)$ a functor.

Define the natural transformation $\eta : T \to LF(T) \circ F$ by requiring that for each object D of $\mathcal{D}$, ηD is the element of the colimiting cocone to $LF(T)(F(D))$ at the object $(D, \mathrm{id}_{FD}, FD)$ of the comma category (F, FD). For each $G : \mathcal{C} \to \mathcal{A}$, the required bijection

$$\mathrm{Nat}(LF(T), G) \to \mathrm{Nat}(T, G \circ F)$$

takes $\lambda : LF(T) \to G$ to $(\lambda F) \circ \eta$. Conversely, given a natural transformation $\mu : T \to GF$ and an object C of $\mathcal{C}$, there is a cocone from $T \circ p$ to GC whose element at an object (D, g, C) of (F, C) is $Gg \circ \mu D : TD \to GFD$. This induces a map $\lambda C : LF(T)(C) \to GC$; these are the components of a natural transformation $\lambda : LF(T) \to G$. The inverse of the bijection just given takes μ to the λ thus constructed.

We have immediately from Theorem 1 (the pointwise construction of adjoints):

Proposition 4. *In the notation of the preceding paragraphs, if every functor $T : \mathcal{D} \to \mathcal{A}$ has a left Kan extension along F, then* $\mathrm{Func}(F, \mathcal{A}) : \mathrm{Func}(\mathcal{C}, \mathcal{A}) \to \mathrm{Func}(\mathcal{D}, \mathcal{A})$ *has a left adjoint.*

Note that by the construction given of Kan extensions, the hypothesis of Proposition 4 will be true if $\mathcal{A}$ is cocomplete.

Right Kan extensions can be defined similarly. A detailed construction of right Kan extensions is given in Mac Lane [1971].

Exercises 1.9

(Diag). Let $\Delta : \mathcal{C} \to \mathcal{C} \times \mathcal{C}$ be the diagonal functor. Find left and right adjoints to Δ. Assume that $\mathcal{C}$ has whatever limits and colimits you need. These are examples of Kan extensions along the (unique) functor $1 + 1 \to 1$.

(CADJ)°. Show that if two composable functors each have a left adjoint, then so does their composite.

(USL2). Show that the functor which takes a set to its set of nonempty finite subsets and a function to its direct image function is the left adjoint to the underlying functor from the category of upper semilattices (see Exercise (USL) of Section 1.2).

(RMOD). Find left *and right* adjoints for the functor which takes an R-module (any fixed ring R) to its underlying Abelian group.

(GPAC). (a) For a fixed group G, let G-Act denote the category of G-actions and equivariant maps. Let U be the forgetful functor. Construct a left adjoint for U. (Hint: it takes a set A to $G \times A$.)
(b) What about a right adjoint?

(TOPA)°. (a) Show that the underlying set functor from $\mathcal{Top}$ to $\mathcal{Set}$ has a left adjoint which takes a set to that set regarded as a discrete topological space.
(b) Show that the underlying set functor in (a) also has a right adjoint.

(ADJCAT)°. Define four functors $\pi_0, U : \mathcal{Cat} \to \mathcal{Set}$ and $D, G : \mathcal{Set} \to \mathcal{Cat}$ as follows:

(a) For any category C, UC is the set of objects of C. For a functor F, UF is what F does to objects.
(b) For a category C, $\pi_0(C)$ is the set of connected components of C—two objects x and y are in the same component if and only if there is a finite sequence $x = x_0, x_1, \cdots, x_n = y$ of objects of C and a finite sequence $a_1, \cdots, a_{n-1}$ of arrows of C with for each $i = 1, \cdots, n-1$, *either* $a_i : x_i \to x_{i+1}$ or $a_i : x_{i+1} \to x_i$. A functor F induces $\pi_0 F$ in the obvious way (it is easy to see that a functor takes a component into a component).
(c) For any set A, DA is the category whose set of objects is A and whose only arrows are identity arrows (DA is the discrete category on A). DF for a function F is forced by the definition for objects.
(d) For a set A, GA is the category whose objects are the elements of A, with exactly one arrow between any two elements. (It follows that every arrow is an isomorphism – i.e., GA is a groupoid). What G does to functions is forced.

Prove that: π_0 is left adjoint to D, D is left adjoint to U, and U is left adjoint to G.

(GRADJ)°. Show that the "underlying graph functor" $U : \mathcal{Cat} \to \mathcal{Grph}$ defined in Section 1.7 has a left adjoint. (Hint: If G is a graph, FG will have the same objects as G, and nonidentity arrows will be "composable sequences" of arrows of G.)

(EQUIII)°. Assume that $L : \mathcal{C} \to \mathcal{D}$ is left adjoint to $R : \mathcal{D} \to \mathcal{C}$. Prove:

(a) R is faithful if and only if the counit εD is epic for every object D of $\mathcal{D}$. (Hint: The map which Yoneda gives from the natural transformation

$$\operatorname{Hom}(-, D) \to \operatorname{Hom}(R(-), RD) \to \operatorname{Hom}(LR(-), D)$$

must be the counit at D. Now reread the definition of epimorphism in Section 1.4).

(b) R is full if and only if the counit is a split monic at every object of $\mathcal{D}$.

(c) L is faithful if and only if the unit is monic for every object of $\mathcal{C}$.

(d) L is full if and only if the unit is a split epic at every object of $\mathcal{C}$.

(e) R is an equivalence of categories if and only if the unit and counit are both natural isomorphisms.

(f) If R is an equivalence of categories, then so is L and moreover then L is also a right adjoint to R.

(SLADJ)°. Show that if $\mathcal{A}$ is a category with finite products and A is an object of $\mathcal{A}$, then the functor from the slice category (see Section 1.1) $\mathcal{A}/A \to \mathcal{A}$ that sends the object $B \to A$ to B—the so-called forgetful functor—has a right adjoint, $B \mapsto B \times A \to A$.

(MONL). Let $\mathcal{M}on$ denote the category of monoids and homomorphisms and $\mathcal{C}at$ the category of categories and functors. Define $L : \mathcal{M}on \to \mathcal{C}at$ as follows: For a monoid M, the objects of LM are the elements of M. An arrow is a pair (k, m) of elements of M; it has domain m and codomain km. Composition is given by the formula

$$(k', km) \circ (k, m) = (k'k, m).$$

If $h : M \to N$ is a homomorphism, $Lh(k, m) = (hk, hm)$. Construct a left adjoint for L.

(UCO)°. (a) Show that if $L : \mathcal{C} \to \mathcal{D}$ is left adjoint to $R : \mathcal{D} \to \mathcal{C}$ with unit η and counit ε, then $R\varepsilon \circ \eta R$ is the identity natural transformation on R and $\varepsilon L \circ L\eta$ is the identity natural transformation on L.

(b) Show that if $L : \mathcal{C} \to \mathcal{D}$ and $R : \mathcal{D} \to \mathcal{C}$ are functors and $\eta : \mathrm{id}_{\mathcal{C}} \to R \circ L$ and $\varepsilon : L \circ R \to \mathrm{id}_{\mathcal{D}}$ are natural transformations satisfying the conclusion of (a), then L is left adjoint to R and η and ε are the unit and counit of the adjunction.

(FRE)°. (a) Show that if L and R are as in (a) of Exercise (UCO), then for any objects A of $\mathcal{C}$ and B of $\mathcal{D}$ and any arrow $f : A \to RB$

of $\mathcal{C}$ there is a unique arrow $g : LA \to B$ of $\mathcal{D}$ for which $Rg \circ \eta A = f$. (This generalizes a well-known property of free groups).

(b) State a dual version of (a) using ε.

(REFL)°. A full subcategory $\mathcal{D}$ of a category $\mathcal{C}$ is **reflective** if the inclusion has a left adjoint, which is called the **reflector.**

(a) Show that $\mathcal{D}$ is a reflective subcategory of $\mathcal{C}$ if and only if for each object A of $\mathcal{C}$ there is an object LA of $\mathcal{D}$ and an arrow $e : A \to LA$ with the property that if $f : A \to B$ is any arrow with B an object of $\mathcal{D}$ then there is a unique arrow $h : LA \to B$ for which $f = h \circ e$.

(b) Show that if $\mathcal{D}$ is a reflective subcategory of $\mathcal{C}$ then the reflector takes an object B of $\mathcal{D}$ to an object isomorphic to B.

(c) If $\mathcal{D}$ is a reflective subcategory of $\mathcal{C}$ with inclusion I and reflector L, show that an object C of $\mathcal{C}$ is isomorphic to an object coming from $\mathcal{D}$ if and only if for each object D' of $\mathcal{D}$ the counit $ILD' \to D'$ induces $\text{Hom}(D', D) \cong \text{Hom}(ILD', D)$.

(d) Show that the category of Abelian groups is a reflective subcategory of the category of groups.

(e) Show that the category of finite sets is not a reflective subcategory of the category of all sets.

(REP). Show that a functor $R : \mathcal{C} \to \mathcal{S}et$ being representable is equivalent to the solution set condition being satisfied uniquely with a singleton solution set.

(GAFT)°. Here is another way of organizing the proof of Freyd's Adjoint Functor Theorem.

(a) Show that $U : \mathcal{D} \to \mathcal{C}$ has an adjoint if and only if for each object C of $\mathcal{C}$, the comma category (C, U) has an initial object.

(b) Show that if $\mathcal{D}$ has and U preserves limits, then (C, U) is complete. (Compare Exercise (CCD) of Section 7.)

(c) Show that the solution set condition is equivalent to (C, U) having a weak initial set, that is a small set of objects with the property that every object is the codomain of at least one morphism whose domain lies in that set.

(d) Show that a category with products and a weak initial set has a weak initial object.

(e) Show that if A is a weak initial object and $E \to A$ the simultaneous equalizer of all the endomorphisms, then E is initial.

(f) Deduce Freyd's adjoint functor theorem (usually known as the G(eneral) A(djoint) F(unctor) T(heorem) to distinguish it from the

SAFT or Special Adjoint Functor Theorem which follows.)

(SAFT)°. A category is said to be **well powered** if every object has only a small set of subobjects. A set $\{Q_i\}$ of objects of a category is said to be a **cogenerating set** if for any pair of objects C and D of the category and any pair of distinct morphisms $f, g : C \to D$, there is at least one Q_i and one morphism $h : D \to Q_i$ with $hf \neq hg$.

(a) Show that a set $\{Q_i\}$ of objects of a complete category is a cogenerating set if and only if every object has a monomorphism into some product of those objects (allowing repetitions).

The Special Adjoint Functor Theorem states that a functor that preserves limits and whose domain is well powered, complete and has a cogenerating set has a left adjoint. Demonstrate this theorem by following the steps below. Assume the hypotheses in each of the steps. The organization of this proof is due to G. M. Kelly.

(b) Show that it is sufficient to prove that a category satisfying the hypotheses has a weak initial set. (See the preceding exercise.)

(c) Show that every object has a unique smallest subobject.

(d) Show that every object which its own smallest subobject can be embedded into a product of members of the cogenerating set in which there are no repetitions. (Hint: Consider a diagram

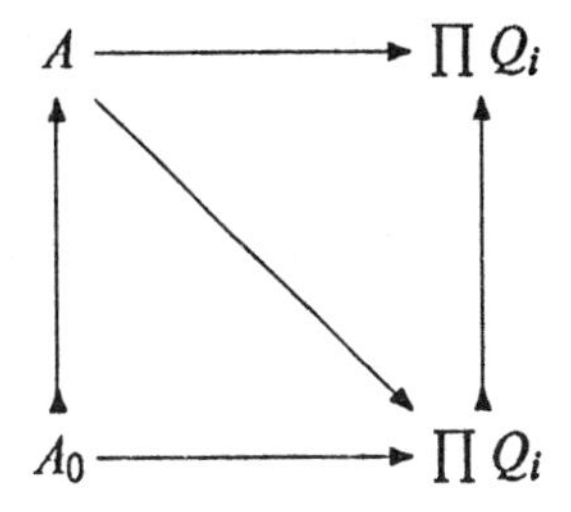

in which the bottom right corner is an irredundant product, i.e., a product with no repetitions.)

(d) Show that the set of all subobjects of irredundant products of members of the cogenerating set forms a weak initial set.

(e) Conclude the SAFT.

1.10. Notes to Chapter 1

Development of category theory

Categories, functors and natural transformations were invented by S. Eilenberg and S. Mac Lane (announced in "The general theory of natural equivalences", [1945]) in order to describe the connecting homomorphism and the long exact sequence in Čech homology and cohomology. The problem was this: homology was defined in the first instance in terms of a cover. If the cover is simple, that is if every non-empty intersection of a finite subset of the cover is a contractible space (as actually happens with the open star cover of a triangulated space), then that homology in terms of the cover is the homology of the space and that is the end of the matter. What is done in Čech theory, in the absence of a simple cover, is to form the direct limit of the homology groups over the set of all covers directed by refinement. This works fine for defining the groups but gives no information on how to define maps induced by, say, the inclusion of a subspace, not to mention the connecting homomorphism. What is missing is the information that homology is natural with respect to refinements of covers as well as to maps of spaces. Fortunately, the required condition was essentially obvious and led directly to the notion of natural tranformation. Only, in order to define natural transformation, one first had to define functor and in order to do that, categories.

The other leading examples of natural transformations were the inclusion of a vector space into its second dual and the commutator quotient of a group. Somewhat surprisingly, in view of the fact that the original motivation came from algebraic topology, is the fact that the Hurewicz homomorphism from the fundamental group of a space to the first homology group of that space was not recognized to be an example until later.

Later, Steenrod would state that no paper had influenced his thinking more than "The general theory of natural equivalences". He explained that although he had been searching for an axiomatic treatment of homology for years and that he of course knew that homology acted on maps (or vice versa, if you prefer) it had never occurred to him to try to base his axiomatics on this fact.

The next decisive step came when Mac Lane [1950] discovered that it was possible to describe the cartesian product in a category by means of a universal mapping property. In fact, he described the direct sum in what was eventually recognized as an additive category by means of two mapping properties, one describing it as a product and the other describing it as a sum. Mac Lane also tried to axiomatize the notion of Abelian category but that was not completely successful. No matter, the universal mapping property described by Mac Lane had shown that it was possible to use categories as an aid to understanding. Later on, Grothendieck succeeded in giving axioms for Abelian categories [1957] and to actually prove something with them – the existence of injectives in an Abelian category with sufficient higher exactness properties. Thus Grothendieck demonstrated that categories could be a tool for actually *doing* mathematics and from then on the development was rapid. The next important step was the discovery of adjoint functors by Kan [1958] and their use as an effective tool in the study of the homotopy theory of abstract simplicial sets.

After that the mainstream of developments in category theory split into those primarily concerned with Abelian categories (Lubkin [1960], Freyd [1964], Mitchell [1964]), which are interesting but tangential to our main concerns here, and those connected with the theories of triples and toposes of which we have more to say later.

Elements

Although the thrust of category theory has been to abstract away from the use of arguments involving elements, various authors have reintroduced one form or another of generalized element in order to make categorical arguments parallel to familiar elementwise arguments; for example, Mac Lane [1971, VIII.4] for Abelian categories and Kock [1981, part II] for Cartesian closed categories. It is not clear whether this is only a temporary expedient to allow older mathematicians to argue in familiar ways or will always form a permanent part of the subject. Perhaps elements will disappear if Lawvere succeeds in his goal of grounding mathematics, both in theory and in practice, on arrows and their composition.

An altogether deeper development has been that of Mitchell [1972] and others of the internal language of a topos (developed thoroughly in a more general setting by Makkai and Reyes [1977]). This allows one to develop arguments in a topos as if the objects were sets, specifically including some use of quantifiers, but with restricted rules of deduction.

Limits

Limits were originally taken over directed index sets—partially ordered sets in which every pair of elements has a lower bound. They were quickly generalized to arbitrary index *categories*. We have changed this to graphs to reflect actual mathematical practice: index categories are usually defined *ad hoc* and the composition of arrows is rarely made explicit. It is in fact totally irrelevant and our replacement of index categories by index graphs reflects this fact. There is no gain—or loss—in generality thereby, only an alignment of theory with practice.

Chapter 2.

Toposes

A topos is, from one point of view, a category with certain properties characteristic of the category of sets. A topos is *not* merely a generalized set theory, but the very elementary constructions to be made in this chapter are best understood, at least at first, by looking at what the constructions mean in $\mathcal{S}et$. From another point of view, a topos is an abstraction of the category of sheaves over a topological space. This latter aspect is described in detail in this chapter.

2.1. Basic Ideas about Toposes

Definition of topos

We will take two properties of the category of sets — the existence of all finite limits and the fact that one can always form the set of subsets of a given set — as the defining properties for toposes.

For a fixed object A of a category $\mathcal{E}$ with finite limits, $- \times A$ is a functor from $\mathcal{E}$ to itself; if $f : B \to B'$, then $f \times A$ is the arrow $(f \circ p_1, p_2) : B \times A \to B' \times A$. By composition, we then have a contravariant functor $\mathrm{Sub}(- \times A) : \mathcal{E} \to \mathcal{S}et$. The **power object** of A (if it exists) is an object $\mathrm{P}A$ which represents $\mathrm{Sub}(- \times A)$, so that $\mathrm{Hom}_{\mathcal{E}}(-, \mathrm{P}A) \simeq \mathrm{Sub}(- \times A)$ naturally. This says precisely that for any arrow $f : B' \to B$, the following diagram commutes, where φ is the natural isomorphism.

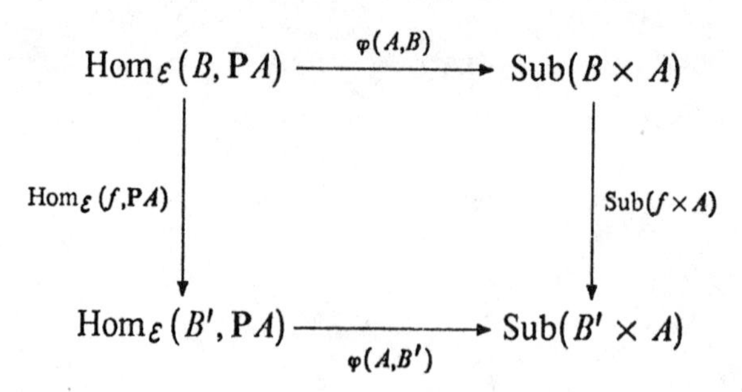

The definition of $\mathbf{P}A$ says that the "elements" of $\mathbf{P}A$ defined on B are essentially the same as the subobjects of $B \times A$. In *Set*, a map f from B to the powerset of A is the same as a relation from B to A (b is related to a if and only if $a \in f(b)$), hence the same as a subset of $B \times A$. When B is the terminal object (any singleton in *Set*), the "elements" of $\mathbf{P}A$ defined on B are the subsets of $A \simeq 1 \times A$; thus $\mathbf{P}A$ is in fact the powerset of A.

In general, if the category has a terminal object 1 and $\mathbf{P}(1)$ exists, then Sub is represented by $\mathbf{P}(1)$, since $1 \times A \cong A$. This object $\mathbf{P}(1)$ is studied in detail in Section 2.3.

Definition. A category $\mathcal{E}$ is a **topos** if $\mathcal{E}$ has finite limits and every object of $\mathcal{E}$ has a power object.

We will assume that $\mathbf{P}A$ is given functionally on $\mathcal{Ob}\,\mathcal{E}$ (it is determined up to isomorphism in any case). This means that for each object A of $\mathcal{E}$, a definite object $\mathbf{P}A$ of $\mathcal{E}$ is given which has the required universal mapping property.

The definition of toposes has surprisingly powerful consequences. (For example, toposes have all finite colimits.) Probably the best analogy elsewhere in mathematics in which a couple of mild-sounding hypotheses pick out a very narrow and interesting class of examples is the way in which the Cauchy-Riemann equations select the analytic functions from all smooth functions of a complex variable.

The properties of toposes will be developed extensively in this Chapter and in Chapters 5 and 6. However, the rest of this section and the next are devoted to examples.

Examples of toposes

(i) The category *Set* is evidently a topos. As we have already pointed out, if X is a set, we can take $\mathbf{P}X$ to be the set of all subsets of X, but that does not determine a unique topos structure on *Set* since we have a choice of φ in diagram (1). The natural choice is to let $\varphi : \mathrm{Hom}(1, \mathbf{P}B) \to \mathrm{Sub}(B)$ be the identity map (thinking of an arrow from 1 to $\mathbf{P}B$ as an element of $\mathbf{P}B$), but we could be perverse and let φ of an element of the powerset be its *complement*.

(ii) To see a more interesting example, let G be a group and let G-*Set* be the category of all sets on which G acts. The morphisms are equivariant G homomorphisms. The existence of finite (in fact, all) limits is an easy exercise. They are calculated "pointwise". If X is a G-set, let $\mathbf{P}X$ denote the set of all subsets of X with G

action given by $gX_0 = \{gx \mid x \in X_0\}$. Note that a global element of PX is a G-invariant subset of X.

Actually, the category of actions by a given monoid, with equivariant maps, is a topos. That will follow from the discussion of functor categories below, since such a category is the same as a $\mathcal{Set}$ -valued functor category from a monoid regarded as a category with one object.

Functor categories

An important example of toposes are $\mathcal{Set}$ -valued functor categories. In order to prove that these categories are toposes, a number of elementary facts about them are needed. A guiding principle in this development is the fact that $\mathrm{Func}(\mathcal{C}, \mathcal{D})$ inherits most of its properties from $\mathcal{D}$ (Exercise (LIMFUN) of Section 1.7).

In this section, $\mathcal{C}$ is a fixed small category and $\mathcal{E} = \mathrm{Func}(\mathcal{C}^{\mathrm{op}}, \mathcal{Set})$. We will outline the proof that $\mathcal{E}$ is a topos. Of course, everything we say in this section is true of $\mathrm{Func}(\mathcal{C}, \mathcal{Set})$, but because of the applications we prefer to state it this way.

Each object C of $\mathcal{C}$ determines an **evaluation map** $\lambda C : \mathcal{E} \to \mathcal{Set}$, where $\lambda C(F) = FC$ and for $\gamma : F \to G$, $\lambda C(\gamma) = \gamma C$.

Proposition 1. *For each object C of $\mathcal{C}$, the evaluation preserves all limits and colimits. I.e., "limits and colimits in $\mathcal{E}$ are computed pointwise". In particular, $\mathcal{E}$ is complete and cocomplete.*

In other words, if $D : \mathcal{I} \to \mathcal{E}$ is a diagram in $\mathcal{E}$, then $(\lim D)(C) = \lim(D(C))$. The proof is in Exercise (LIMFUN) of Section 1.7.

Corollary 2. *For a fixed object E, the functor $- \times E : \mathcal{E} \to \mathcal{E}$ commutes with all colimits.*

Proof. The property claimed for this functor is valid when $\mathcal{E} = \mathcal{Set}$ by Exercise (CCL), and the Proposition allows one to extend it to an arbitrary functor category.

The fact that Sub is representable in a topos (by $P(1)$) means that it takes colimits to limits. (Exercise (REPLIM) of Section 1.7). In particular, $\mathrm{Sub}(\sum A_i) = \prod \mathrm{Sub}(A_i)$ and if

$$A \overset{p_1}{\underset{p_2}{\rightrightarrows}} B \xrightarrow{\quad c \quad} C \tag{2}$$

is a coequalizer in a topos, then

$$\text{(3)} \qquad \mathrm{Sub}(A) \underset{\mathrm{Sub}(p_2)}{\overset{\mathrm{Sub}(p_1)}{\leftleftarrows}} \mathrm{Sub}(B) \xleftarrow{\mathrm{Sub}(c)} \mathrm{Sub}(C)$$

is an equalizer. The first is easy to see in $\mathcal{S}et$, but a direct proof of the second fact in $\mathcal{S}et$, not using the fact that Sub is representable, is surprisingly unintuitive.

As a step toward proving that $\mathcal{E}$ is a topos, we prove the fact just mentioned for $\mathcal{E}$.

Proposition 3. *If $D : \mathcal{I} \to \mathcal{E}$ is a diagram in $\mathcal{E}$, then*

$$\mathrm{Sub}(\operatorname{colim} D) \simeq \lim(\mathrm{Sub}(D)).$$

Proof. We use repeatedly the fact that the result is true in $\mathcal{S}et$ because there Sub is representable by the two-element set. Let $F = \operatorname{colim} FD$. For an object i of $\mathcal{I}$, the cocone $FDi \to F$ gives a cone $\mathrm{Sub}(F) \to \mathrm{Sub}(FDi)$ which in turn gives $\mathrm{Sub}(F) \to \lim \mathrm{Sub}(FDi)$. Now to construct an arrow going the other way, let $EDi \subseteq FDi$ be a compatible family of subobjects, meaning that whenever $i \to j$

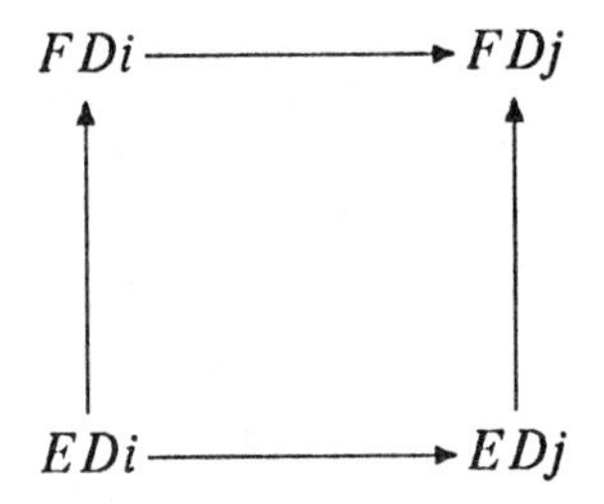

is a pullback. Let $E = \operatorname{colim} EDi$. Since colimits preserve monos (by Proposition 1, since they do so in $\mathcal{S}et$), E is a subfunctor of F. This gives a map $\lim \mathrm{Sub}(FDi) \to \mathrm{Sub}(F)$. Finally, to see that both composites are the identity, it suffices to see that all those constructions are identical to the ones carried out in $\mathcal{S}et^{\mathrm{Ob}(\mathcal{C})}$, for which the result follows from Proposition 1 since it is true in $\mathcal{S}et$.

Theorem 4. *$\mathcal{E}$ is a topos.*

Proof. Finite limits, indeed all limits, exist by Proposition 1. As for $\mathbf{P}$, for a functor E, let $\mathbf{P}E$ be defined by letting $\mathbf{P}E(A)$ be the set of subfunctors of $\mathrm{Hom}(-, A) \times E$. It is straightforward to verify that $\mathbf{P}E$ is a functor.

We must show that subfunctors of $F \times E$ are in natural one to one correspondence with natural transformations from F to $\mathbf{P}E$. We show this first for F representable, say $F = \mathrm{Hom}(-, A)$. We have

$$\begin{aligned}\mathrm{Nat}(F, \mathbf{P}E) &= \mathrm{Nat}(\mathrm{Hom}(-, A), \mathbf{P}E) \simeq \mathbf{P}E(A) \\ &= \mathrm{Sub}(\mathrm{Hom}(-, A) \times E) = \mathrm{Sub}(F) \times E).\end{aligned}$$

The case for general F follows from Proposition 3 and the fact that F is a colimit of representable functors (Exercise (FCR), Section 1.8).

Exercises 2.1

(PTTP)°. Prove that the product of two toposes is a topos.

(EPS)°. Let $B = \mathbf{P}A$ in Diagram (1); the subobject of $\mathbf{P}A \times A$ corresponding to $\mathrm{id}_{\mathbf{P}A}$ is denoted $\in A$ (it is the "element of" relation in $\mathcal{S}et$). Prove that for any subobject $U \to A \times B$ there is a unique arrow $\Phi_U : A \to \mathbf{P}B$ which makes the following diagram a pullback.

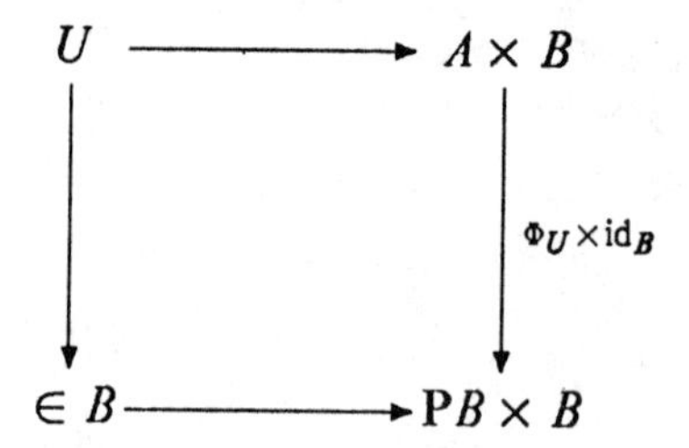

(INJ)°. An object B of a category is **injective** if for any subobject $A_0 \to A$ and arrow $f : A_0 \to B$ there is an arrow $\tilde{f} : A \to B$ extending f. Prove that in a topos any power object is injective. (Hint: Every subobject of $A_0 \times B$ is a subobject of $A \times B$. Now use Exercise (EPS).)

(SCQ). Show that in diagram (2) in $\mathcal{S}et$, if X is an equivalence relation and c is the class map, then $\mathrm{Sub}\, c$ in diagram (3) takes a set of equivalence classes to its union. What are $\mathrm{Sub}\, p_1$ and $\mathrm{Sub}\, p_2$?

(SUBLIM)°. Complete the proof of Proposition 3.

(CCL)°. Show that if X is a set, the functor $- \times X : \mathcal{S}et \to \mathcal{S}et$ commutes with colimits. (Hint: Show that $- \times X$ is left adjoint to $(-)^X$ and use Corollary 2 of Section 1.9).

2.2. Sheaves on a Space

Categories of sheaves were the original examples of toposes. In this section we will consider sheaves over topological spaces in some detail and prove that the category of sheaves over a fixed space is a topos. In Section 5.1, we give Grothendieck's generalization of the concept of sheaf. He invented it for use in algebraic geometry, but we will use it as the fundamental tool in building the connection between toposes and theories.

An excellent introduction to sheaf theory may be found in Tennison [1975].

Let X be a topological space and $\mathcal{O}(X)$ the category of open sets of X and inclusions. As we have seen, the category $\mathrm{Func}(\mathcal{O}(X)^{\mathrm{op}}, \mathcal{S}et)$ is a topos. An object of this category is called a **presheaf** on X, and the topos is denoted $\mathcal{P}sh(X)$. If the open set V is contained in the open set U, the induced map from FU to FV is denoted $F(U, V)$ and is called a **restriction map.** In fact, we often write $x|V$ instead of $F(U, V)x$ for $x \in FU$. This terminology is motivated by the example of rings of continuous functions mentioned in Section 1.4.

A presheaf is called a **sheaf** if it satisfies the following "local character" condition: If $\{U_i\}$ is an open cover of U and $x_i \in FU_i$ is given for each i in such a way that $x_i|U_i \cap U_j = x_j|U_i \cap U_j$ for all i and j, then there is a unique $x \in FU$ such that $x|U_i = x_i$.

The full subcategory of $\mathcal{P}sh(X)$ whose objects are the sheaves on X is denoted $\mathcal{S}h(X)$.

Examples. (i) For each topological space Y, the functor which assigns to each open set U of X the set of continuous functions from U to Y can easily be seen to be a sheaf.

(ii) Given a topological space Y and continuous map $p : Y \to X$, for each open U in X let $\Gamma(U, Y)$ denote the set of all continuous maps $s : U \to Y$ such that $p \circ s$ is the inclusion of U in X. These are called **sections** of p. Then $\Gamma(-, Y) : O(X)^{\mathrm{op}} \to \mathcal{S}et$ is a sheaf, called the **sheaf of sections** of p. We will see below (Theorem 3) that every sheaf arises this way.

The definition of sheaf is expressible by an exactness condition:

Proposition 1. *$F : \mathcal{O}(X)^{\mathrm{op}} \to \mathcal{S}et$ is a sheaf if and only if for every open set U and for every open cover $\{U_i\}$ of U, the following diagram is an equalizer.*

$$FU \longrightarrow \prod(FU_i) \underset{d_1}{\overset{d_0}{\rightrightarrows}} \prod F(U_i \cap U_j)$$

In this diagram, the left arrow is induced by restrictions. As for d_0 and

d_1, *they are the unique arrows for which the diagrams*

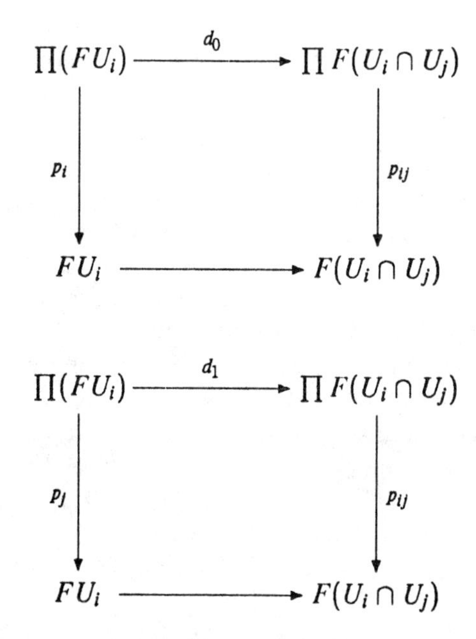

commute. The bottom arrows are the restriction maps.

Proof. Exercise.

Sheaf categories are toposes

Theorem 2. $Sh(X)$ *is a topos.*

Proof. We know Func($\mathcal{O}(X)^{\text{op}}$, Set) has limits, so to see that $Sh(X)$ has finite limits, it is sufficient to show that the limit of a diagram of sheaves is a sheaf. This is an easy consequence of Proposition 1 and is omitted.

The method by which we proved that Set-valued functor categories are toposes suggests that we define $\mathbf{P}(F)$ to be the functor whose values at U is the set of subsheaves (i.e., subobjects in $Sh(X)$) of $F \times \text{Hom}(-, U)$. Since $\mathcal{O}(X)$ is a partially ordered set, the sheaf $G = F \times \text{Hom}(-, U)$ has a particularly simple form, namely $G(V) = F(V)$ if $V \subseteq U$ and $G(V)$ is empty otherwise. Thus we write $F|U$ for $F \times \text{Hom}(-, U)$. Hence $(\mathbf{P}F)U$ is the set of subsheaves of $F|U$. It is necessary to show that this defines a sheaf (it is clearly a presheaf).

Let U_i be a cover of U. Suppose for each i we have a subsheaf G_i of $F|U_i$ such that $G_i|U_i \cap U_j = G_j|U_i \cap U_j$. Define G so that for all $V \subseteq U$,

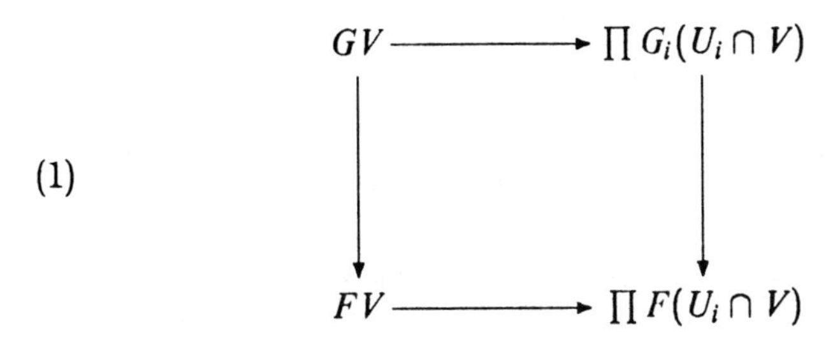

is a pullback. For other open sets V, GV is of course empty. Restriction maps are induced by the pullback property. It is clear that G is a subfunctor of F.

We first show that $G|U_j = G_j$. For $V \subseteq U_j$, the fact that $G_i(U_i \cap U_j) = G_j(U_i \cap U_j)$ implies that $G_i(V \cap U_j) = G_j(V \cap U_j)$. Thus we have the commutative diagram

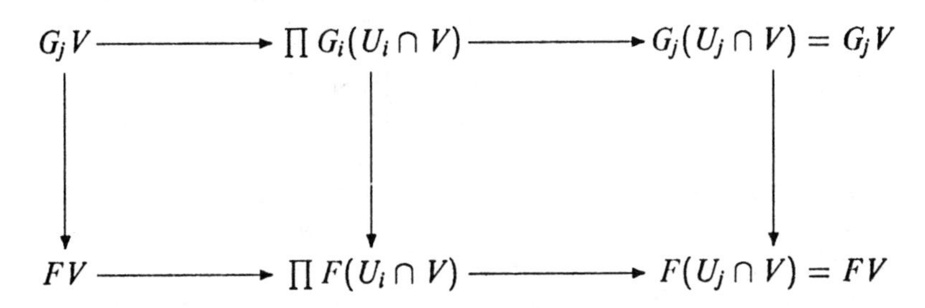

in which the top middle node is also $\prod G_i(U_j \cap V)$, the outer rectangle is a pullback and the middle arrow is a mono as G_i is a subfunctor of F. It follows from Exercise (PBCC) that the left square is a pullback too. But that pullback is $G(V)$ by definition.

To see that G is a subsheaf, let V_k be a cover of V. By Proposition 1 we need to show that the top row of the following diagram is an equalizer. By Exercise (EQPB)(a) it is sufficient to show that the left square in the diagram is a pullback.

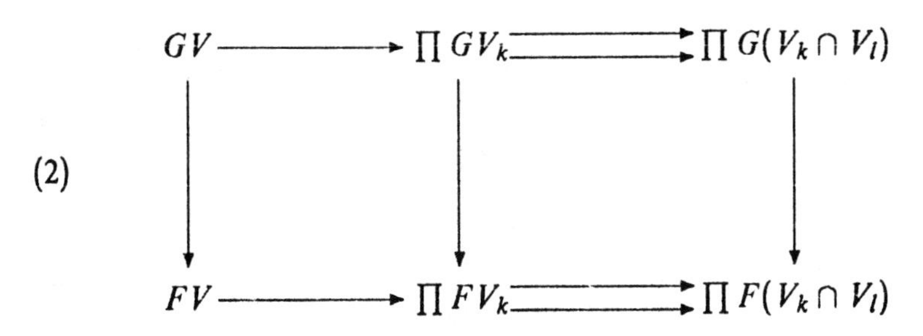

In the following commutative cube,

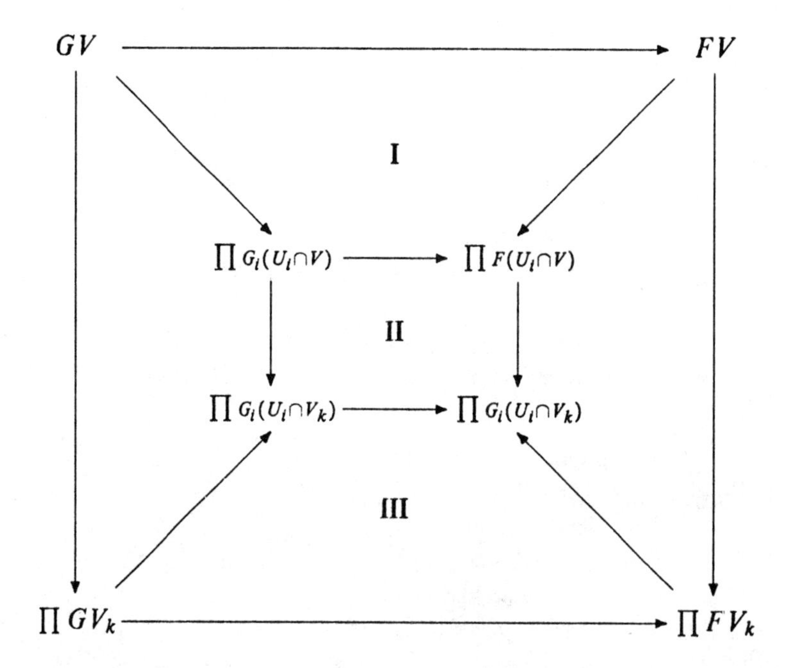

the square labelled I is a pullback by the definition of G and because $G|U_i = G_i$. Number III is a product of squares which are pullbacks from the definition of G. Finally II is a product of squares, each of which is the left hand square in a diagram of type (2) above with G_i replacing G and is a pullback because G_i is a sheaf (see Exercise (EQPB)(b). It follows from Exercise (PBBC) that the outer square is a pullback, which completes the proof.

Sheafification

In the rest of this section, we outline some functorial properties of toposes of sheaves. These results follow from more general results to be proved later when we discuss Grothendieck topologies, and so may be skipped. However, you may find considering this special case first helpful in understanding and motivating the ideas introduced later. We give only the constructions involved in the proofs; the verifications are left as exercises.

We saw in example (ii) that any space over X defines a sheaf on X. Given a presheaf F on X we will construct a space LF and a continuous map (in fact local homeomorphism) $p : LF \to F$ for which, for an open set U of X, the elements of FU are sections of p (when F is a sheaf).

Thus we need somehow to find the points of LF that lie over a particular point x of X. We construct the set of such points by using colimits.

Form the diagram of sets and functions consisting of all those sets FU for all open U which contain x and all the restriction maps between them. This diagram is actually a directed system. The colimit in $\mathcal{S}et$ of this diagram is denoted F_x and called the **stalk** or **fiber** of F at x. An element of F_x is an equivalence class of pairs (U, s) where $x \in U$ and $s \in FU$. The equivalence relation is defined by requiring that $(U, s) \cong (V, t)$ if and only if there is an open set W such that $x \in W \subseteq U \cap V$ and $s|W = t|W$. The equivalence class determined by (U, s) is denoted s_x and is called the **germ** determined by s at x. Thus s determines a map $\hat{s}$ from U to F_x which takes x to s_x.

Now let LF be the disjoint union of all the stalks of F. We topologize LF with the topology generated by the images of all the maps $\hat{s}$ for all open U in X and all sections $s \in FU$. These images actually form a basis (Exercise (IBAS)). LF comes equipped with a projection $p : LF \to X$ which takes s_x to x. This projection p is continuous and is in fact a local homeomorphism, meaning that for any y in LF there is an open set V of LF containing y for which the image of $p|V$ is an open set in X and $p|V$ is a homeomorphism from V to $p(V)$. To see this, let U be an open set containing $p(y)$ and define $V = \hat{s}(U)$, where y is the equivalence class containing (U, s). Note that LF is not usually Hausdorff, even when X is.

If $f : F \to G$ is a map of presheaves, then f induces (by the universal property of colimits) a map $f_x : F_x \to G_x$ for each $x \in X$, and so a map $Lf : LF \to LG$. It is a nice exercise to see that Lf is continuous and that this makes L a functor from $\mathcal{P}sh(X)$ to $\mathcal{T}op/X$.

LF is called the **total space** of F (or just the "space" of F). (Many people follow the French in calling the total space the "espace étalé". It is wrong to call it the "étale space" since "étale" is a different and also mathematically significant word.) We will denote by LH/X the category of spaces (E, p) over X with p a local homeomorphism.

The function Γ defined in Example (ii) above also induces a functor (also called Γ) from the category $\mathcal{T}op/X$ of spaces over X to $\mathcal{P}sh(X)$: If E and E' are spaces over X and $u : E \to E'$ is a map over X, then $\Gamma(u)$ is defined to take a section s of the structure map of E to $u \circ s$. It is easy to see that this makes Γ a functor. Note that for E over X, $\Gamma(E)$ is a presheaf, hence a functor from $\mathcal{O}(X)^{\mathrm{op}}$ to $\mathcal{S}et$; its value $\Gamma(E)(U)$ at an open set U is customarily written $\Gamma(U, E)$.

Theorem 3. *L is left adjoint to Γ. Moreover, L is a natural equivalence between $\mathcal{S}h(X)$ and LH/X.*

Proof. We will construct natural transformations $\eta : \mathrm{id} \to \Gamma \circ L$ and $\varepsilon : L \circ \Gamma \to \mathrm{id}$ for which $\Gamma\varepsilon \circ \eta\Gamma = \mathrm{id}$ and $\varepsilon L \circ L\eta = \mathrm{id}$, from which the adjointness will follow (Exercise (Uco) of Section 1.9).

Let F be a presheaf on X. On an open set U of X, define the natural transformation ηF by requiring that $(\eta F)U$ take an element s of FU to the section $\hat{s}$. On the other hand, for a space E over X, the continuous map εE is defined to take an element s_x of $L(\Gamma(E))$ to $s(x)$. The necessary verifications are left to the reader, as is the proof that when F is a sheaf, ηF is an isomorphism, and when $p : E \to X$ is a local homeomorphism, εE is a homeomorphism over X. The latter two facts prove that L is a natural equivalence between $\mathcal{Sh}(X)$ and LH/X.

The functor $\Gamma \circ L : \mathcal{Psh}(X) \to \mathcal{Sh}(X)$ is called the **sheafification functor**.

Corollary 4. *For any topological space X, LH/X is a topos.*

Corollary 5. *$\mathcal{Sh}(X)$ is a reflective subcategory of $\mathcal{Psh}(X)$.*

Proof. The reflector is the sheafification functor.

Change of base space

Any continuous function between topological spaces induces a pair of functors between the sheaf categories which are adjoint.

Given a continuous function $f : X \to Y$ and a sheaf F on X, the **direct image** functor $f_* : Sh(X) \to Sh(Y)$ is defined to be the restriction of

$$\mathrm{Func}(f^{-1}, \mathcal{Set}) : \mathrm{Func}(\mathcal{O}(X)^{\mathrm{op}}, \mathcal{Set}) \to \mathrm{Func}(\mathcal{O}(Y)^{\mathrm{op}}, \mathcal{Set}).$$

Note that f^{-1} is a functor from $\mathcal{O}(Y)^{\mathrm{op}}$ to $\mathcal{O}(X)^{\mathrm{op}}$. Thus f_* is composition with f^{-1}.

On the other hand, given a local homeomorphism $p : E \to Y$, $f^*(E)$ is defined to be the pullback

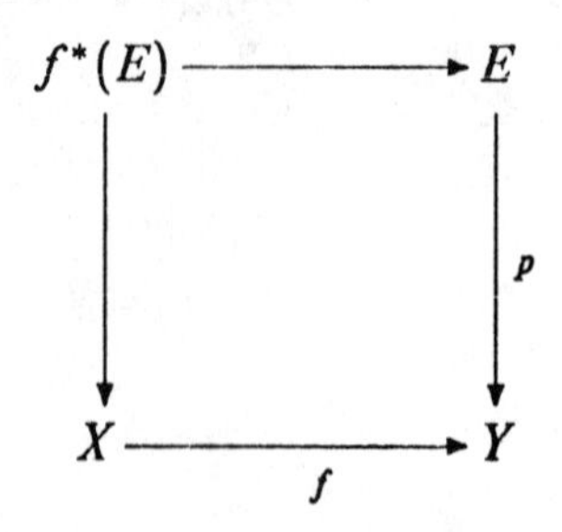

so that $f^*(E) = \{(x,e) \mid fx = pe\}$. It is easy to see that the map $(x,e) \to x$ is a local homeomorphism. On sections, f^* takes a section s of p defined on an open set V of Y to the map which takes $x \in f^{-1}(V)$ to (x, sfx).

Proposition 6. *f^* is left adjoint to f_*. Moreover, f^* preserves all finite limits.*

Note that f^* perforce preserves all colimits since it has a right adjoint.

Proof. f^* is the restriction of the pullback functor from Top/Y to Top/X, which has as a left adjoint composing with f (the proof is easy). Thus it preserves limits in Top/Y; but finite limits in LH/Y are the same as in Top/Y (again easy).

There is a natural map from E to $f_* \circ f^*(E)$ whose component on an open set V of Y takes a section s in $\Gamma(V,E)$ to the function $s \circ f : f^{-1}(V) \to E$, which by definition is an element of $\Gamma(V, f_* \circ f^*(E))$.

On the other hand, let F be a sheaf on X and let $x \in X$. For any open V of Y for which $x \in f^{-1}(V)$, we have a map from $\Gamma(f^{-1}(V)), F)$ to the stalk F_x, hence a map from $\Gamma(V, f_*(F))$ (which is, by definition, the same as $\Gamma(f^{-1}(V)))$ to F_x. This directed system defines a map t_x from the stalk of $f_*(F)$ at fx to the stalk of F at x. We then define a natural transformation from f^*f_* to the identity which takes a section s of $f^*f_*(F)$ defined on U to the section of F which takes $x \in U$ to $t_x(s(x))$. These two natural transformations satisfy the hypotheses of Exercise (Uco) of Section 1.9, so that f^* is left adjoint to f_*. The detailed verifications, which get a bit intricate, are left to the reader. This is a special case of adjoints of functors induced by maps of theories. The general situation will be dealt with in Chapter 8.

A **geometric morphism** between toposes is a functor $f : \mathcal{E} \to \mathcal{E}'$ with a left adjoint f^* which preserves finite limits. The functor f is usually written f_*, and f^* is called its "inverse image". Thus a continuous map $f : X \to Y$ of topological spaces induces a geometric morphism from $Sh(X)$ to $Sh(Y)$. We will study geometric morphisms in detail in Chapter 5.

Exercises 2.2

(CONT). (a) Verify that Examples (i) and (ii) really are sheaves.

(b) Show that Example (i) is a special case of Example (ii). (Hint: Consider the projection from $Y \times X$ to X.)

(EQL)°. Prove Proposition 1.

(LFU)°. Show that L is a functor.

(SFSH). Show that a subfunctor of a sheaf need not be a sheaf.

(IBAS). Let F be a presheaf on a topological space X. Show that the images $\hat{s}(U)$ for all open sets U in X and all sections $s \in FU$ form a basis for a topology on LF.

(LGNT)°. Carry out the verifications that prove Theorem 3. (You have to prove that for each F, ηF is a natural transformation, and for each p, εp is a continuous map; that η and ε are natural transformations; and that they satisfy the requirements in (a) of Exercise (Uco) of Section 1.9.)

(IFFS). Using the notation of the preceding exercise, prove that εE is an isomorphism if and only if $p : E \to X$ is a local homeomorphism, and ηF is an isomorphism if and only if F is a sheaf.

(STK). Show that two sheaves over the same space can have the same stalks at every point without being the same sheaf. (Hint: Look at a double covering of a circle versus two single circles lying over a circle.)

(LOC)°. Prove that the pullback of a local homeomorphism is a local homeomorphism.

(PT)°. Show that every point of a topological space X induces a geometric morphism from $\mathcal{S}et$ to $\mathcal{S}h(X)$ which takes a sheaf over X to its stalk over the point.

(EQPB)°. Consider the diagram

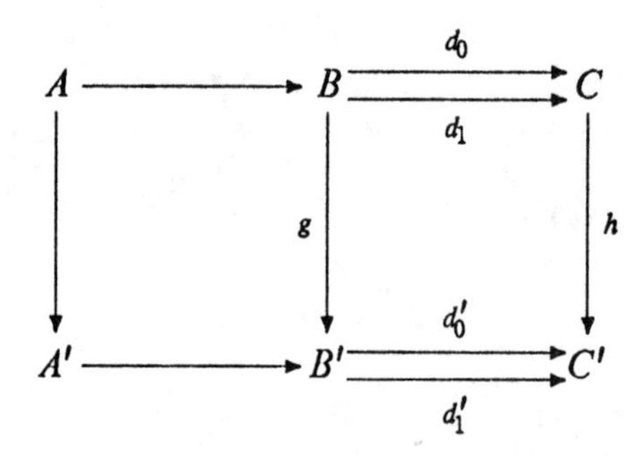

in which we assume the left square commutes, and $h \circ d_i = d_i' \circ g$, $i = 1, 2$. (We often say that such a square commutes serially, a notion which we will use a lot in later chapters.)

(a) Show that if the bottom row is an equalizer and the left square is a pullback, then the top row is an equalizer.
(b) Show that if the top row is an equalizer, the bottom row has both composites the same, and f is monic, then the left square is pullback.

(PBCC)°. Consider the diagram

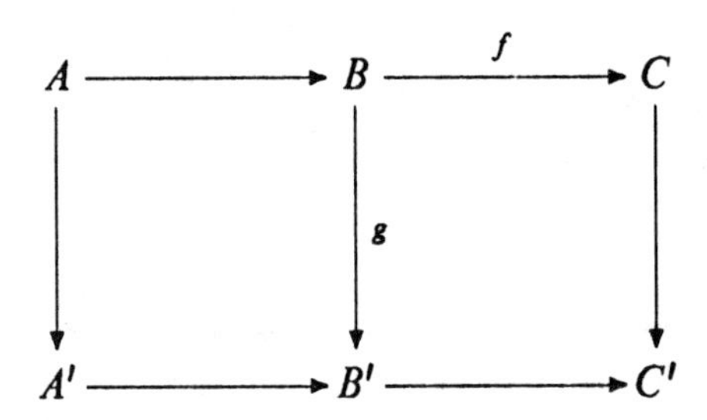

(a) Show that if both squares are pullbacks then so is the outer rectangle.

(b) Show that if the outer rectangle is a pullback and f and g are jointly monic, then the left square is a pullback. (f and g are jointly monic if $f(x) = f(y)$ and $g(x) = g(y)$ implies that $x = y$. Such a square is called a **mono square.**)

2.3. Properties of Toposes

In this section and the next, we will state and prove those basic properties of toposes which can conveniently be proved without using triple theory.

In the following, $\mathcal{E}$ is a topos with power-object function $\mathbf{P}$.

Functoriality of P

Proposition 1. *Let $\mathcal{A}$ and $\mathcal{B}$ be categories and $\Phi : \mathcal{A}^{\mathrm{op}} \times \mathcal{B} \to \mathbf{Set}$ be a functor. Let $F : \mathrm{Ob}\,\mathcal{A} \to \mathrm{Ob}\,\mathcal{B}$ be a function such that for each object A of $\mathcal{A}$ there is a natural (in B) equivalence*

$$\Phi(A, B) \simeq \mathrm{Hom}(FA, B).$$

Then there is a unique way of extending F to a functor (also denoted F) $\mathcal{A} \to \mathcal{B}$ in such a way that the above equivalence is a natural equivalence in both A and B.

Proof. Fix a morphism $f : A' \to A$. We have a diagram of functors on

$\mathcal{B}$

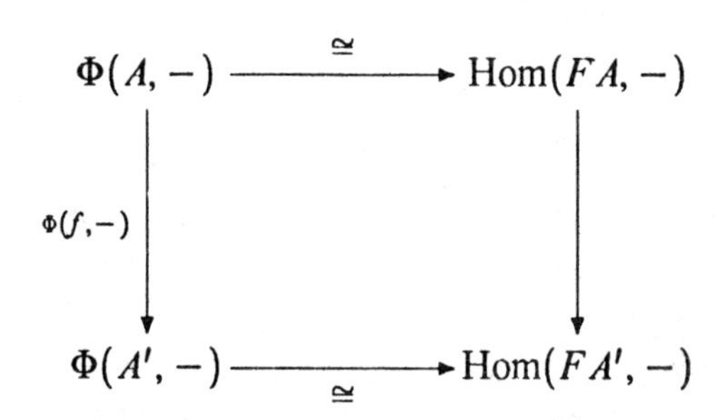

in which the right arrow is defined by the indicated isomorphisms to make the diagram commute. The result is a natural (in $\mathcal{B}$) transformation from $\mathrm{Hom}(FA, -)$ to $\mathrm{Hom}(FA', -)$ which is induced by a morphism we denote $Ff : FA' \to FA$. The naturality in $\mathcal{A}$ and the functoriality of F are clear.

Proposition 2. $\mathbf{P}$ *has a unique extension to a functor from* $\mathcal{E}^{\mathrm{op}}$ *to* $\mathcal{E}$ *with the property that for any arrow* $g : A' \to A$*, the following diagram commutes. Here,* φ *is the natural isomorphism of Section 2.1.*

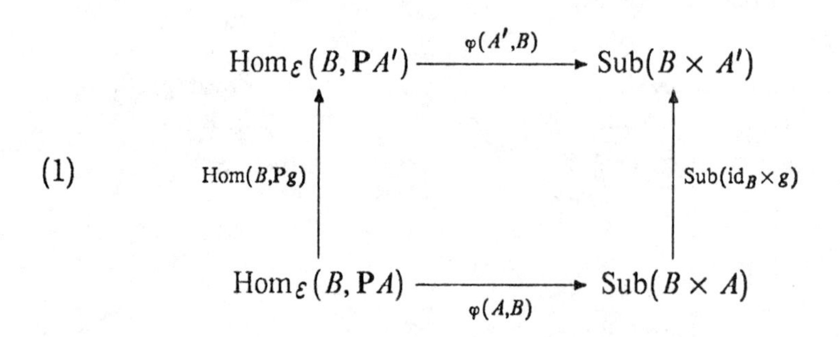

Proof. Apply Proposition 1 above with $\mathcal{A} = \mathcal{E}$, $\mathcal{B} = \mathcal{E}^{\mathrm{op}}$, and $F = \mathbf{P}^{\mathrm{op}}$.

It is worthwhile to restate what we now know about $\mathbf{P}$ in view of the definition of Sub. If $s : U \to B \times A$ is a subobject and $[s] = \Phi^{-1}(s) : B \to \mathbf{P}A$ is the corresponding element of $\mathbf{P}A$, then diagram (1) of Section 2.1 says that for an arrow $f : B' \to B$, the element $[s]f$ of $\mathbf{P}A$ defined on B' corresponds via the adjunction to $\mathrm{Sub}(f \times \mathrm{id}_B)(s)$ which is the pullback of s along $f \times \mathrm{id}_B$. On the other hand, given $g : A' \to A$, then according to diagram (1) of this section, the element $\mathbf{P}g(s)$ of $\mathbf{P}A'$ corresponds to $\mathrm{Sub}(\mathrm{id}_B \times g)(s)$, which is the pullback of s along $\mathrm{id}_B \times g$.

Proposition 3. $\mathbf{P}^{\mathrm{op}} : \mathcal{E} \to \mathcal{E}^{\mathrm{op}}$ *is left adjoint to* $\mathbf{P} : \mathcal{E}^{\mathrm{op}} \to \mathcal{E}$.

Proof. The arrow $A \times B \to B \times A$ which switches coordinates induces a natural isomorphism from the bifunctor whose value at (B, A) is $\mathrm{Sub}(B \times A)$ to the bifunctor whose value at (B, A) is $\mathrm{Sub}(A \times B)$. This then induces a natural isomorphism

$$\mathrm{Hom}_{\mathcal{E}}(B, \mathbf{P}A) \simeq \mathrm{Hom}_{\mathcal{E}}(A, \mathbf{P}B) = \mathrm{Hom}_{\mathcal{E}^{\mathrm{op}}}(\mathbf{P}B, A)$$

which proves the Proposition.

We sometimes say, "$\mathbf{P}$ is adjoint to itself on the left."

The subobject classifier

Since $A \simeq A \times 1$, the subobject functor is represented by $\mathbf{P}(1)$. This object is so important in a topos that it deserves its own name, which is traditionally Ω. It follows from the Yoneda Lemma that Ω has a representative subobject true : $\Omega_0 \to \Omega$ with the property that for any object A and any subobject $a : A_0 \to A$ there is a unique map $\chi a : A \to \Omega$ such that a is the pullback of true along χa. This means that there is a map $A_0 \to \Omega_0$ (whose nature will be clarified by Proposition 4) for which the following diagram is a pullback:

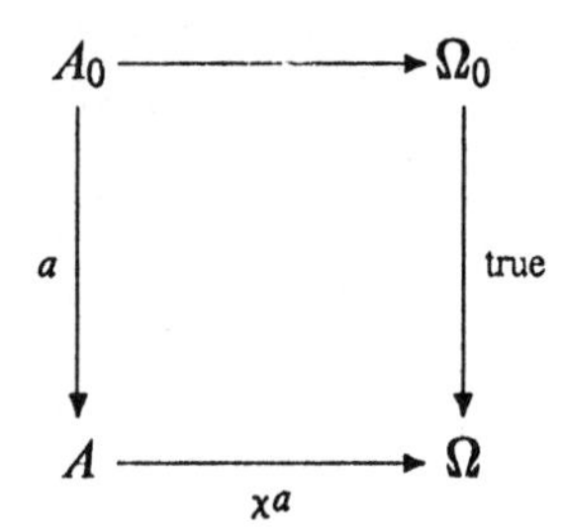

Proposition 4. Ω_0 *is the terminal object.*

Proof. For a given object A, there is at least one map from A to Ω_0; namely the map u given by the following pullback:

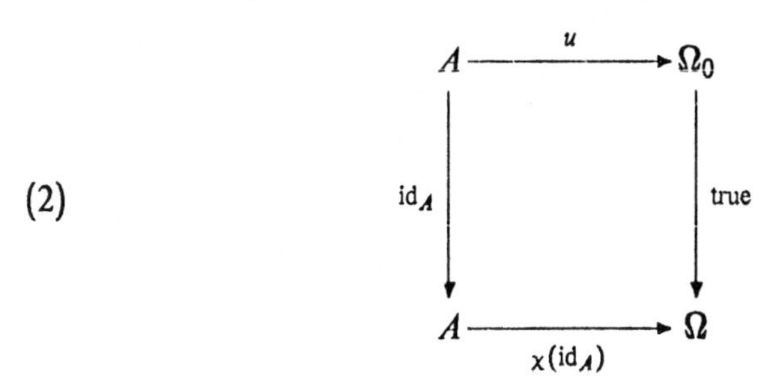

To see that u is the only map, suppose $v : A \to \Omega_0$ is another map. Then this diagram is a pullback (see Exercise (PBM) of Chapter 1.6):

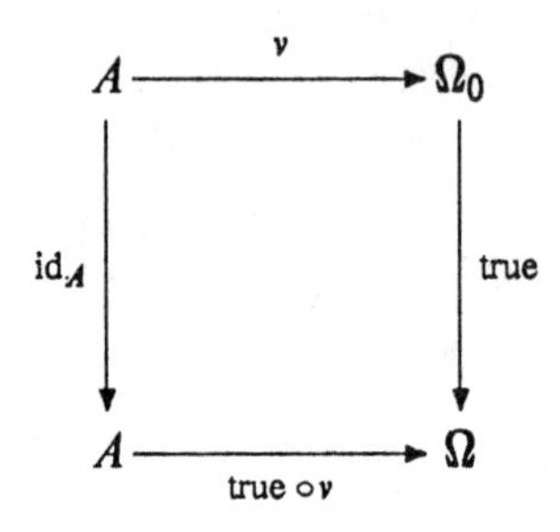

The uniqueness part of the universal mapping property of Ω says that $\text{true} \circ v = \chi(\text{id}_A)$, which is $\text{true} \circ u$ by diagram (2). Since true is mono, this means $u = v$. It follows that every object has exactly one map to Ω_0.

Ω is called the **subobject classifier.** In *Set*, any two element set is a subobject classifier in two different ways, depending on which of the two elements you take to be true. If for each set A you take $\mathbf{P}A$ to be the actual set of subsets of A, then the preceding construction makes the subobject classifier the set of subsets of a singleton set with true taking the value of the nonempty subset. The subobject classifier in a category of G-sets is a two element set with the trivial action, and comments similar to those just made apply here too. As we will see, in most toposes Ω is not nearly so simple.

Recall from Exercise (REGMON) of Section 1.7 that a monomorphism is regular if it is the equalizer of two arrows. In that exercise you were asked to show that every morphism whose domain is 1 is a regular mono and that the pullback of a regular mono is a regular mono. Thus:

Corollary 5. *Every monomorphism in a topos is regular.*

The singleton map

We will define a special arrow $\{\} : A \to \mathbf{P}A$ which in the case $\mathcal{E} = \mathit{Set}$ is the map taking x to the singleton set containing x. Its importance lies in the fact that composing with $\{\}$ internalizes the construction of the graph of a function, for if $f : B \to A$, then $\{\}f : B \to \mathbf{P}A$ corresponds to the subobject $[(b, a) \mid a = f(b)]$ of $B \times A$, which in *Set* is in fact the graph of f. (Thus by this definition, the graph of f in *Set* is the set of ordered pairs $(b, f(b))$ *regarded as a subobject of* $B \times A$, so that the graph carries with it the information about the codomain of f as well as its domain.)

The fact that $\{\}f$ should correspond to the graph of f suggests the way to construct $\{\}$. Let γ be the natural transformation from $\mathrm{Hom}_{\mathcal{E}}(-,A)$ to $\mathrm{Sub}(-\times A)$ defined by having γB take $f : B \to A$ to the subobject $(\mathrm{id}_B,f) : B \to B\times A$. Observe in the first place that if $(\mathrm{id}_B,f)\circ u = (\mathrm{id}_B,f)\circ v$, then $(u,fu) = (v,fv)$ so that (id_B,f) is indeed monic. To show that γ is a natural transformation translates into showing that for any arrow $g : B' \to B$,

$$\mathrm{Sub}(g\times A)(\mathrm{id}_B,f) = (\mathrm{id}_{B'},fg).$$

By definition of Sub, that requires showing that the following diagram is a pullback, which is easy.

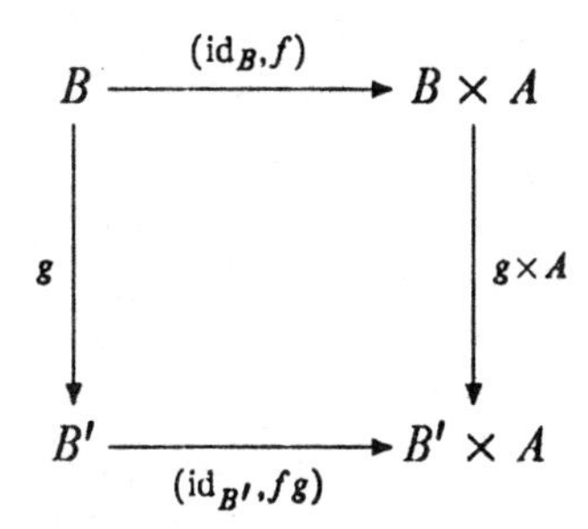

Now let $\overline{\gamma}$ be the natural transformation

$$\overline{\gamma} : \mathrm{Hom}_{\mathcal{E}}(-,A) \to \mathrm{Sub}(-\times A) \simeq \mathrm{Hom}_{\mathcal{E}}(-,PA).$$

Let $\{\} : A \to PA$ be the corresponding arrow given by the Yoneda Lemma. Recall that according to the proof of the Yoneda Lemma, for $f : B \to A$, $\overline{\gamma}B(f) = \{\}f$.

The following proposition just says that a morphism is determined by its graph.

Proposition 6. $\{\}$ *is monic.*

Proof. If $f,f' : B \to A$ are two morphisms for which $(\mathrm{id}_B,f) : B \to B\times A$ and $(\mathrm{id}_B,f') : B \to B\times A$ give equivalent subobjects, then there is an isomorphism j of B for which $(j,fj) = (\mathrm{id}_B,f')$, whence $f = f'$. Since by construction $\{\}f$ corresponds by adjunction to the subobject $(\mathrm{id}_B,f) : B \to B\times A$, $\{\}$ must be monic.

Equivalence relations

As observed in Exercise (EQC) of Section 1.8, the kernel pair of a regular epimorphism is an effective equivalence relation. In a topos, the converse is true:

Theorem 7. *In a topos, every equivalence relation is effective.*

Proof. Let E be an equivalence relation on A. E is a subobject of $A\times A$, so corresponds to an arrow $[\ ]_E : A \to \mathbf{P}A$ (which in $\mathcal{S}et$ is the class map). An element of A defined on T is sent to the subobject of $T\times A$ (element of $\mathbf{P}A$ defined on T) which is the pullback of the diagram

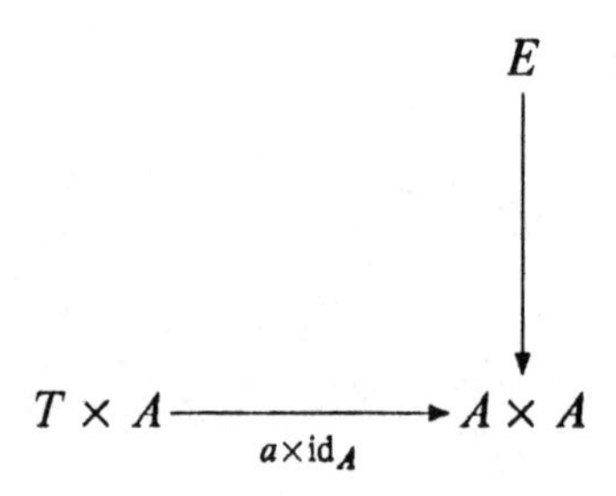

Thus if $a \in^T A$ and $(t, a') \in^V T\times A$, then $(t, a') \in^V [a]_E$ if and only if $(a\circ t, a') \in^V E$. This fact is used twice in the proof below.

To show that E is the kernel pair of $[\ ]_E$, we must show that if a_1 and a_2 are elements of A defined on T then $[a_1]_E = [a_2]_E$ if and only if $(a_1, a_2) \in E$. To see this, let $(t, a) : V \to T\times A$ be an element of $T\times A$ defined on V. The corresponding subset of $\mathrm{Hom}(V, T\times A)$ is

$$[a_1]_E = \{(t,a) \mid (a_1\circ t, a) \in E\};$$

$[a_2]_E$ is defined similarly. If $[a_1]_E = [a_2]_E$, let $V = T$, $t = \mathrm{id}_T$, and $a = a_2$. By reflexivity, $(\mathrm{id}_T, a_2) \in [a_2]_E$, hence belongs to $[a_1]_E$. Therefore $(a_1, a_2) = (a_1\circ \mathrm{id}_T, a_2) \in E$.

For the converse, suppose $(a_1, a_2) \in E$. Then for all $t : V \to T$, $(a_1\circ t, a_2\circ t) \in E$. Suppose $(t, a) \in [a_1]_E$ (defined on V), then $(a_1\circ t, a) \in E$ and therefore by symmetry and transitivity, $(a_2\circ t, a) \in E$. Hence $[a_1]_E \subseteq [a_2]_E$. The other inclusion follows by symmetry. This shows that E is the kernel pair of $[\ ]_E$.

Exercises 2.3

(PCA). Let $G : \mathcal{B} \to \mathcal{A}$ be a functor and $F : \mathcal{O}b\,\mathcal{A} \to \mathcal{O}b\,\mathcal{B}$ be a function such that for each $A \in \mathcal{A}$ there is a natural (in B) equivalence

$$\mathrm{Hom}(A, GB) \simeq \mathrm{Hom}(FA, B).$$

Use Proposition 1 to show that F has a unique extension to a functor left adjoint to G. (This gives a second proof of the pointwise construction of adjoints, Section 1.9).

(BAL)°. Prove that in a topos an arrow which is both a monomorphism and an epimorphism is an isomorphism. (Hint: Use Corollary 5.)

(TF). (Interchanging true and false). Describe how to define **P** in *Set* so that the subobject classifier is the set of subsets of a one-element set and the value of true is the empty set.

(OMT)°. Let X be a topological space. If U is open in X, define $\Omega(U)$ to be the set of open subsets of U. If $V \subseteq U$, let $\Omega(U, V)(W) = W \cap V$.

(a) Show that Ω is a sheaf, and is the subobject classifier in $Sh(X)$.
(b) What is $\{\}$?

(TPPB)°. Suppose that for each object A of a topos there is a map $\mathrm{j}A : \mathrm{Sub}(A) \to \mathrm{Sub}(A)$ with the property that whenever

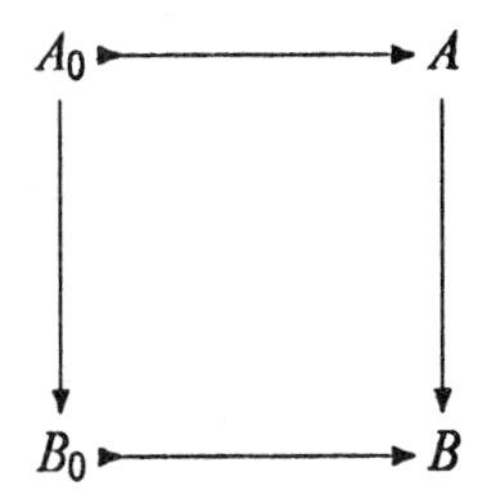

is a pullback, then there is a pullback

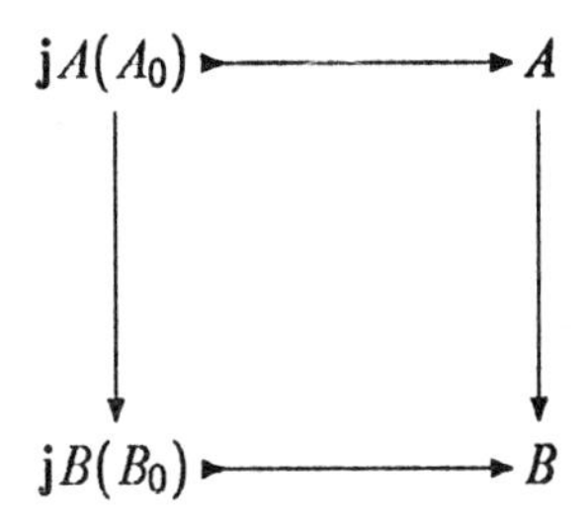

where the top arrow is the inclusion.

Use the Yoneda lemma to show that show that these functions constitute a natural endomorphism of **P**.

(LMA). (a) Show that if M is a monoid and $\mathcal{E}$ is the topos of left M actions and equivariant maps, then Ω is the set of left ideals of M, with action $mL = \{n \mid nm \in L\}$.
(b) Show that in $\mathcal{E}$, Ω has exactly two global elements.

2.4. The Beck Conditions

The Beck conditions are useful technical conditions concerning inverse images and forward images induced by inclusions.

A mono $a : A_0 \to A$ induces a set function $a \circ - : \mathrm{Sub}(A_0) \to \mathrm{Sub}(A)$ by composition, taking the subobject determined by $u : A_1 \to A_0$ to the subobject determined by $a \circ u$.

Proposition 1 (Beck Condition, external version). *Let*

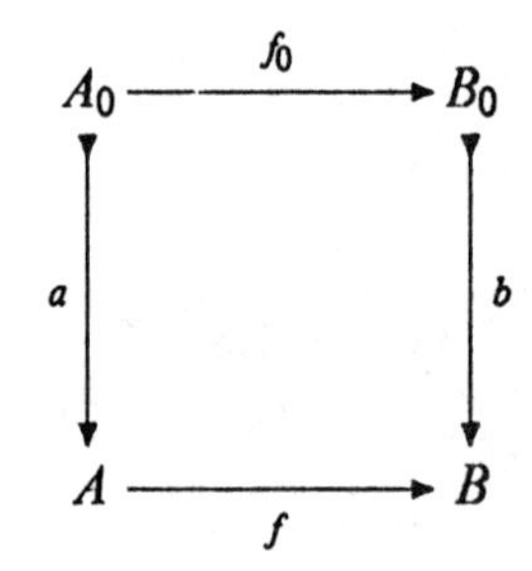

be a pullback. Then

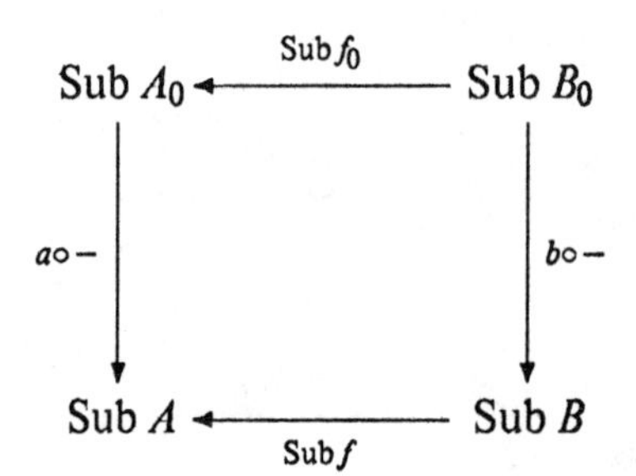

commutes.

Proof. This translates into proving that if both squares in the following diagram are pullbacks, then so is the outer rectangle. That is Exercise (Pbcc)(a) of Section 2.3.

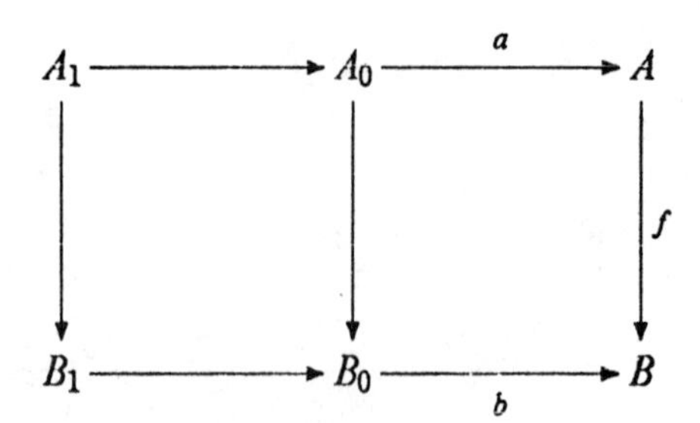

Observe that in *Set*, A_0 is the inverse image of B_0 along f.

The object $\mathbf{P}A$ is said to "internalize" $\mathrm{Sub}(A)$. For a given monic $a : A_0 \to A$, there is an arrow $\exists a : \mathbf{P}A_0 \to \mathbf{P}A$ which internalizes $a \circ -$ in the same sense. To construct $\exists a$, we first observe that $a \circ -$ induces an arrow

$$(B \times a) \circ - : \mathrm{Sub}(B \times A_0) \to \mathrm{Sub}(B \times A)$$

for any object B.

Proposition 2. $(- \times a)$ *is a natural transformation from* $\mathrm{Sub}(- \times A_0)$ *to* $\mathrm{Sub}(- \times A)$.

Proof. Suppose $f : B' \to B$ is given. Then the diagram

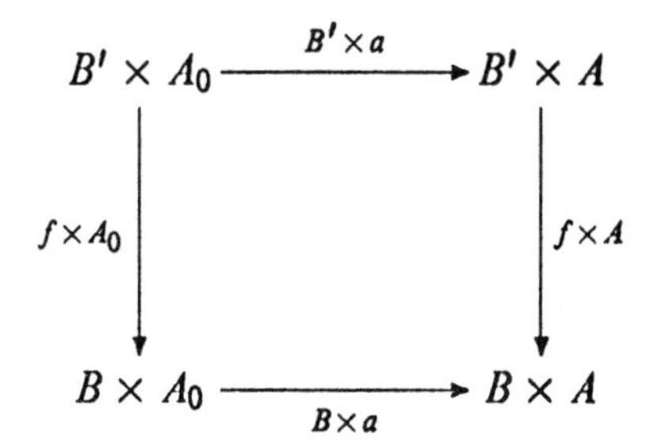

is a pullback (easy exercise), so that by Proposition 1,

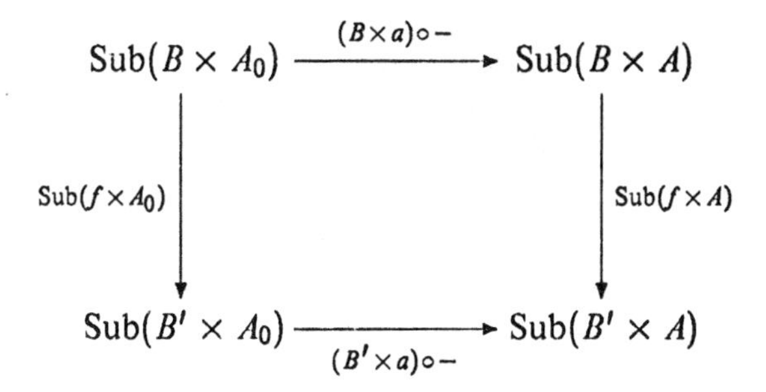

commutes as required for $(- \times a)$ to be a natural transformation.

Definition. If $a : A_0 \to A$ is monic, $\exists a : \mathbf{P}A_0 \to \mathbf{P}A$ is the arrow induced by the natural transformation

$$\mathrm{Hom}_{\mathcal{E}}(-, \mathbf{P}A_0) \simeq \mathrm{Sub}(- \times A_0) \to \mathrm{Sub}(- \times A) \simeq \mathrm{Hom}_{\mathcal{E}}(-, \mathbf{P}A).$$

$\exists a$ takes an element of $\mathbf{P}A_0$ to the element of $\mathbf{P}A$ regarded as the same subobject.

Proposition 3 (The Beck condition, internal version). *If*

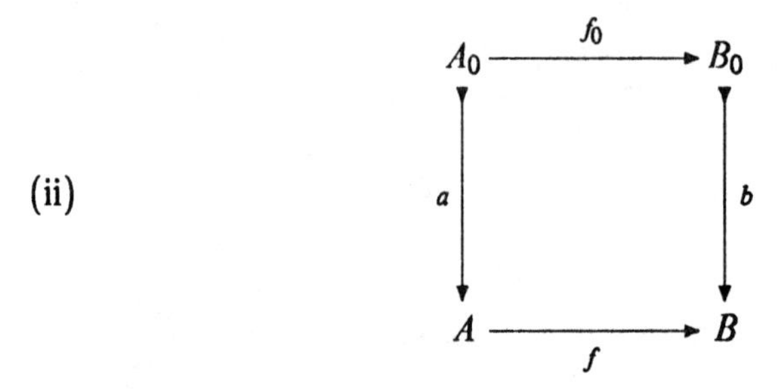

is a pullback, then

(iii)
$$\begin{array}{ccc} PA_0 & \xleftarrow{Pf_0} & PB_0 \\ {\scriptstyle \exists a}\downarrow & & \downarrow{\scriptstyle \exists b} \\ PA & \xleftarrow[Pf]{} & PB \end{array}$$

commutes.

What the Beck condition really says is that if X is a subobject of B_0, $f^{-1}(X)$ is unambiguously defined.

Proof. If (ii) is a pullback, then so is

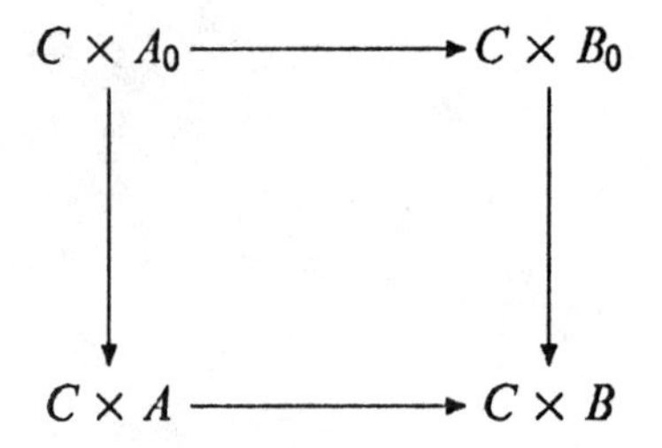

so by Proposition 1,

$$\begin{array}{ccc} \mathrm{Sub}(C \times A_0) & \longleftarrow & \mathrm{Sub}(C \times B_0) \\ \downarrow & & \downarrow \\ \mathrm{Sub}(C \times A) & \longleftrightarrow & \mathrm{Sub}(C \times B) \end{array}$$

commutes. Hence because diagram (1) of Section 2.1 commutes,

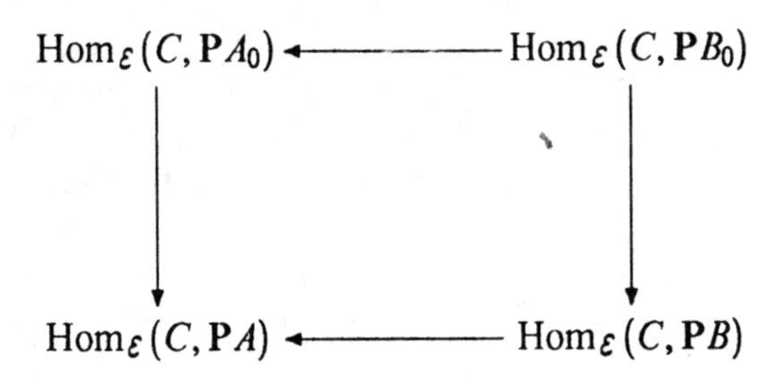

commutes, so that (iii) commutes by definition of $\exists a$.

Exercises 2.4

(ExID)°. Show that for any object A in a topos, $\exists \mathrm{id}_A = \mathrm{id}_{\mathbf{P}A}$.

(ExINV)°. Show that for any monic $a : A_0 \to A$ in a topos, $\mathbf{P}a \circ \exists a = \mathrm{id}_{A_0}$.

2.5. Notes to Chapter 2

The development of topos theory resulted from the confluence of two streams of mathematical thought from the sixties. The first of these was the development of an axiomatic treatment of sheaf theory by Grothendieck and his school of algebraic geometry. This axiomatic development culminated in the discovery by Giraud that a category is equivalent to a category of sheaves for a Grothendieck topology if and only if it satisfies the conditions for being what is now called a Grothendieck topos (section 6.8).The main purpose of the axiomatic development was to be able to define sheaf cohomology. This purpose was amply justified by Deligne's proof of the Weil conjectures [1974].

The second stream was Lawvere's continuing search (which, it is probably only a slight exaggeration to state, had characterized his career to that date) for a natural way of founding mathematics (universal algebra, set theory, category theory, etc.) on the basic notions of morphism and composition of morphisms. All formal (and naive) presentations of set theory up to then had taken as primitives the notions of elements and sets with membership as the primitive relation. What Lawvere had in mind for set theory was to take sets and functions as the primitives (and you don't really need the sets if you are interested in reducing the number of primitives to a minimum—see Exercise (SGRPOID) of Chapter 1.1) and the partial operation of composition as the basic relation.

In a formal way it is clear that this can always done by defining the terminal object 1 and then an element as a morphism with domain 1. Subobjects and membership can be readily defined and it is clear that set theory can be recovered. However, Lawvere did not have in mind a slavish translation of Zermelo-Fraenkel set theory into categorical language, but rather a treatment in which functions were clearly the fundamental notion. See Lawvere [1965] for an example of this. The closest he had come prior to 1969 was the notion of a hyperdoctrine which is similar to that of a topos except that PA is a category rather than an object of the ambient category.

The foundation of mathematics on the concept of function or arrow as primitive is revolutionary, but no more revolutionary than the introduction of set theory was early in the century. The idea of constructing a quotient space without having to have an ambient space including it, for example, was made possible by the introduction of set theory, in particular by the advent of the rather dubious idea that a set can be an element of another set. There is probably nothing in the introduction of topos theory as foundations more radical than that.

In the fall of 1969, Lawvere and Tierney arrived together at Dalhousie University and began a research project to study sheaf theory axiomatically. To be a possible foundation for set theory, the axioms had to be elementary — which Giraud's axioms were not. The trick was to find enough elementary consequences of these axioms to build a viable theory with.

The fact that a Grothendieck topos has arbitrary colimits and a set of generators allows free use of the special adjoint functor theorem to construct adjoints to colimit-preserving functors. Lawvere and Tierney began by assuming explicitly that some of these adjoints existed and they and others pared this set of hypotheses down to the current set.

They began by defining a topos as a category with finite limits and colimits such that for each $f : A \to B$ the functor

$$f^* : \mathcal{E}/A \to \mathcal{E}/B$$

gotten by pulling back along f has a right adjoint and that for each object A , the functor $\mathcal{E}^{\mathrm{op}} \to \mathit{Set}$ which assigns to B the set of partial maps B to A is representable. During the year at Dalhousie, these were reduced to the hypothesis that $\mathcal{E}$ be cartesian closed (i.e. that

$$-- \times A : \mathcal{E} \to \mathcal{E}/A$$

have a right adjoint) and that partial functions with codomain 1 (that is, subobjects) be representable. Later Mikkelsen showed that finite colimits could be constructed and Kock that it was sufficient to assume finite limits and power objects.

The resulting axioms, even when the axioms for a category are included, form a much simpler system on which to found mathematics than the Zermelo-Fraenkel axioms. Moreover, they have many potential advantages, for example in the treatment of variability.

It has been shown that the topos axioms augmented by axioms of two-valuedness and choice give a model of set theory of power similar to that of Zermelo-Fraenkel, but weaker in that all the sets appearing in any axiom of the replacement schema must be quantified over sets rather than over the class of all sets. See Mitchell [1972] and Osius [1974, 1975].

Chapter 3.

Triples

From one point of view, a triple is an abstraction of certain properties of algebraic structures. From another point of view, it is an abstraction of certain properties of adjoint functors (Theorem 1 of Section 3.1). Triple theory has turned out to be an important tool for studying toposes. In this chapter, we develop those parts of the theory we need to use in developing topos theory. In Chapter 9, we present additional topics in triple theory.

3.1. Definition and Examples

A **triple** $\mathsf{T} = (T, \eta, \mu)$ on a category $\mathcal{C}$ is an endofunctor $T : \mathcal{C} \to \mathcal{C}$ together with two natural transformations $\eta : \mathrm{id}_{\mathcal{C}} \to T$, $\mu : TT \to T$ subject to the condition that the following diagrams commute.

(1)

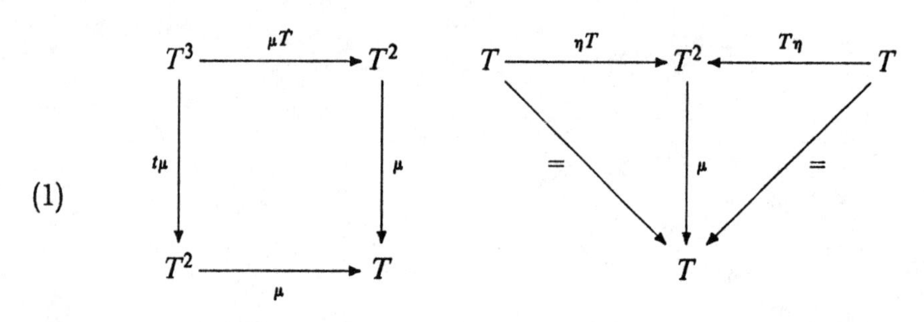

associative identity left and right unitary identities

In these diagrams, T^n means T iterated n times. As explained in Section 1.3, the component of μT at an object X is the component of μ at TX, whereas the component of $T\mu$ at X is $T(\mu X)$; similar descriptions apply to η.

The terms "monad", "triad", "standard construction" and "fundamental construction" have also been used in place of "triple".

Examples

The reader will note the analogy between the identities satisfied by a triple and those satisfied by a monoid. In fact, the simplest example of a triple involves monoids:

(i) Let M be a monoid and define $T : \mathit{Set} \to \mathit{Set}$ by $TX = M \times X$. Let $\eta X : X \to M \times X$ take x to $(1_M, x)$ and $\mu X : M \times M \times X \to M \times X$ take (m, n, x) to (mn, x). Then the associative and unitary identities follow from those of M.

(ii) In a similar way, if R is a commutative ring and A an associative unitary R-algebra, there is a triple on the category of R-modules taking M to $A \otimes M$. The reader may supply η and μ.

(iii) A third example is obtained by considering an object C in a category $\mathcal{C}$ which has finite sums, and defining $T : \mathcal{C} \to \mathcal{C}$ by $TX = X + C$. Take $\eta X : X \to X + C$ to be the injection into the sum and $\mu X : X + C + C \to X + C$ to be $\mathrm{id}_X + \nabla$, where ∇ is the codiagonal—the map induced by id_C.

(iv) If $\mathcal{C}$ is a category with arbitrary products and D is an object of $\mathcal{C}$, we can define a triple $\mathrm{T} = (T, \eta, \mu)$ on $\mathcal{C}$ by letting $TC = D^{\mathrm{Hom}(C,D)}$. To define T on arrows, as well as to define η and μ, we establish some notation which will be very useful later. For $u : C \to D$, let $\langle u \rangle : TC \to D$ be the corresponding projection from the product. Then for $f : C' \to C$, we must define

$$Tf : D^{\mathrm{Hom}(C',D)} \to D^{\mathrm{Hom}(C,D)}.$$

The universal mapping property of the product is such that the map Tf is uniquely determined by giving its projection on every coordinate. So if $v : C \to D$, define $\langle v \rangle \circ Tf = \langle v \circ f \rangle$. The proof of functoriality is trivial. We define $\eta C : C \to TC$ by $\langle u \rangle \circ \eta C = u$ and $\mu : T^2C \to TC$ by $\langle u \rangle \circ \mu C = \langle \langle u \rangle \rangle$. We could go into more detail here, but to gain an understanding of the concepts, you should work out the meaning of the notation yourself. Once you have facility with the notation, the identities are trivial to verify, but they were mind-bogglingly hard in 1959 using elements. You might want to try to work these out using elements to see the difficulty, which comes in part because the index set is a set of functions.

(v) More generally, if $\mathcal{C}$ is a category with arbitrary products and $\mathcal{D}$ is a *set* of objects of $\mathcal{C}$, let

$$TC = \prod\{D \mid D \in \mathcal{D}, f : C \to D\}.$$

This defines T on objects. The remainder of the construction is similar to the one above and is left to the reader.

(vi) An example of a different sort is obtained from the free group construction. Let $T : \mathcal{Set} \to \mathcal{Set}$ take X to the underlying set of the free group generated by X. Thus TX is the set of equivalence classes of words made up of symbols x and x^{-1} for all $x \in X$; the equivalence relation is that generated by requiring that any word containing a segment of the form xx^{-1} or of the form $x^{-1}x$ be equivalent to the word obtained by deleting the segment. We will denote the equivalence class of a word w by $[w]$, and we will frequently say "word" instead of "equivalence class of words". The map ηX takes x to $[x]$, whereas μX takes a word of elements of TX, i.e., a word of words in X, to the word in X obtained by dropping parentheses. For example, if x, y and z are in X, then $[xy^2z^{-1}]$ and $[z^2x^2]$ are in TX, so $w = [[xy^2z^{-1}][z^2x^2]] \in TTX$, and $\mu X(w) = [xy^2zx^2] \in TX$. There are many similar examples based on the construction of free algebraic structures of other sorts. Indeed, every triple in $\mathcal{Set}$ can be obtained in essentially that way, provided you allow infinitary operations in your algebraic structures.

Sheaves

In this section we describe the first triple ever explicitly considered. It was produced by Godement, who described the construction as the standard construction of an embedding of a sheaf into a "flabby" sheaf (*faisceau flasque*).

In Example (2) of Section 2.2, we described how to construct a sheaf Γ given any continuous map $p : Y \to X$ of topological spaces. If each fiber of p (that is, each set $p^{-1}(x)$ for $x \in X$) is endowed with the structure of an Abelian group in such a way that all the structure maps $+ : Y \times_X Y \to Y$, $0 : X \to Y$ (which assigns the 0 element of $p^{-1}(x)$ to x), and $- : Y \to Y$ are continuous, then the sheaf of sections becomes in a natural way an Abelian group. In the same way, endowing the fibers with other types of algebraic structure (rings or R-modules are the examples which most often occur in mathematical practice) in such a way that all the structure maps are continuous makes the sheaf of sections become an algebraic structure of the same kind. In fact a sheaf of Abelian groups is an Abelian group at three levels: fibers, sections, and as an Abelian group object in the category $\mathcal{Sh}(X)$. (See Exercise (SHAB)).

If Y is retopologized by the coarsest topology for which p is continuous (so all fibers are indiscrete), then every section of p is continuous. In general, given any set map $p : Y \to X$, using the coarsest topology this way produces a sheaf Rp which in fact is the object part of a functor $R : \mathcal{Set} / |X| \to \mathcal{Sh}(X)$, where $|X|$ is the discrete space with the same points as X. (See Exercise (RFUN)). The resulting sheaf has the property that all its restriction maps are surjective. Such a sheaf is called **flabby,** and Godement was interested in them for the purpose of constructing resolutions of objects to compute homology groups with.

For each sheaf F, Godement constructed a sheaf TF (which turns out to be flabby) which defines the object map of the functor part of a triple, as we will describe. Let Y be the disjoint union of the stalks of F, and let $p : Y \to X$ take the stalk F_x to x. Topologize Y by the coarsest topology for which p is continuous and let TF be the sheaf of sections of Y. (Compare the construction of LF in Section 2.2).

Evidently, $TFU = \sum\{F_x \mid x \in U\}$. Define $\eta F : F \to F$ by requiring that $\eta F(s)$ be the equivalence class containing s in the stalk at x. Then ηF is monic because of the uniqueness condition in the definition of sheaf (Exercise (UNMN)). Defining μ is complicated. We will postpone that until we have shown how adjoint pairs of functors give rise to triples; then we will factor T as the composite of a pair of adjoints and get μ without further work.

If F is an R-module, then so is TF (Exercise (RUL)). In this case, iterating T gives Godement's standard resolution.

Adjunctions give triples

The difficulty in verifying the associative identity for μ in examples like the group triple, as well as the fact that every known triple seemed to be associated with an adjoint pair, led P. Huber [1961] to suspect and prove:

Theorem 1. *Let* $U : \mathcal{B} \to \mathcal{C}$ *have a left adjoint* $F : \mathcal{C} \to \mathcal{B}$ *with adjunction morphisms* $\eta : \mathrm{id} \to UF$ *and* $\epsilon : FU \to \mathrm{id}$*. Then* $\mathsf{T} = (UF, \eta, U\epsilon F)$ *is a triple on* $\mathcal{C}$*.*

Proof. The unitary identities

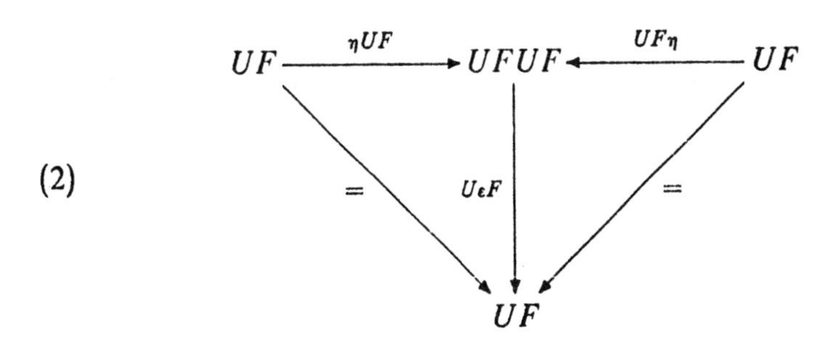

are just $\eta U \circ U\epsilon = \mathrm{id}$ evaluated at F (see Exercise (UCO) of Section 1.9) and U applied to $F\eta \circ \varepsilon F = \mathrm{id}$, respectively. The associative identity

(3)

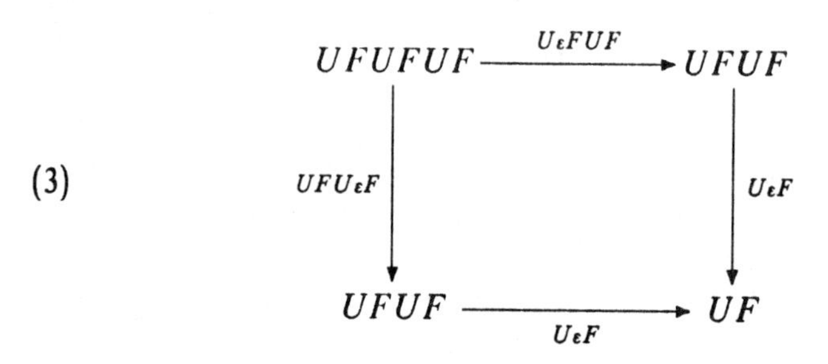

is U applied to the following diagram, which is then evaluated at F:

(4)

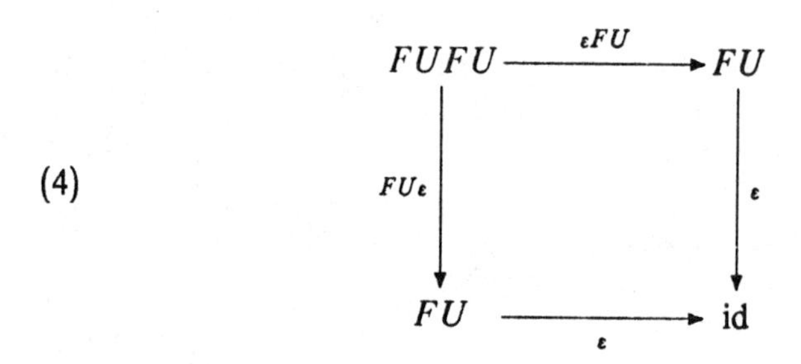

At an object Y, this last diagram has the form

(5)

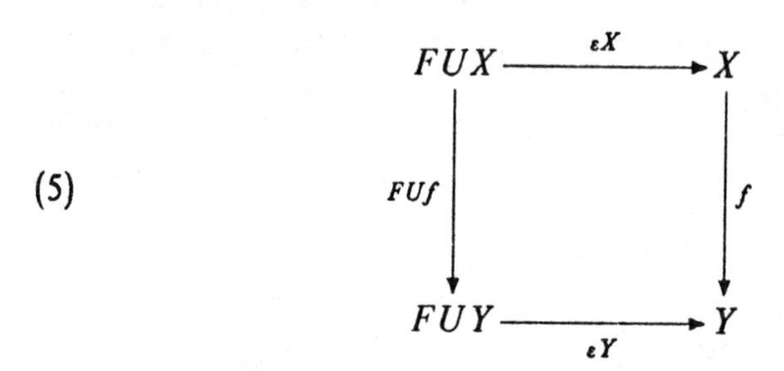

(for $X = FUY$ and $f = \varepsilon Y$) which commutes because ε is natural. (This is an example of part (a) of Exercise (GOD) of Section 1.3).

The group triple of example (vi) of course arises from the adjunction of the underlying set functor and the free group functor. We will see in Section 3.2 that in fact every triple arises from an adjoint pair (usually many different ones).

The factorization in the case of the triple for sheaves we began to construct above is $T = RU$, where $U : \mathcal{Sh}(X) \to \mathcal{Set}\,/\,|X|$ has $U(F) = \bigcup F_x$ and $R : \mathcal{Set}\,|X| \to \mathcal{Sh}(X)$ is the functor defined previously. Then R is left adjoint to U and produces a triple $\mathsf{T} = (T, \eta, U\varepsilon R)$.

Cotriples

A **cotriple** in a category $\mathcal{B}$ is a triple in $\mathcal{B}^{\text{op}}$. Thus $\mathsf{G} = (G, \varepsilon, \delta)$ (this is standard notation) is a cotriple in $\mathcal{B}$ if G is an endofunctor of $\mathcal{B}$, and $\varepsilon : G \to \text{id}$, $\delta : G \to G^2$ are natural transformations satisfying the duals to the diagrams (1) above. (Thus a cotriple is the opposite of a triple, not the dual of a triple. The dual of a triple — in other words, a triple in $\mathcal{C}at^{\text{op}}$ — is a triple.)

Proposition 2. *Let $U : \mathcal{B} \to \mathcal{C}$ have a left adjoint $F : \mathcal{C} \to \mathcal{B}$ with adjunction morphisms $\eta : \text{id} \to UF$ and $\varepsilon : FU \to \text{id}$. Then $\mathsf{G} = (FU, \varepsilon, F\eta U)$ is a cotriple on $\mathcal{C}$.*

Proof. This follows from Theorem 1 and the observation that U is left adjoint to F as functors between $\mathcal{B}^{\text{op}}$ and $\mathcal{C}^{\text{op}}$ with unit ε and counit η.

Exercises 3.1

(PTRP). Let $\mathbf{P}$ denote the functor from $\mathcal{S}et$ to $\mathcal{S}et$ which takes a set to its powerset and a function to its direct image function (Section 1.2). For a set X, let ηX take an element of X to the singleton containing x, and let μX take a set of subsets of X (an element of $\mathbf{P}X$) to its union. Show that $(\mathbf{P}, \eta, \mu)$ is a triple in $\mathcal{S}et$. (Hint: compare Exercise (USL2) of Section 1.9).

(POLY). Let R be any commutative ring. For each set X, let TX be the set of polynomials in a finite number of variables with the variables in X and coefficients from R. Show that T is the functor part of a triple (μ is defined to "collect terms").

(TRE). An ordered binary rooted tree (OBRT) is a binary rooted tree (assume trees are finite in this problem) which has an additional linear order structure (referred to as left/right) on each set of siblings. An X-labeled OBRT (LOBRT/X) is one together with a function from the set of terminal nodes to X. Show that the following construction produces a triple in $\mathcal{S}et$: For any set X, TX is the set of all isomorphism classes of LOBRT/X. If $f : X \to Y$, then Tf is relabeling along f (take a tree in TX and change the label of each node labeled x to $f(x)$). ηX takes $x \in X$ to the one-node tree labeled x, and μX takes a tree whose labels are trees in TX to the tree obtained by attaching to each node the tree whose name labels that node.

(TRE2). Let $\mathcal{B}$ be the category of sets with one binary operation (subject to no conditions) and functions which preserve the binary operation.

(a) Show that the triple of Exercise (TRE) arises from the underlying set functor $\mathcal{B} \to \mathit{Set}$ and its left adjoint.
(b) Give an explicit description of the cotriple in $\mathcal{B}$ induced by the adjoint functors in (a).

(GRCO). Give an explicit description of the cotriple in $\mathcal{Grp}$ induced by the underlying set functor and the free group functor.

(MONCO). Let M be a monoid and $G = \mathrm{Hom}(M, -) : \mathit{Set} \to \mathit{Set}$. If X is a set and $f : M \to X$, let $\varepsilon X(f) = f(1)$ and $[\delta X(f)](m)(n) = f(mn)$ for $m, n \in M$. Show that δ and ε are natural transformations making (G, ε, δ) a cotriple in Set.

(ETAMON)°. Show that if T is any triple on $\mathcal{C}$ and A is an object of $\mathcal{C}$, and there is at least one mono $A \to TA$, then ηA is monic. (Hint: If m is the monic, put Tm into a commutative square with η and use a unitary identity.)

(UNMN). Prove that ηF as defined in the section on sheaves is monic.

(RFUN). Complete the proof that R, defined in the section on sheaves, is a functor.

(RUL). Show that the functors R and U constructed in the section on sheaves have the properties that R is left adjoint to U and that if F is an R-module in the category of sheaves of sets over the underlying space, then so is TF.

(SHAB). Show that the following three statements about a sheaf F on a topological space X are equivalent:

(i) Every fiber of F is an Abelian group in such a way that the structure maps (addition, negation and picking out 0) are continuous in the total space of F.
(ii) For every open U of X, FU is an Abelian group in such way that all the restriction maps are homomorphisms.
(iii) F is an Abelian group object in the category $\mathcal{Sh}(X)$. (See Section 1.7.)

3.2. The Kleisli and Eilenberg-Moore Categories

After Huber proved Theorem 1 of Section 3.1, P. J. Hilton conjectured that *every* triple arises from an adjoint pair. The answer was provided more or less simultaneously, using two distinct constructions, by Eilenberg and Moore [1965] and by H. Kleisli [1965].

Theorem 1. *Let* $\mathsf{T} = (T, \eta, \mu)$ *be a triple on* $\mathcal{C}$. *Then there is a category* $\mathcal{B}$ *and an adjoint pair* $F : \mathcal{C} \to \mathcal{B}$, $U : \mathcal{B} \to \mathcal{C}$, *such that* $T = UF$, $\eta : \mathrm{id} \to UF = T$ *is the unit and* $\mu = U\varepsilon F$ *where* ε *is the counit of the adjunction.*

Proof. Construction 1 (Kleisli). The insight which makes this construction work is that *if* a category $\mathcal{B}'$ and an adjoint pair $F : \mathcal{C} \to \mathcal{B}'$, $U : \mathcal{B}' \to \mathcal{C}$ exist with $T = UF$, then the full subcategory $\mathcal{B}$ of $\mathcal{B}'$ of objects of the form FA for A an object of $\mathcal{C}$ must, by definition, have the property that

$$\mathrm{Hom}_{\mathcal{B}}(FA, FB) \cong \mathrm{Hom}_{\mathcal{C}}(A, UFB) = \mathrm{Hom}_{\mathcal{C}}(A, TB).$$

This is the clue that enables us to define $\mathcal{B}$ in terms of the given data $\mathcal{C}$ and T.

The category $\mathcal{B}$ will have the same objects as $\mathcal{C}$. For arrows, set $\mathrm{Hom}_{\mathcal{B}}(A, B) = \mathrm{Hom}_{\mathcal{C}}(A, TB)$. If $f : A \to TB \in \mathrm{Hom}_{\mathcal{B}}(A, B)$ and $g : B \to TC \in \mathrm{Hom}_{\mathcal{B}}(B, C)$, then we let $g \circ f \in \mathrm{Hom}_{\mathcal{B}}(A, C)$ be the composite

$$A \xrightarrow{\ f\ } TB \xrightarrow{\ Tg\ } T^2C \xrightarrow{\ \mu C\ } TC.$$

The identity arrow on an object A is ηA. It is an elementary exercise, using the associative and unitary identities (and naturality) to see that these definitions make $\mathcal{B}$ a category.

The functor $U : \mathcal{B} \to \mathcal{C}$ is defined by $UA = TA$; if $f : A \to B \in \mathrm{Hom}_{\mathcal{B}}(A, B)$, then Uf is defined to be

$$TA \xrightarrow{\ Tf\ } T^2B \xrightarrow{\ \mu B\ } TB\,.$$

F is defined by $FA = A$; if $f : A \to B \in \mathrm{Hom}_{\mathcal{C}}(A, B)$, Ff is the composite

$$A \xrightarrow{\ \eta A\ } TA \xrightarrow{\ Tf\ } TB$$

which is the same as

$$A \xrightarrow{\ f\ } B \xrightarrow{\ \eta B\ } TB\,.$$

The required equivalence $\mathrm{Hom}_{\mathcal{C}}(A, UB) \cong \mathrm{Hom}_{\mathcal{B}}(FA, B)$ is the same as $\mathrm{Hom}_{\mathcal{C}}(A, TB) \cong \mathrm{Hom}_{\mathcal{B}}(A, B)$, which is true by definition. The rest is left as an exercise.

Observe that Kleisli's category is in some sense as small as it could be because it makes F surjective on objects. This observation can be made precise (Proposition 2 below and Exercise (KEM)).

The **Kleisli category** is denoted $\mathcal{K}(\mathsf{T})$.

Construction 2 (Eilenberg-Moore). The category constructed by Eilenberg and Moore is in effect all the possible quotients of objects in Kleisli's category. Of course, we have to say this using only the given ingredients, so the definition is in terms of a map $a : TA \to A$ which is to be thought of as underlying the quotient map:

A T**-algebra** is a pair (A, a) where A is an object of $\mathcal{C}$ and $a : TA \to A$ an arrow of $\mathcal{C}$ subject to the condition that these two diagrams commute:

(1)

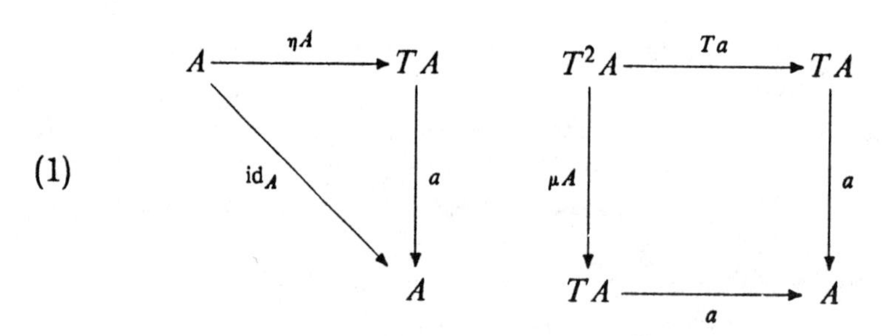

The arrow a is the **structure map** of the algebra.

A map $f : (A, a) \to (B, b)$ of $\mathcal{B}$ is a map $f : A \to B$ of $\mathcal{C}$ for which

(2)

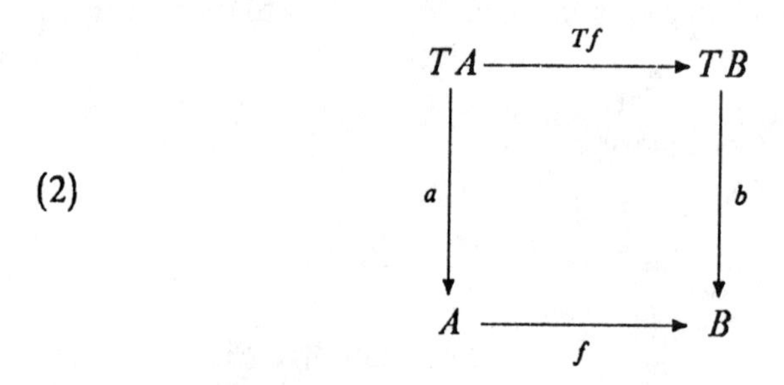

commutes.

The category of T-algebras and T-algebra maps is denoted $\mathcal{C}^{\mathsf{T}}$. Define $U^{\mathsf{T}} : \mathcal{C}^{\mathsf{T}} \to \mathcal{C}$ by $U^{\mathsf{T}}(A, a) = A$ and $U^{\mathsf{T}}f = f$, and $F^{\mathsf{T}} : \mathcal{C} \to \mathcal{C}^{\mathsf{T}}$ by $F^{\mathsf{T}}A = (TA, \mu A)$, $F^{\mathsf{T}}f = Tf$. Most of the proof that this produces a pair of adjoint functors for which the conditions of the theorem hold is straightforward. The required adjunction

$$\alpha : \mathrm{Hom}_{\mathcal{C}^{\mathsf{T}}}((UFC, \mu C), (C', c')) \to \mathrm{Hom}_{\mathcal{C}}(C, C')$$

takes a morphism $h : UFC \to C'$ of algebras to $h \circ \eta C$ and its inverse takes $g : C \to C'$ to $c' \circ UFg$.

It is worthwhile to examine the definition when T is the group triple. Here, a is a set map from the underlying set of the free group on A to A. The definition $xy = a([xy])$, where x and y are elements of A, so that $[xy]$ is an element of TA, gives a multiplication on A which in fact makes A a group—this follows with some diagram-chasing from the diagrams in (1) (Exercise (GRPT)). The identity element is $a([\ ])$. Observe that associativity does *not* follow from the right hand diagram in (1): associativity is built into the definition of TA as consisting of strings of elements. The right diagram in (1) says that if you take a string of free-group elements, in other words a string of strings, then you can either multiply each string out using the multiplication of the group A, then multiply the resulting elements together, or you can first erase parentheses, making one long string, and then multiply it out—either way gives you the same result. In other words, *substitution commutes with evaluation*, which is the essence of algebraic manipulation.

Conversely, in the case of the group triple, every group can be represented as an algebra for the triple: For a group G, use the quotient map $UFG \to G$ which takes a string to its product in G.

As suggested by the discussion in the proof of the theorem, we have a simple description of the Kleisli category. In this proposition we use the common convention of describing as an embedding any functor which is faithful and takes non-isomorphic objects to non-isomorphic objects (a much stronger condition than merely reflecting isomorphisms; the underlying functor from groups to sets reflects isomorphisms without having that property). This convention exemplifies the categorical imperative that the actual identity of the objects is irrelevant.

Proposition 2. *$\mathcal{K}(\mathsf{T})$ is embedded in $\mathcal{C}^{\mathsf{T}}$ as the full subcategory generated by the image of F.*

Proof. The embedding is $\Phi : \mathcal{K}(\mathsf{T}) \to \mathcal{C}^{\mathsf{T}}$ defined by $\Phi(A) = (TA, \mu A)$, and for $f : A \to TB$, $\Phi(f)$ is the composite

$$TA \xrightarrow{Tf} T^2B \xrightarrow{\mu B} TB\,.$$

The Eilenberg-Moore comparison functor

Proposition 2 is only part of the story. In fact, $\mathcal{K}(\mathsf{T})$ is initial and $\mathcal{C}^{\mathsf{T}}$ is final among all ways of factoring T as an adjoint pair. We will describe how this works with $\mathcal{C}^{\mathsf{T}}$ and leave the other part to you (Exercise (KEM)).

Suppose we have $F : \mathcal{C} \to \mathcal{B}$, $U : \mathcal{B} \to \mathcal{C}$, with F left adjoint to U and unit and counit $\eta : \mathrm{id} \to UF$, $\varepsilon : FU \to \mathrm{id}$, with $\mathsf{T} = (T = UF, \eta, U\varepsilon F)$. Let $U^{\mathsf{T}} : \mathcal{C}^{\mathsf{T}} \to \mathcal{B}$, $F^{\mathsf{T}} : \mathcal{B} \to \mathcal{C}^{\mathsf{T}}$ be the adjoint pair given by Construction 2 of the proof of Theorem 1. The **Eilenberg-Moore comparison functor** is the functor $\Phi : \mathcal{B} \to \mathcal{C}^{\mathsf{T}}$ which takes B to $(UB, U\varepsilon B)$ and f to Uf. It is easy to see that this really is a functor, and in fact the only functor for which $U^{\mathsf{T}} \circ \Phi = U$ and $\Phi \circ F = F^{\mathsf{T}}$. This says $\mathcal{C}^{\mathsf{T}}$ is the terminal object in the category of adjoint pairs which induce T (Exercise (KEM)).

The Eilenberg-Moore functor is in many important cases an isomorphism or equivalence of categories, a topic which is pursued in sections 3.3 and 3.4.

Coalgebras for a cotriple

If $\mathsf{G} = (G, \varepsilon, \delta)$ is a cotriple in a category $\mathcal{C}$, the construction of the Eilenberg Moore category of algebras of G regarded as a triple in $\mathcal{C}^{\mathrm{op}}$ yields, when all arrows are reversed, the category $\mathcal{C}_{\mathsf{G}}$ of **coalgebras of** G. Precisely, a G**-coalgebra** is a pair (A, α) with $\alpha : A \to GA$ for which the following diagrams commute:

(3)

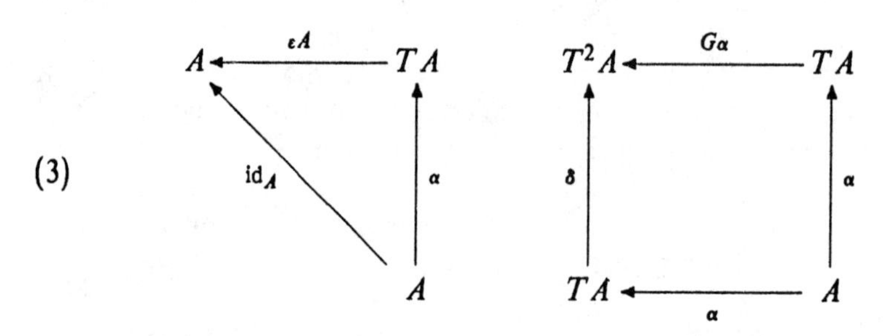

A morphism $f : (A, \alpha) \to (B, \beta)$ is an arrow $f : A \to B$ for which $Gf \circ \alpha = \beta \circ f$.

We will prove in Section 3.5 that when a functor has both a left and a right adjoint, the corresponding categories of algebras and coalgebras are isomorphic. Our major use of cotriples in this book will be based on the fact that the category of coalgebras for a left exact cotriple (meaning the functor is left exact) in a topos is itself a topos.

Exercises 3.2

(FRTR)°. Let $\mathsf{T} = (T, \eta, \mu)$ be a triple, A and B be objects of the underlying category.

(a) Show that $(TA, \mu A)$ is an algebra for T. (Such algebras are called **free**.)

(b) Show that for any $f : A \to B$, Tf is an algebra map from $(TA, \mu A)$ to $(TB, \mu B)$.
(c) Show that μA is an algebra morphism from $(TTA, \mu TA)$ to $(TA, \mu A)$.

(STAR). (Manes) Let $\mathcal{C}$ be a category. Show that the following data:

(i) A function $T : \mathcal{Ob}(\mathcal{C}) \to \mathcal{Ob}(\mathcal{C})$;
(ii) for each pair of objects C and D of $\mathcal{C}$ a function $\mathrm{Hom}(C, TD) \to \mathrm{Hom}(TC, TD)$, denoted $f \mapsto f^*$;
(iii) for each object C of $\mathcal{C}$ a morphism $\eta C : C \to TC$;

subject to the conditions:

(i) For $f : C \to TD$, $f = \eta TD \circ f^*$;
(ii) for any object C, $(\eta C)^* = \mathrm{id}_{TC}$;
(iii) for $f : C \to TD$ and $g : D \to TE$, $(g^* \circ f)^* = g^* \circ f^*$;

are equivalent to a triple on $\mathcal{C}$. (Hint: An elegant way to attack this exercise is to use the data to define the Kleisli category for the triple, using the pointwise adjunction construction (Theorem 1 of Section 1.9) to get the adjoint pair whose corresponding triple is the one sought.)

(GRPT). Show that if T is the group triple, the Eilenberg-Moore comparison functor $\Phi : \mathcal{Grp} \to \mathcal{Set}^{\mathsf{T}}$ is an isomorphism of categories.

(SUBALG). Let T be a triple in a category $\mathcal{C}$. Let (C, c) be an algebra for T and let B be a subobject of C. Show that a map $b : TB \to B$ is an algebra structure on B for which inclusion is an algebra map if and only if

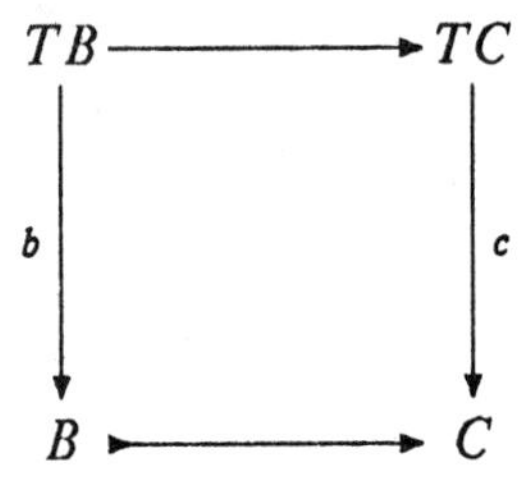

commutes, and that there cannot be more than one such map b. (This says in effect that B "is" a subalgebra if and only if it is "closed under the operations"—in other words, $c(TB) \subseteq B$. In Section 6.4, we give a condition for a subobject of a coalgebra of a left exact cotriple in a topos to be a subcoalgebra.)

(KEM). For a given triple T in a category $\mathcal{C}$, let $\mathcal{E}$ be the category in which an object is a category $\mathcal{B}$ together with an adjoint pair of functors $F : \mathcal{C} \to \mathcal{B}$, $U : \mathcal{B} \to \mathcal{C}$ which induces the triple T via Theorem 1 of Section 3.1, and in which an arrow from $(\mathcal{B}, F, U)$ to $(\mathcal{B}', F', U')$ is a functor $G : \mathcal{B} \to \mathcal{B}'$ for which $U \circ G = U'$ and $G \circ F' = F$. Show that $\mathcal{K}(\mathsf{T})$ is the initial object in $\mathcal{E}$ and $\mathcal{C}^{\mathsf{T}}$ is the terminal object.

(MONCO2). Show that the coalgebras for the cotriple defined in Exercise (MONCO) of Section 3.1 form a category isomorphic to the category of sets acted on on the right by M.

(KCTW). Show that for a triple T in a category $\mathcal{C}$ each of the following constructions give a category $\mathcal{K}$ isomorphic to the Kleisli category.

(a) $\mathcal{K}$ is the full subcategory of $\mathcal{C}^{\mathsf{T}}$ whose objects are the image of F^{T}, i.e. all objects of the form $(F^{\mathsf{T}}C, \mu C)$, C an object of $\mathcal{C}$.
(b) $\mathcal{K}^{\mathrm{op}}$ is the full subcategory of $\mathrm{Func}(\mathcal{C}^{\mathsf{T}}, \mathcal{S}et)$ whose objects are of the form $\mathrm{Hom}_{\mathcal{C}}(C, U^{\mathsf{T}}(-))$, C an object of $\mathcal{C}$.

Linton used the second definition, in which F^{T} does not appear, to study algebraic theories in the absence of a left adjoint to the underlying functor.

(ECMP). (Linton) Let $\mathsf{T} = (T, \eta, \mu)$ be a triple in $\mathcal{C}$. Let $\mathcal{K}$ be the Kleisli category of T and $F_{\mathsf{T}} : \mathcal{C} \to \mathcal{K}$ be the left adjoint to $U_{\mathsf{T}} : \mathcal{K} \to \mathcal{C}$. Let $H : \mathcal{C}^{\mathsf{T}} \to \mathrm{Func}(\mathcal{K}^{\mathrm{op}}, \mathcal{S}et)$ denote the functor which takes (C, c) to the restriction of $\mathrm{Hom}_{\mathcal{C}^{\mathsf{T}}}(-, (C, c))$ to $\mathcal{K}$, where $\mathcal{K}$ is regarded as a subcategory of $\mathcal{C}^{\mathsf{T}}$ as in Exercise (KCTW)(a) above. Prove that the diagram

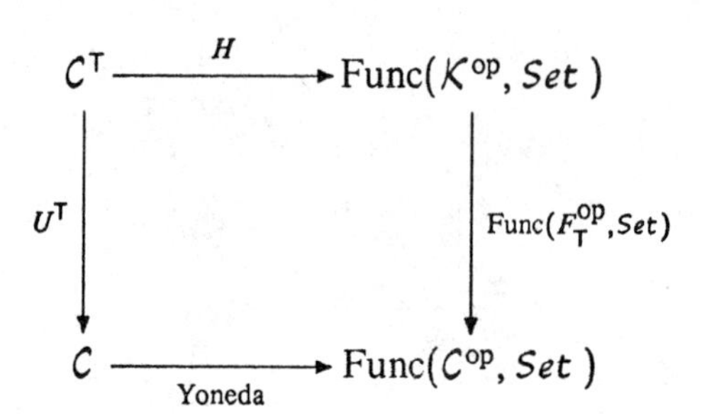

is a pullback.

3.3. Tripleability

In this section we will be concerned with the question of deciding when a functor $U : \mathcal{B} \to \mathcal{C}$ with a left adjoint has the property that the Eilenberg-Moore category for the corresponding triple is essentially the same as $\mathcal{B}$.

To make this precise, a functor U which has a left adjoint for which the corresponding Eilenberg-Moore comparison functor Φ is an equivalence of categories is said to be **tripleable**. If Φ is full and faithful, we say that U is of **descent type** (and if it is tripleable, that it is of **effective descent type**).

We will often say $\mathcal{B}$ is tripleable over $\mathcal{C}$ if there is a well-understood functor $U : \mathcal{B} \to \mathcal{C}$. Thus $\mathcal{G}rp$ is tripleable over $\mathcal{S}et$, for example (Exercise (GRPT) of section 3.2).

In this section we state and prove a theorem due to Beck giving conditions on a functor $U : \mathcal{B} \to \mathcal{C}$ which insure that it is tripleable. Variations on this basic theorem will be discussed in Sections 3.4 and 9.1. Before we can state the main theorem, we need some background.

Reflecting isomorphisms

A functor U **reflects isomorphisms** if whenever Uf is an isomorphism, so is f. For example, the underlying functor $U : \mathcal{G}rp \to \mathcal{S}et$ reflects isomorphisms — that is what you mean when you say that a group homomorphism is an isomorphism if and only if it is one to one and onto. (Warning: That U reflects isomorphisms is *not* the same as saying that if UX is isomorphic to UY then X is isomorphic to Y — for example, two groups with the same number of elements need not be isomorphic). Observe that the underlying functor $U : \mathcal{T}op \to \mathcal{S}et$ does not reflect isomorphisms.

Proposition 1. *Any tripleable functor reflects isomorphisms.*

Proof. Because equivalences of categories reflect isomorphisms, it is sufficient to show that, for any triple $\mathbf{T}$ in a category $\mathcal{C}$, the underlying functor $U : \mathcal{C}^{\mathbf{T}} \to \mathcal{C}$ reflects isomorphisms. So let $f : (A, a) \to (B, b)$ have the property that f is an isomorphism in $\mathcal{C}$. Let $g = f^{-1}$. All we need to show is that

(1)
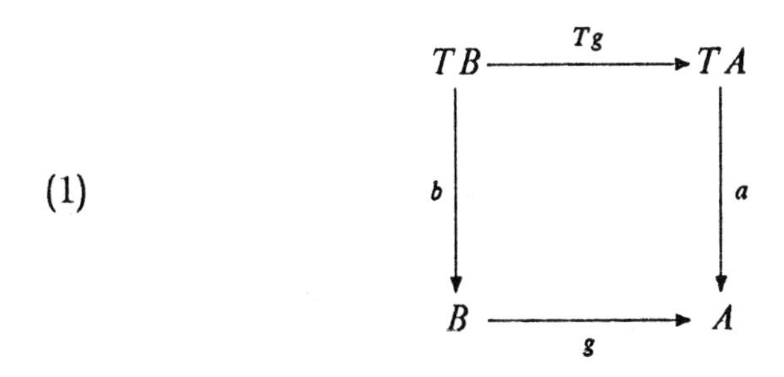

commutes. This calculation shows that:

$$\begin{aligned} a \circ Tg &= g \circ f \circ a \circ Tg = g \circ b \circ Tf \circ Tg \\ &= g \circ b \circ T(f \circ g) = g \circ b \circ T(\mathrm{id}) = g \circ b. \end{aligned}$$

Contractible coequalizers

A **parallel pair** in a category is pair of maps with the same domain and codomain:

(2)
$$A \underset{d^1}{\overset{d^0}{\rightrightarrows}} B$$

The parallel pair above is **contractible** (or **split**) if there is an arrow $t : B \to A$ with

$$d^0 \circ t = \mathrm{id},$$

and

$$d^1 \circ t \circ d^0 = d^1 \circ t \circ d^1.$$

A **contractible coequalizer** consists of objects and arrows

(3)
$$A \underset{t}{\overset{d^0,\ d^1}{\rightleftarrows}} B \underset{s}{\overset{d}{\rightleftarrows}} C$$

for which

(i) $d^0 \circ t = \mathrm{id}$,
(ii) $d^1 \circ t = s \circ d$,
(iii) $d \circ s = \mathrm{id}$, and
(iv) $d \circ d^0 = d \circ d^1$.

We will eventually see that any Eilenberg-Moore algebra is a coequalizer of a parallel pair which becomes contractible upon applying U^{T}.

Proposition 2. (a) *A contractible coequalizer is a coequalizer.*
(b) *If* (3) *is a contractible coequalizer in a category* $\mathcal{C}$ *and* $F : \mathcal{C} \to \mathcal{D}$ *is any functor, then*

(4)
$$FA \underset{Ft}{\overset{Fd^0,\ Fd^1}{\rightleftarrows}} FB \underset{Fs}{\overset{Fd}{\rightleftarrows}} FC$$

is a contractible coequalizer.

(c) *If*

$$A \overset{d^0}{\underset{d^1}{\rightrightarrows}} B \xrightarrow{\ d\ } C \qquad (5)$$

is a coequalizer, then the existence of t making

$$A \rightrightarrows B,\quad t: B \to A \qquad (6)$$

a contractible pair forces the existence of s making (5) *a contractible coequalizer.*

Proof. To show (a), let $f : B \to D$ with $f \circ d^0 = f \circ d^1$. The unique $g : C \to D$ required by the definition of coequalizer is $f \circ s$ ("To get the induced map, compose with the contraction s"). It is straightforward to see that $f = g \circ d$, and g is unique because d is a split epimorphism. Statement (b) follows from the fact that a contractible coequalizer is defined by equations involving composition and identities, which functors preserve. A coequalizer which remains a coequalizer upon application of any functor is called a **absolute coequalizer**; thus (a) and (b) together say that a contractible coequalizer is an absolute coequalizer.

As for (c), if t exists, then by assumption, $d^1 \circ t$ coequalizes d^0 and d^1, so there is a unique $s : C \to B$ with $s \circ d = d^1 \circ t$. But then

$$d \circ s \circ d = d \circ d^1 \circ t = d \circ d^0 \circ t = d$$

and d is epi, so $d \circ s = \mathrm{id}$.

If $U : \mathcal{B} \to \mathcal{C}$, a U**-contractible coequalizer pair** is a pair of morphisms as in (2) above for which there is a contractible coequalizer

$$UA \overset{U(d^0)}{\underset{U(d^1)}{\rightrightarrows}} UB \xrightarrow{\ d\ } C,\quad t: UB \to UA,\quad s: C \to UB \qquad (7)$$

in $\mathcal{C}$.

Proposition 3. *Let $U : \mathcal{B} \to \mathcal{C}$ be tripleable and*

$$B' \overset{d^0}{\underset{d^1}{\rightrightarrows}} B \qquad (8)$$

be a U-contractible coequalizer pair. Then (8) has a coequalizer $d : B \to B''$ in $\mathcal{B}$ and

(9)
$$UB' \underset{U(d^1)}{\overset{U(d^0)}{\rightrightarrows}} UB \xrightarrow{Ud} UB''$$

is a coequalizer in $\mathcal{C}$.

Before beginning the proof, we need some terminology for commutative diagrams. A diagram like

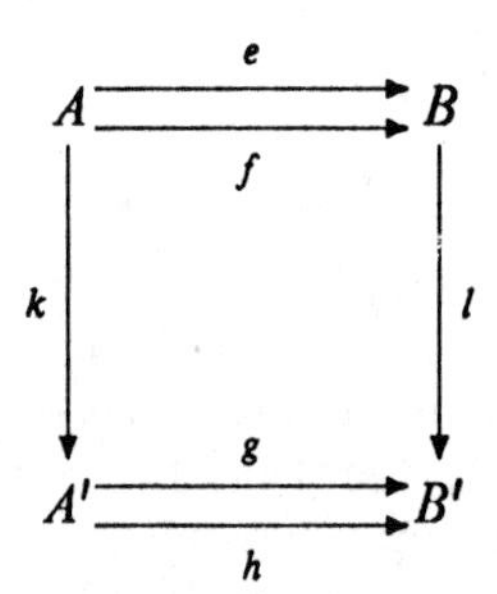

is said to **commute serially** if $g \circ k = l \circ e$ and $h \circ k = l \circ f$. (The arrows are matched up in accordance with the order they occur in the diagram). In more complicated diagrams the analogous convention will be understood. For example,

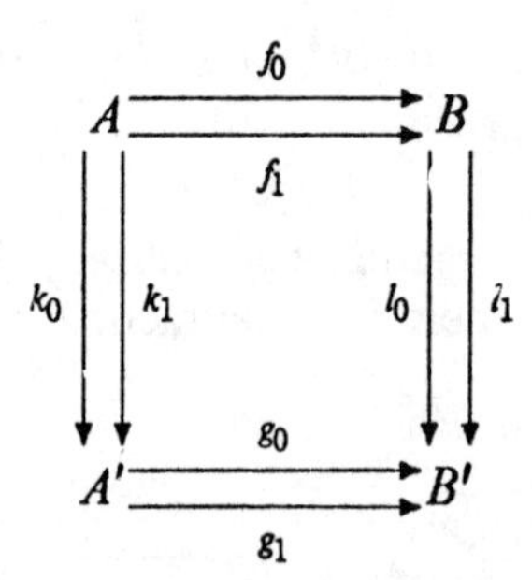

will be said to commute serially if $g_i \circ k_j = l_j \circ f_i$ for all four possible choices of $i, j = 0, 1$.

Proof of Proposition 3. (This proof was suggested by J. Beck). As in Proposition 1, let $\mathcal{B} = \mathcal{C}^{\mathsf{T}}$. Thus we are given a pair

(10)
$$(C', c') \underset{d^1}{\overset{d^0}{\rightrightarrows}} (C, c)$$

and we suppose

$$(11)\qquad C' \underset{t}{\overset{d^0,\ d^1}{\rightrightarrows}} C \underset{s}{\overset{d}{\rightleftarrows}} C''$$

is a contractible coequalizer in $\mathcal{C}$. Then by Proposition 2(b), all three rows of the following diagram (in which we have not yet defined c'') are contractible coequalizers.

(12)

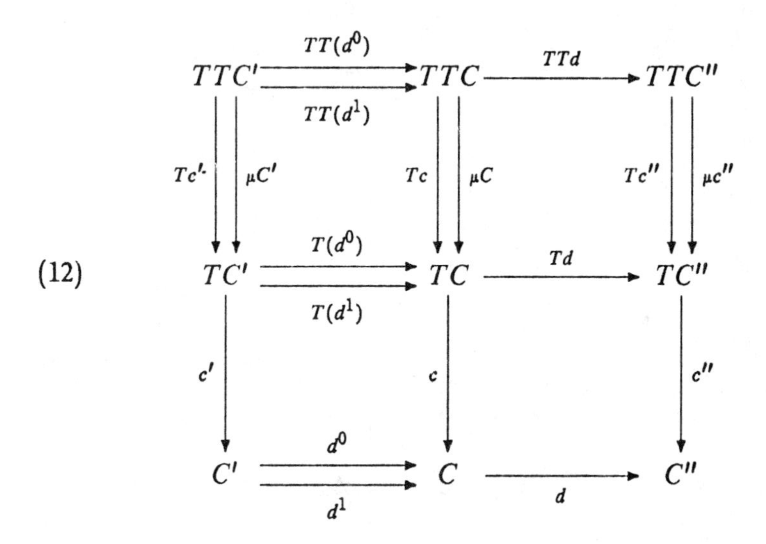

The lower left square commutes serially because d^0 and d^1 are algebra maps by assumption. This implies that $d \circ c$ coequalizes $T(d^0)$ and $(T(d^1)$. Using the fact that the middle row is a coequalizer, we define c'' to be the unique arrow making the bottom right square commute. By the proof of Proposition 2(a) ("To get the induced map, compose with the contraction") we have

$$c'' = d \circ c \circ Ts$$

a fact we will need in the proof of Proposition 4.

We first prove that c'' is a structure map for an algebra. The upper right square commutes serially, the square with the μ's because μ is a natural transformation, and the one with Tc and Tc'' because it is T of the bottom right square. Using this and the fact that c is a structure map, we have

$$c'' \circ \mu c'' \circ T^2 d = c'' \circ Tc'' \circ T^2 d$$

so, since T^2d is an epimorphism (it is split),

$$c'' \circ Tc'' = c'' \circ \mu c''.$$

The unitary law for algebras is obtained by replacing the top row of (12) by the bottom row and the vertical arrows by $\eta c'$, ηc, and $\eta c''$, and using a similar argument. We know d is an algebra map because the bottom right square commutes.

Finally, we must show that (C'', c'') is a coequalizer, so suppose $f : (C, c) \to (E, e)$ coequalizes d^0 and d^1. Then because C'' is their coequalizer in $\mathcal{C}$, there is a unique arrow $u : C'' \to E$ for which $f = u \circ d$ We need only to prove that the right square in

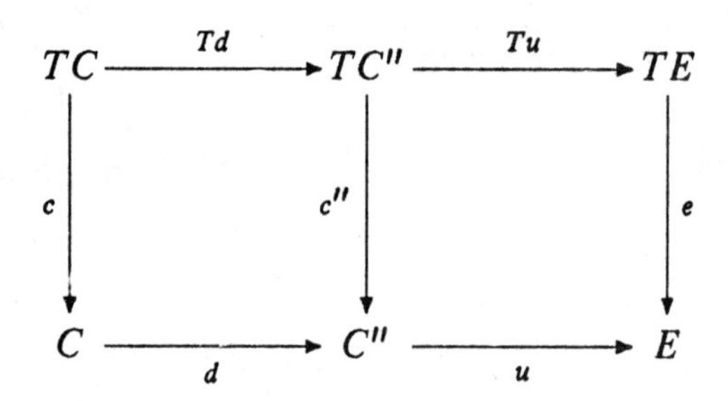

commutes. This follows from the fact that the left square and the outer rectangle commute and Td is epi.

Note that we have actually proved that U^{T} creates coequalizers of U^{T}-contractible pairs.

Algebras are coequalizers

A parallel pair is said to be **reflexive** if the two arrows have a common right inverse. A **reflexive coequalizer** diagram is a coequalizer of a reflexive parallel pair.

Proposition 4. *Let* T *be a triple on* $\mathcal{C}$. *Then for any* (A, a) *in* $\mathcal{C}^{\mathsf{T}}$,

$$(TTA, \mu TA) \underset{Ta}{\overset{\mu A}{\rightrightarrows}} (TA, \mu A) \tag{13}$$

is a reflexive U*-contractible coequalizer pair whose coequalizer in* $\mathcal{C}$ *is* (A, a).

Proof. The associative law for μ implies that μA is an algebra map, and the naturality of μ implies that Ta is an algebra map. It follows from the identities for triples and algebras that

$$(14)\qquad TTA \underset{\eta TA}{\overset{\mu A}{\underset{Ta}{\rightrightarrows}}} TA \underset{\eta A}{\overset{a}{\rightleftarrows}} A$$

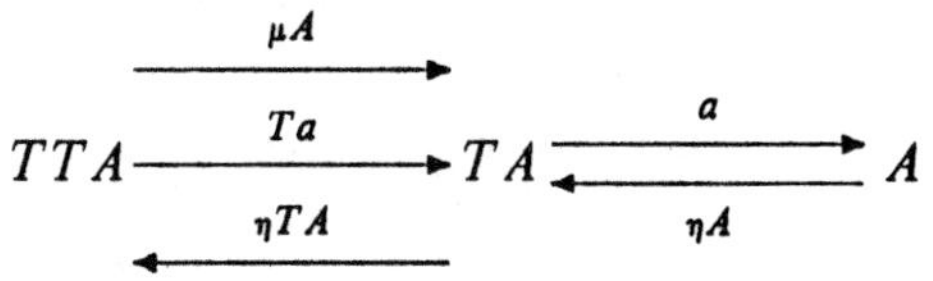

is a contractible coequalizer. Thus by Proposition 3, there is an algebra structure on A which coequalizes μA and Ta in C^{T}. As observed in the proof of that proposition, the structure map is

$$a \circ \mu A \circ T\eta A = a,$$

as claimed.

The common right inverse for Ta and μA is $T\eta A$:

$$Ta \circ T\eta A = T(a \circ \eta A) = T(\mathrm{id}) = \mathrm{id}$$

by the unitary law for algebras and

$$\mu A \circ T(\eta A) = U(\varepsilon FA \circ F\eta A) = U(\mathrm{id}) = \mathrm{id}$$

by Exercise (Uco), Section 1.9. The naturality of μ implies that $T\eta$ is an algebra map.

Corollary 5. *If, given*

$$(15)\qquad \mathcal{B} \underset{F}{\overset{U}{\rightleftarrows}} \mathcal{C}$$

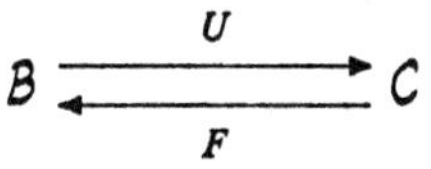

F is left adjoint to U, then for any object B of $\mathcal{B}$,

$$(16)\qquad FUFUB \underset{FU\varepsilon B}{\overset{\varepsilon FUB}{\rightrightarrows}} FUB$$

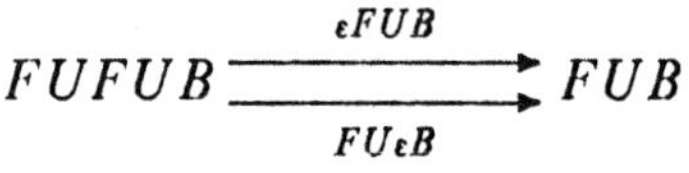

is a reflexive U-contractible coequalizer pair.

Proof. $U\varepsilon F$ is an algebra structure map by Exercise (FRTR) of Section 3.2 and the definition of C^{T}. Thus by Proposition 4, it is the coequalizer of the diagram underlying (16). The common right inverse is $F\eta U$.

Beck's Theorem

We will now prove a number of lemmas culminating in two theorems due to Beck. Theorem 9 characterizes functors U for which the comparison functor is full and faithful, and Theorem 10 characterizes tripleable functors.

In the lemmas, we speak of an adjoint pair (15) with associated triple T in $\mathcal{C}$ and cotriple G in $\mathcal{B}$.

Lemma 6. *The diagram*

$$(17) \qquad C \xrightarrow{\eta C} TC \underset{T\eta C}{\overset{\eta TC}{\rightrightarrows}} T^2C$$

is an equalizer for every object C of $\mathcal{C}$ if and only if ηC is a regular mono for all objects C of $\mathcal{C}$.

Proof. If (17) is an equalizer then obviously ηC is regular mono. Suppose ηC is regular mono; then we have *some* equalizer diagram of the form

$$(18) \qquad C \xrightarrow{\eta C} TC \underset{d^1}{\overset{d^0}{\rightrightarrows}} C'$$

It is sufficient to show that for any element w of TC, d^0 and d^1 agree on w if and only if ηTC and $T\eta C$ agree on w.

(i) If $d^0 \circ w = d^1 \circ w$, then $w = \eta C \circ g$ for some g, so

$$\eta TC \circ w = \eta TC \circ \eta C \circ g = T\eta C \circ \eta C \circ g = T\eta C \circ w.$$

(ii) If $\eta TC \circ w = T\eta C \circ w$, then $T\eta TC \circ Tw = T^2\eta C \circ Tw$. But

$$(19) \qquad TC \xrightarrow{\eta C} T^2C \underset{T^2\eta C}{\overset{T\eta TC}{\rightrightarrows}} T^3C$$

is a contractible equalizer, with contractions μC and μTC. Thus $Tw = T\eta C \circ h$ for some h, so that $Td_0 \circ Tw = Td_1 \circ Tw$. Hence in the diagram below, which commutes because η is a natural transformation, the two composites are equal.

(20)

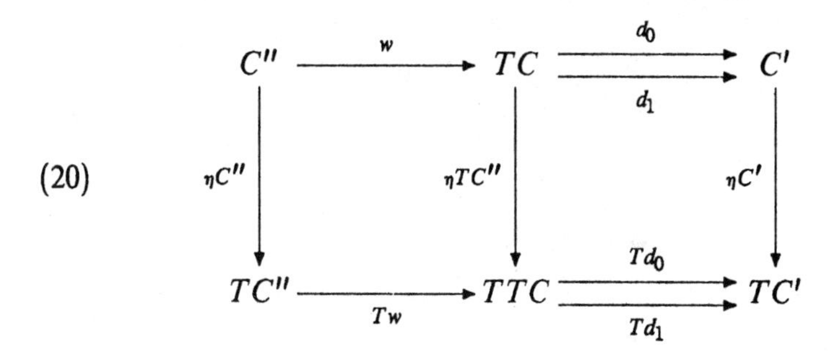

It follows from the fact that $\eta C'$ is (regular) mono that $d_0 \circ w = d_1 \circ w$.

Dually, we have

Corollary 7. *εB is a regular epi for every object B of $\mathcal{B}$ if and only if*

(21)
$$FUFUB \underset{\varepsilon FUB}{\overset{FU\varepsilon B}{\rightrightarrows}} FUB \xrightarrow{\varepsilon B} B$$

is a coequalizer for every object B of $\mathcal{B}$.

Lemma 8. *For all objects B and B' of $\mathcal{B}$,*

$$\mathrm{Hom}_{\mathcal{B}}(FUB, B') \cong \mathrm{Hom}_{\mathcal{C}^{\mathrm{T}}}(\Phi FUB, \Phi B'),$$

where $\Phi : \mathcal{B} \to \mathcal{C}^{\mathrm{T}}$ is the comparison functor. (In other words, "Φ is full and faithful on arrows out of free objects").

Proof. We have,

$$\begin{aligned}
\mathrm{Hom}_{\mathcal{B}}(FUB, B') &\cong \mathrm{Hom}_{\mathcal{C}}(UB, UB') \\
&\cong \mathrm{Hom}_{\mathcal{C}}(UB, U^{\mathrm{T}}(UB', U\varepsilon B')) \\
&\cong \mathrm{Hom}_{\mathcal{C}^{\mathrm{T}}}(F^{\mathrm{T}}(UB), (UB', U\varepsilon B')) \\
&\cong \mathrm{Hom}_{\mathcal{C}^{\mathrm{T}}}((TUB, \mu UB), (UB', U\varepsilon B')) \\
&\cong \mathrm{Hom}_{\mathcal{C}^{\mathrm{T}}}(\Phi FUB, \Phi B').
\end{aligned}$$

Theorem 9 (Beck). *Φ is full and faithful if and only if εB is a regular epi for all objects B of $\mathcal{B}$.*

Proof. $U\varepsilon B$ is the structure map of an algebra in $\mathcal{C}^{\mathrm{T}}$ by Exercise (FRTR), and so by Proposition 4 is a coequalizer of a parallel pair with domain in the image of F^{T}, hence in the image of Φ. Since $\Phi(\varepsilon B) = U\varepsilon B$, it follows from Exercise (FFC) that if Φ is full and faithful, then εB is a regular epi.

Conversely, suppose εB is a regular epi. If $f, g : B \to B'$ in $\mathcal{B}$ and $Uf = Ug$, then $FUf = FUg$ and this diagram commutes:

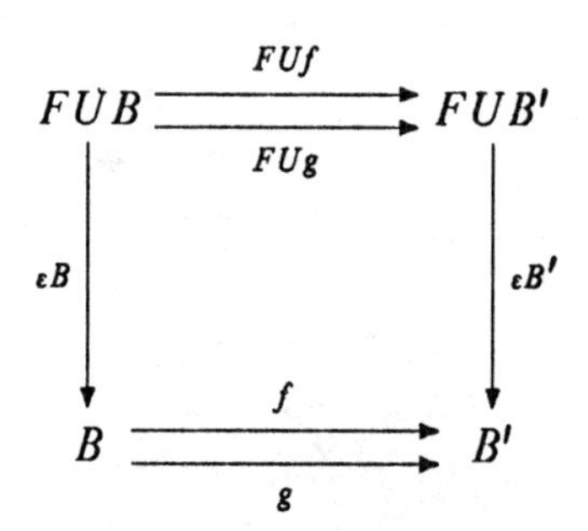

Thus since εB is epi, $f = g$. Hence Φ is faithful.

Under the hypothesis, (21) is a U-contractible coequalizer diagram by Corollaries 5 and 7. Since $U^{\mathsf{T}} \circ \Phi = U$, applying Φ to (21) gives a U^{T}-contractible coequalizer diagram. It follows that the horizontal edges of the diagram below are equalizers; the top row homsets are computed in $\mathcal{B}$ and the bottom row in $\mathcal{C}^{\mathsf{T}}$. The vertical arrows are those induced by Φ; by Lemma 8, the middle and right one are isomorphisms. Thus the left one is an isomorphism, too, proving the Proposition.

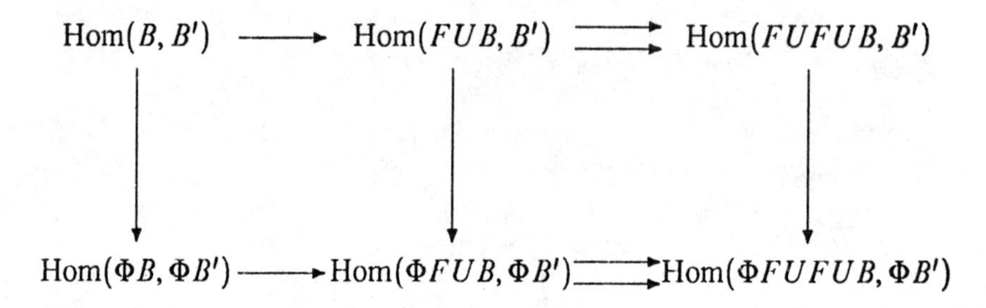

Theorem 10 (Beck's Precise Tripleability Theorem). *$U : \mathcal{B} \to \mathcal{C}$ is tripleable if and only if*

(i) *U has a left adjoint.*
(ii) *U reflects isomorphisms.*
(iii) *$\mathcal{B}$ has coequalizers of reflexive U-contractible coequalizer pairs and U preserves them.*

Proof. If U is tripleable it has a left adjoint F by definition and it satisfies (ii) and (iii) by Propositions 1 and 3. (Note that in fact $\mathcal{B}$ has and U preserves the coequalizers of all U-contractible parallel pairs, not merely reflexive ones—that is the way Beck originally stated the theorem).

Now, to do the other direction, we know (16) is a reflexive U-contractible coequalizer pair, so by (iii) it has a coequalizer B'. Since εB coequalizes (16) (because ε is a natural transformation), we know that there is an arrow f making

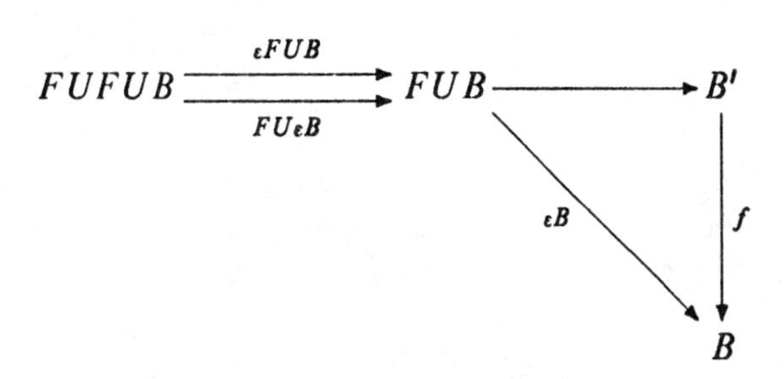

commute. However, as observed in the proof of Corollary 5, $U\varepsilon B$ is coequalizer of U of this diagram, so Uf is an isomorphism. Hence f is an isomorphism, so εB is a regular epi, so that Φ is full and faithful by Theorem 9.

The argument just given that f is an isomorphism can easily be used to show that in fact any functor U satisfying (ii) and (iii) must reflect coequalizers of reflexive U-contractible coequalizer pairs (Exercise (URFL)).

For any object (C, c) of $\mathcal{C}^{\mathsf{T}}$, we must find an object B of $\mathcal{B}$ for which $\Phi(B) \cong (C, c)$. Now Φ of the following diagram is (13),

$$FUFC \underset{\varepsilon FC}{\overset{Fc}{\rightrightarrows}} FC$$

so it is a reflexive U-contractible coequalizer. Thus by assumption there is a coequalizer B for which the sequence underlying

$$FUFC \underset{\varepsilon FC}{\overset{Fc}{\rightrightarrows}} FC \longrightarrow B$$

is

$$UFUFC \underset{U\varepsilon FC}{\overset{UFc}{\rightrightarrows}} UFC \longrightarrow UB$$

By Proposition 4, this last diagram is U^{T} of a coequalizer diagram in $\mathcal{C}^{\mathsf{T}}$ with coequalizer (C, c). Since U^{T} reflects such coequalizers and $U^{\mathsf{T}} \circ \Phi = U$, it follows that $\Phi(B) \cong (C, c)$, as required.

Theorem 10 is the *precise* tripleability theorem as distinct from certain theorems to be discussed in Section 3.5 which give conditions for tripleability which are sufficient but not necessary. Theorem 10 is commonly known by its acronym "PTT".

Other conditions sufficient for tripleability are discussed in Section 9.1.

We extract from the last paragraph of the proof of the PTT the following proposition, which we need later. This proposition can be used to provide an alternate proof of the equivalence of $\mathcal{C}^{\mathsf{T}}$ and $\mathcal{B}$ in the PTT (see Exercise (EQUIII) of Section 1.9).

Proposition 11. *If $U : \mathcal{B} \to \mathcal{C}$ has a left adjoint and $\mathcal{B}$ has coequalizers of reflexive U-contractible coequalizer pairs, then the comparison functor Φ has a left adjoint.*

Proof. The object B constructed in the last paragraph of the proof of Theorem 10 requires only the present hypotheses to exist. Define $\Psi(C,c)$ to be B. If $g : (C,c) \to (D,d)$, then the diagram

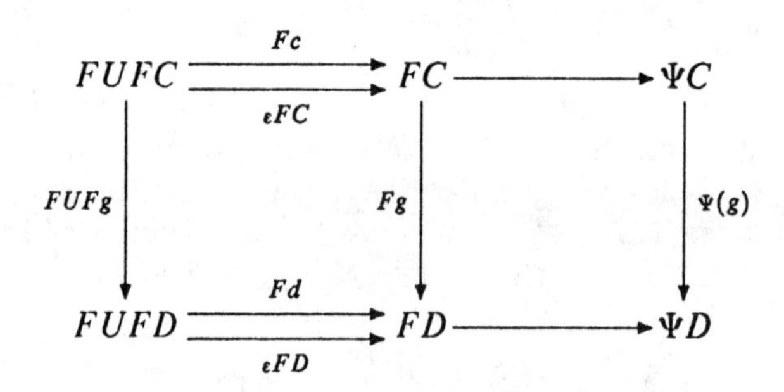

commutes serially, the top square because c and d are structure maps and the bottom one because ϵ is a natural transformation. Thus it induces a map $\Psi(g)$. It is straightforward to check that this makes Ψ a functor which is left adjoint to the comparison functor Φ.

Compact Hausdorff spaces

We illustrate the use of the PTT by proving that compact Hausdorff spaces are tripleable over Set. This fact was actually proved by F. E. J. Linton before Beck proved the theorem, using an argument which, after suitable generalization, became Duskin's Theorem of Section 9.1.

The underlying set functor U from the category $\mathit{CptHaus}$ of compact Hausdorff spaces and continuous maps to Set has a left adjoint β. βX is the Stone-Čech compactification of the set X considered as a discrete space.

Proposition 12. *U is tripleable.*

Proof. The statement that U reflects isomorphisms is the same as the statement that a bijective continuous map between compact Hausdorff spaces is a homeomorphism, which is true. A pair $d^0, d^1 : C' \to C$ of continuous maps between compact Hausdorff spaces has a coequalizer in $\mathcal{Top}$ which is necessarily preserved by the underlying set functor since that functor has a right adjoint (the functor which puts the indiscrete topology on a set X).

The quotient is compact and will be Hausdorff if and only if the kernel pair is closed. Thus we will be finished if we show that the kernel pair of this coequalizer is closed. That kernel pair is the equivalence relation generated by R (the relation which is the image of C' in $C \times C$). By Exercise (SPO)(c) below the kernel pair is $R \circ R^{\text{op}}$. Now R is closed in $C \times C$ (it is the image of a map of a compact space into a Hausdorff space), and so is compact Hausdorff. Hence the fiber product $R \times_C R^{\text{op}}$ is compact Hausdorff, so is closed in $C \times C \times C$. $R \circ R^{\text{op}}$ is the image of that space in the Hausdorff space $C \times C$ and so is closed, as required.

The functor part of this triple takes a set to the set of ultrafilters on it. The functor which takes a set to the set of filters on it is also part of a triple, the algebras for which are continuous lattices (Day [1975], Wyler [1981]). Continuous lattices are also algebras for a triple in the category of topological spaces and elsewhere. Another example of this last phenomenon of being the category of algebras for triples in different categories is the category of monoids, which is tripleable over $\mathcal{Set}$ and also (in three different ways) over $\mathcal{Cat}$ (Wells [1980]).

Exercises 3.3

(URFL). Show that a functor $U : \mathcal{B} \to \mathcal{C}$ which reflects isomorphisms has the property that if a diagram D in $\mathcal{B}$ has a colimit and X is a cocone from D in $\mathcal{B}$ for which UX is a colimit of UD, then X is a colimit of D. Do the same for limits. (These facts are summarized by the slogan: "A functor that reflects isomorphisms reflects all limits and colimits it preserves." This slogan exaggerates the matter slightly: The limits and colimits in question have to be assumed to exist.)

(SPO)°. (Suggested in part by Barry Jay.) A parallel pair $d^0, d^1 : X' \to X$ in $\mathcal{Set}$ determines a relation R on X, namely the image of $(d^0, d^1) : X' \to X \times X$. Conversely, a relation R on a set S defines a parallel pair from R to S (the two projection maps).

(a) Show that a relation R in $\mathcal{Set}$ determines a reflexive parallel pair if and only if it contains the diagonal.

(b) Show that a relation R in $\mathcal{S}et$ determines a contractible parallel pair if and only if each equivalence class $[x]$ in the equivalence relation E generated by R contains an element x_* with the property that xEx' if and only if xRx_* and $x'Rx_*$.

(c) Show that the parallel pair determined by an ordering on a set is contractible if and only if every connected component of the ordered set has a maximum element.

(d) Show that an equivalence relation in $\mathcal{S}et$ is always contractible.

(e) Show that if R is the relation determined by any contractible parallel pair in $\mathcal{S}et$, then $E = R \circ R^{\text{op}}$ is the equivalence relation generated by R.

(CRCO). Show that the algebra map c'' constructed in the proof of Proposition 3 is the only structure map $TC'' \to C''$ which makes d an algebra map.

(NTN)°. Prove that (19) is a contractible coequalizer.

(UPHI). Show that the Eilenberg-Moore comparison functor is the only functor $\Phi : \mathcal{B} \to \mathcal{C}^{\mathsf{T}}$ for which $U^{\mathsf{T}} \circ \Phi = U$ and $\Phi \circ F = F^{\mathsf{T}}$. (Hint: Show that for $B \in \mathcal{O}b(\mathcal{B})$, $\Phi(B)$ must be (UB, b) for some arrow $b : UFUB \to UB$), and then consider this diagram:

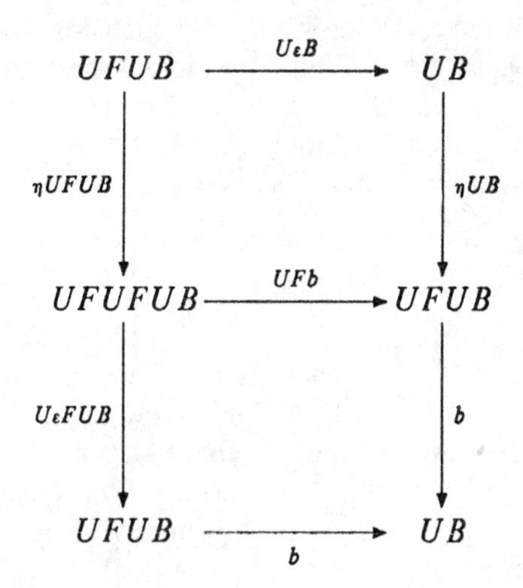

(PPTT). Let U be a functor with left adjoint F. Then the comparison functor for the induced triple is an isomorphism of categories (not merely an equivalence) if and only if U reflects isomorphisms and creates coequalizers for reflexive U-contractible coequalizer pairs. (Compare Exercise (CRRF) of Section 1.7.)

(REFL). A subcategory $\mathcal{C}_0$ of a category $\mathcal{C}$ is **reflective** (or **reflexive**) if the inclusion functor has a left adjoint. Show that the inclusion functor of a reflective subcategory is tripleable.

(EQRF). Show that an equivalence of categories reflects isomorphisms.

(FFC). Show that if H is a full and faithful functor and Hf is the coequalizer of a parallel pair with domain in the image of H, then f is a coequalizer.

3.4. Properties of Tripleable Functors

In this section we describe various properties a tripleable functor must have. Some of these are useful in the development of topos theory, and as necessary conditions for tripleability they are also useful in showing that certain functors are not tripleable, as we will illustrate.

Completeness of categories of algebras

If $\mathsf{T} = (T, \eta, \mu)$ is a triple in a category $\mathcal{C}$, then the category $\mathcal{C}^{\mathsf{T}}$ of algebras is "as complete as $\mathcal{C}$ is", in the following sense:

Theorem 1. *Let* T *be a triple in* $\mathcal{C}$. *Then* $U^{\mathsf{T}} : \mathcal{C}^{\mathsf{T}} \to \mathcal{C}$ *creates limits. Hence any tripleable functor reflects limits.*

Proof. In the following, we write U for U^{T} for simplicity. Let $D : \mathcal{I} \to \mathcal{C}^{\mathsf{T}}$ be a diagram, and let C be the limit of $U \circ D$ in $\mathcal{C}$. We must find an algebra structure map $c : TC \to C$ making (C, c) the limit of D, and that structure must be unique.

Let n be an object of $\mathcal{I}$. Then Dn is a structure $(UD(n), \alpha n)$. Let $\beta n : C \to U(D(n))$ be the element of the limit cone corresponding to n. Then the elements $(\alpha n) \circ (T\beta n)$ form a cone from TC to UD, and so induce an arrow $c : TC \to C$ which is the required algebra structure.

The fact that the resulting structure is a T-algebra follows from the jointly monic nature of the elements βn. For example, associativity follows from the diagram

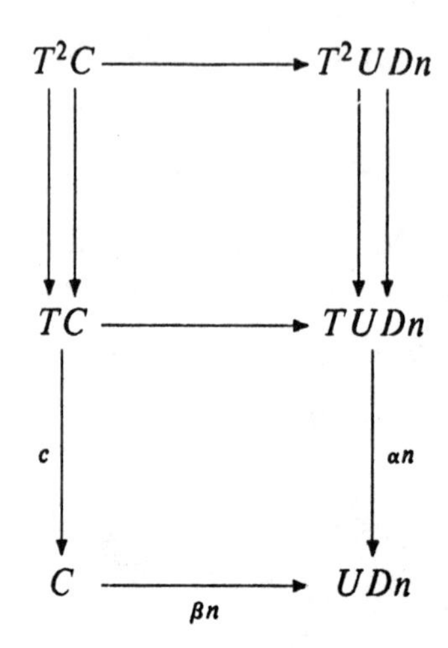

To see that (C, c) is the limit, suppose we have a cone with elements

$$\gamma n : (C', c') \to Dn = (UDn, \alpha n).$$

Applying U, this gives a cone from C' to UDn, which induces an arrow $C' \to C$ by the fact that C is a limit. The left square in the following diagram must then commute because the βn are a jointly monic family. That means that the map $C' \to C$ is an algebra map as required.

(1) 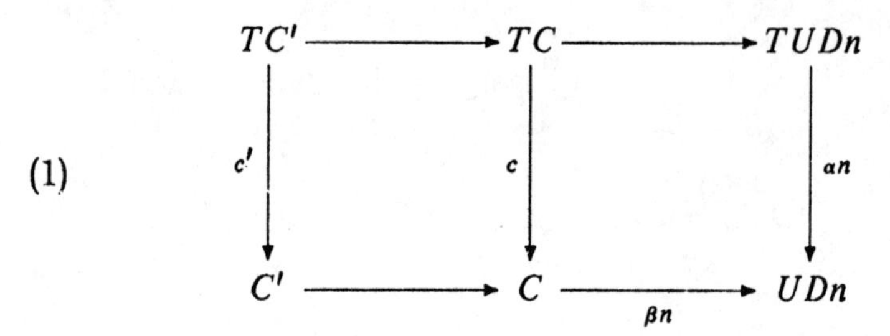

Corollary 2. *If $U : \mathcal{B} \to \mathcal{C}$ is tripleable and $\mathcal{C}$ is complete, then so is $\mathcal{B}$.*

Note that U in the preceding corollary need not create limits, since $\mathcal{B}$ might be equivalent to but not isomorphic to $\mathcal{C}^{\mathsf{T}}$.

In Section 9.3 we will describe sufficient conditions for proving $\mathcal{B}$ cocomplete.

Banach spaces

The fact a subobject is a subalgebra if and only if it is closed under the operations (Exercise (SUBALG) of 3.2) is a useful necessary condition for tripleability. We apply this criterion to Banach spaces.

Let Ban denote the category whose objects are real Banach spaces and whose morphisms are linear maps which do not increase the norm. Let $U : Ban \to Set$ be the functor which takes a space B to its unit ball $\{b \in B \mid |B| \leq 1\}$, and maps to their restrictions. We will show that U reflects isomorphisms and has a left adjoint, but is nevertheless not tripleable.

U reflects isos: Suppose $f : B \to C$ is such that Uf is an isomorphism. The proof repeatedly uses the fact that for any $b \in B$, $b/|b|$ is in UB. Thus f is injective (if $f(b) = 0$ then Uf must take $b/|b|$ to 0) and surjective (if $c \in C$ then $c/|c|$ is the image of some $b \in UB$, so c is the image of $b|c|$). It is also necessary to show that f^{-1} preserves the norm, which will follow if we show that f preserves the norm. Let $b \in B$ and $n = |f(b/|b|)|$. Then

$$f(b/(n|b|)) = (1/n)f(b/|b|)$$

must be in UC because its norm is 1. Since f is injective and Uf is surjective onto UC, $b/(n|b|)$ is in UB, whence $n \geq 1$. Since f does not increase norms, $n \leq 1$, so $n = 1$. Thus f preserves norms, as required.

The left adjoint to U assigns to a set X the set FX of all functions $f : X \to R$ for which $\sum_{(x \in X)} |f(x)| \leq \infty$. The norm $|f(x)| = \sum |f(x)|$ makes FX a Banach space. It is in fact $l^1(X)$, where X is regarded as a measure space with atomic measure.

The identities for an algebra for the triple induced by F and U imply by a somewhat long but straightforward argument that for a given Banach space C, the induced algebra structure $c = U\varepsilon C : UFUC \to UC$ is defined for $f \in FUC$ by

$$c(f) = \sum_{(x \in X)} x|f(x)|.$$

Then if C is the closed interval $[-1, 1]$, the only $f \in FUC$ for which $c(f) = 1$ is the function with value 1 at 1 and 0 elsewhere, and the only f with $c(f) = -1$ are the functions with value -1 at 1 and 0 elsewhere, and with value 1 at -1 and 0 elsewhere. Letting $B = (-1, 1)$, this implies that $c(FUB) \subseteq B$, which means by Exercise (SUBALG) of 3.2 that there is an algebra structure on the *open* interval which agrees with c. But the open interval $(-1, 1)$ with the usual addition and scalar multiplication is not the closed ball of any Banach space, so U is not tripleable.

Even worse is the three-point algebra $[0, 1]/(0, 1)$, which is the coequalizer of the inclusion of $(0, 1)$ in $[0, 1]$ and the zero map. It is instructive to work out the algebra structure on this algebra.

Tripleability over Set

A useful necessary condition for tripleability over $\mathcal{S}et$ is the following proposition.

Proposition 3. *If $U : \mathcal{B} \to \mathcal{S}et$ is a tripleable functor, then the pullback of a regular epi in $\mathcal{B}$ is a regular epi.*

Proof. If c is a regular epi in $\mathcal{B}$, then it is the coequalizer of its kernel pair (Corollary 2 and Exercise (EQC) of Section 1.8). This kernel pair induces an equivalence relation in $\mathcal{S}et$ which, like any such, is split. Hence by Proposition 2(c) of Section 3.3, c is a U-contractible coequalizer. Hence Uc is a regular epi in $\mathcal{S}et$. Let b be the pullback of c along an arrow f. Then Ub is a regular epi in $\mathcal{S}et$ since U preserves pullbacks and the pullback of a regular epi is a regular epi in $\mathcal{S}et$. Ub is then the class map of an equivalence relation and so b is a U-contractible coequalizer in $\mathcal{B}$. Since U is tripleable, the PTT implies that b is regular.

Corollary 4. *The category $\mathcal{C}at$ of small categories and functors is not tripleable over $\mathcal{S}et$.*

(Note that this means there is *no* functor from $\mathcal{C}at$ to $\mathcal{S}et$ which is tripleable—in particular, neither of the functors which take a category to its set of objects or its set of arrows.)

Proof. Let **1** denote the category with one object and one arrow, and **2** the category with two objects and exactly one non-identity morphism f going from one object to the other. There are two functors from **1** to **2**, and to form their coequalizer is to identify the domain and codomain of f. In this coequalizer we must have the arrows f^2, f^3, and so on, and there is no reason for any equalities among them, so the coequalizer is the monoid $(\mathsf{N}, +)$ regarded as a category with one object. Thus the arrow $\mathbf{2} \to \mathsf{N}$ is a regular epi. See Exercise (SURJ).

Now let $\mathsf{M} \rightarrowtail \mathsf{N}$ denote the submonoid of even integers. Let 2 (as opposed to **2**) denote $1 + 1$, that is the category with two objects and no nonidentity arrows. It is easy to see that

(2)

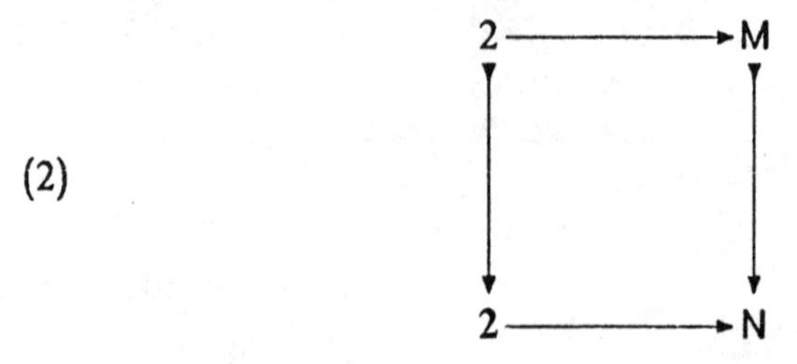

is a pullback and that the top arrow is not a regular epi.

(It is a notable phenomenon that a functor may merge objects and therefore make arrows compose with each other that never dreamed of composing before the functor was applied. This makes colimits in $\mathcal{C}at$ hard to understand and is probably the main reason why the oft-expressed notion that a category is a "monoid with many objects" has only limited fruitfulness.)

On the other hand, $\mathcal{C}at$ *is* tripleable over the category $\mathcal{G}rph$ of graphs. The left adjoint L to the underlying functor $U : \mathcal{C}at \to \mathcal{G}rph$ is defined by making LG (for a graph G) the category whose nonidentity arrows are all composable paths of arrows of G. (Exercise (GRADJ) of Section 1.9). Composition is concatenation of paths. Identity arrows have to be included separately.

That U reflects isomorphisms is the familiar fact that the inverse of a bijective functor is a functor.

Finally, suppose

$$A \rightrightarrows B \tag{3}$$

is a pair of functors for which

$$UA \rightrightarrows UB \xrightarrow{d} G \tag{4}$$

is a contractible coequalizer. In particular, it is an absolute coequalizer, so applying UL gives a coequalizer

$$ULUA \rightrightarrows ULUB \xrightarrow{ULd} ULG \tag{5}$$

If $b : LUB \to B$ is the map corresponding to the given category structure on B (taking a composable path to its composite), then by the coequalizer property there is a map g making the following diagram commute.

(6)

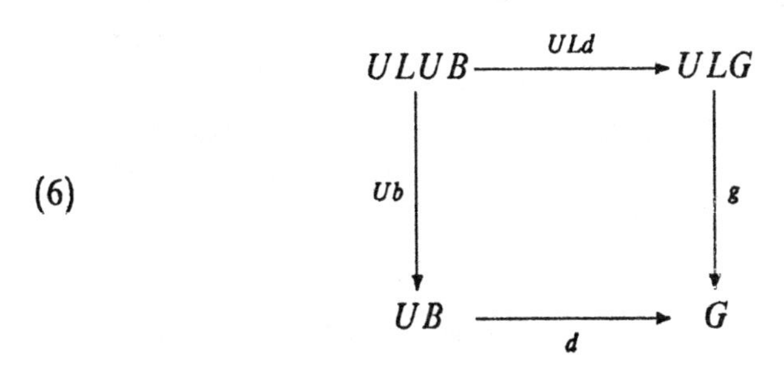

and it is easy to see from the universal property of coequalizers that both horizontal arrows are surjective. A straightforward check using that surjectivity then shows that defining composition of arrows a and b of ULG by $a \circ b = g(a, b)$ and identity arrows as images of identity arrows in UB makes G a category, d a functor which is the required coequalizer.

Grph is also tripleable over sets; this follows from Exercise (GRMN) below and Exercise (MAT) in Section 3.5. Thus the composite of two tripleable functors need not be tripleable. For another example, see Exercise (TABT) below. In the next section we will state theorems giving circumstances under which the composite of tripleable functors is tripleable.

Exercises 3.4

(GRMN). Show that *Grph* is isomorphic to the category of setvalued actions of the monoid M with multiplication table

	i	s	t
i	i	s	t
s	s	s	t
t	t	s	t

(Hint: If M acts on X, X is the set of arrows of a graph and sx and tx are its source and target).

(TABT). (a) Show that the inclusion of the category of torsion-free Abelian groups into the category of Abelian groups is tripleable. (Hint: Try Exercise (REFL) of Section 3.3).

(b) Show that the category of Abelian groups is tripleable over *Set* .

(c) Show that the underlying set functor from the category of torsion-free Abelian groups to *Set* has a left adjoint (Hint: restrict the free-Abelian-group functor).

(d) Show that the functor in (c) reflects isomorphisms.

(e) Show that the functor in (c) is *not* tripleable. (Hint: The two maps $(m, n) \to m + 2n$ and $(m, n) \to m$ from $\mathbf{Z} \oplus \mathbf{Z}$ to $\mathbf{Z}$ determine a U-split equivalence relation but are not the kernel pair of anything).

(SURJ). Using the notation of Corollary 4, we have the commutative

diagram

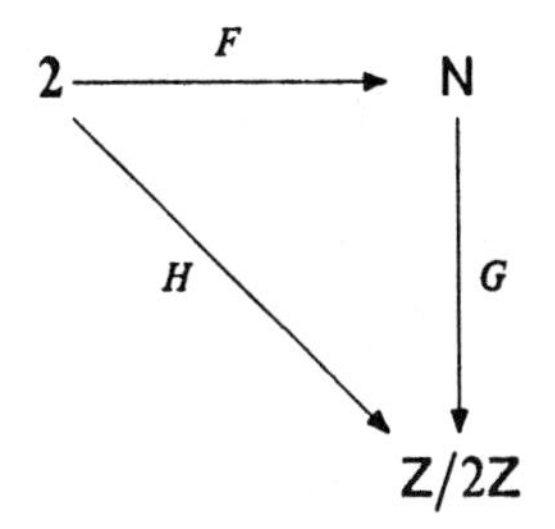

We have already seen that F is regular epi. Show that G is regular epi and H is not so that the composite of regular epis need not be one. This destroys a very reasonable sounding conjecture.

3.5. Sufficient Conditions for Tripleability

We define two useful sufficient conditions on a functor which make it tripleable, which in addition allow us to give circumstances under which the composite of tripleable functors is tripleable.

Such a composite can fail to be tripleable if the first functor applied fails to lift contractible coequalizers to *contractible* coequalizers. However, the composite might still be tripleable if the second functor lifts *all* coequalizers. This motivates the following two definitions.

A functor $U : \mathcal{B} \to \mathcal{C}$ satisfies the hypotheses of the "VTT" (Vulgar Tripleability Theorem) if it has a left adjoint, reflects isomorphisms, and if any reflexive U-contractible coequalizer pair is already a contractible coequalizer in $\mathcal{B}$. Since contractible coequalizers are preserved by any functor, it follows that if U satisfies the VTT it must be tripleable.

U satisfies the hypotheses of the "CTT" (Crude Tripleability Theorem) if U has a left adjoint and reflects isomorphisms, $\mathcal{B}$ has coequalizers of those reflexive pairs (f, g) for which (Uf, Ug) is a coequalizer and U preserves those coequalizers. Such a functor is clearly tripleable.

Proposition 1. *Suppose $U_1 : \mathcal{A} \to \mathcal{B}$ and $U_2 : \mathcal{B} \to \mathcal{C}$.*

(a) *If U_1 and U_2 both satisfy CTT (respectively VTT) then so does $U_1 \circ U_2$.*

(b) *If U_1 satisfies CTT, U_2 satisfies PTT and $U_3 : \mathcal{C} \to \mathcal{D}$ satisfies VTT, then $U_3 \circ U_2 \circ U_1$ is tripleable.*

Proof. Easy.

The following proposition is a different sort, imposing hypotheses on the categories involved which imply the existence of an adjoint. A **pointed** category is a category with an object 0 which is both initial and terminal, which implies that for any two objects A and B there is a (necessarily unique) arrow from A to B which factors through 0. The category of groups, for example, is pointed.

Proposition 2. *Let $\mathcal{A}$ be a complete, cocomplete pointed category, let $U : \mathcal{B} \to \mathcal{A}$ be tripleable, and let $\mathcal{C}$ be a small category. Define $V : \mathrm{Func}(\mathcal{C}, \mathcal{B}) \to \mathcal{A}$ by taking a functor Ψ to the product of all objects $U\Psi(C)$ of $\mathcal{A}$ over all objects C of $\mathcal{C}$. Then V is tripleable.*

Outline of Proof. We will refer repeatedly to the following diagram, in which S is the set of objects of $\mathcal{C}$ and i is the inclusion. In the direction from $\mathrm{Func}(\mathcal{C}, \mathcal{B})$ to $\mathcal{A}$, both routes represent a factorization of V, and the square commutes.

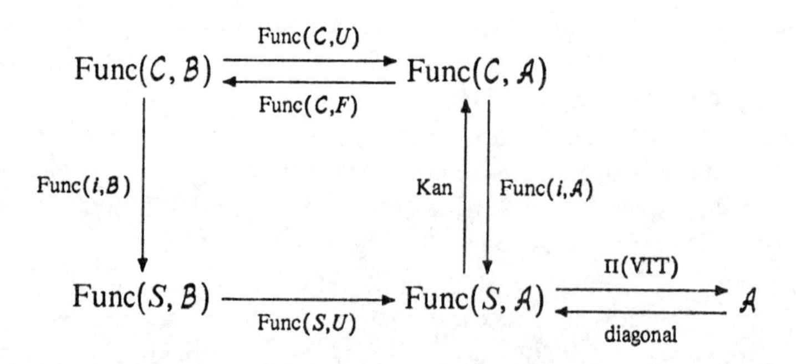

F is the left adjoint of U, so $\mathrm{Func}(\mathcal{C}, F)$ is the left adjoint of $\mathrm{Func}(\mathcal{C}, U)$. The left adjoint of $\mathrm{Func}(i, \mathcal{A})$ exists by Kan extensions because $\mathcal{A}$ is cocomplete (see Section 1.9). The functor $\prod$ takes a functor $F : S \to \mathcal{A}$ to the product of all its values; it is easy to see the diagonal map (takes an object of $\mathcal{A}$ to the constant functor with than object as value) is the left adjoint. Thus by composition, V has a left adjoint.

The left vertical arrow preserves colimits because colimits (like limits) are computed pointwise. Thus it satisfies the coequalizer condition of CTT. Note that it might not have a left adjoint, although it will using Kan extensions if $\mathcal{B}$ is also cocomplete. $\mathrm{Func}(S, U)$ satisfies the coequalizer condition of PTT by a similar argument. The pointedness of $\mathcal{A}$ implies that $\prod$ satisfies VTT. This works as follows: If $T : S \to \mathcal{A}$ and C is an object of $\mathcal{C}$ (element of S), then there is a canonical embedding $TC \to \prod T$ (which is the product of all the objects TC') induced by the identity map on TC and the point maps from TC to all the objects TC'. This makes the composite

$$TC \longrightarrow \prod T \xrightarrow{\text{proj}} TC$$

the identity. This enables one to transfer the maps involved in a U-split coequalizer in $\mathcal{A}$ up to Func$(S, \mathcal{A})$, verifying VTT. Thus V factors into a composite of functors satisfying the coequalizer conditions of CTT, PTT and VTT in that order, so it satisfies the PTT.

The observation using pointedness above, applied to V instead of $\prod$, yields the proof that V reflects isomorphisms. It follows that V satisfies the CTT, hence is tripleable.

Exercises 3.5

(Mat). Show that if M is any monoid, the underlying functor to $\mathcal{S}et$ from the category of actions by M on sets and equivariant maps is tripleable.

(ModT). Same as the preceding exercise for the category of R-modules for any fixed ring R.

(MonTrp). Show that the functor L of Exercise (MonL) of Section 1.9 is tripleable.

(Ptd). (a) Show that any map $1 \to 0$ in a category is an isomorphism.
(b) Show that if a category has 1 and equalizers and if $\text{Hom}(1, -)$ is never empty, then 1 is initial.

3.6. Morphisms of Triples

In this section we define a notion of morphism of triples on a given category in such a way that functors between Eilenberg-Moore categories which commute with the underlying functors correspond bijectively with morphisms of triples. In the process of proving this, we will describe (Proposition 1) a method of constructing morphisms of triples which will be used in Section 3.7.

Let $\mathsf{T} = (T, \eta, \mu)$ and and $\mathsf{T}' = (T', \eta', \mu')$ be triples on a category $\mathcal{C}$. A morphism morphism of triples $\alpha : \mathsf{T} \to \mathsf{T}'$ is a natural transformation $\alpha : T \to T'$ making diagrams (1) and (2) below commute. In (2), the notation α^2 denotes $\alpha\alpha$ in the sense of Exercise (God), Section 1.3. Thus

α^2 is the morphism $T'\alpha \circ \alpha T$ which, because α is a natural transformation, is the same as $\alpha T' \circ T\alpha$.

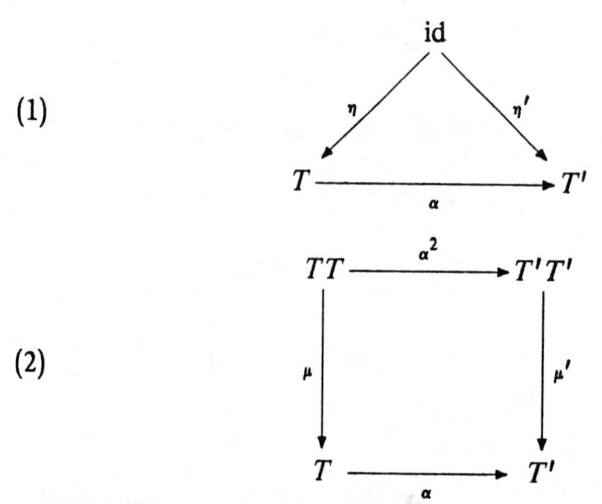

The following proposition gives one method of constructing morphisms of triples.

Proposition 1. *In the notation of the preceding paragraphs, let* $\sigma : TT' \to T'$ *be a natural transformation for which*

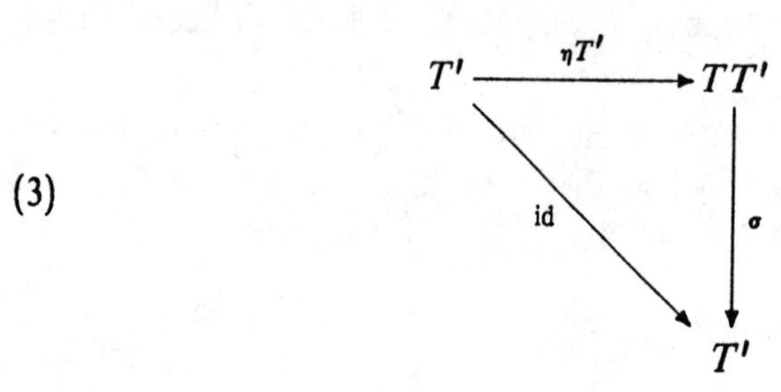

and

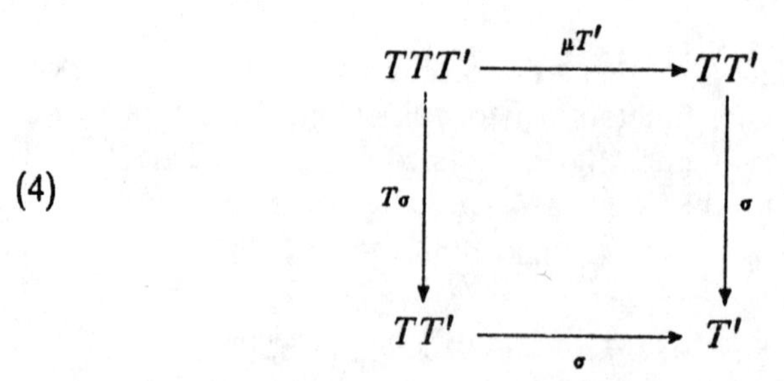

commute. Let $\alpha = \sigma \circ T\eta' : T \to T'$. *Then* α *is a morphism of triples.*

Proof. That (1) commutes follows from the commutativity of

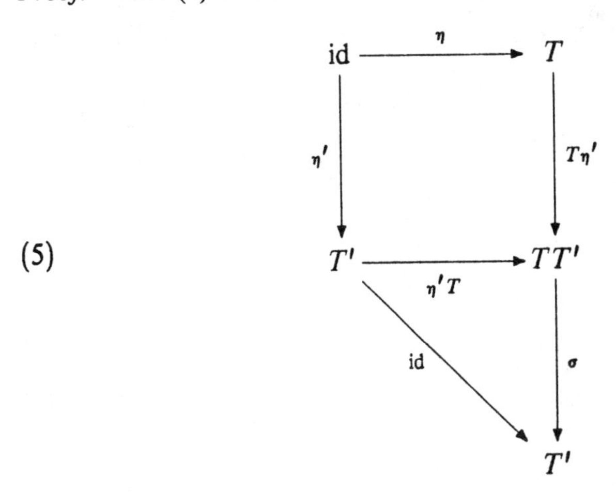

In this diagram, the square commutes because η is a natural transformation and the triangle commutes by (3).

The following diagram shows that (2) commutes.

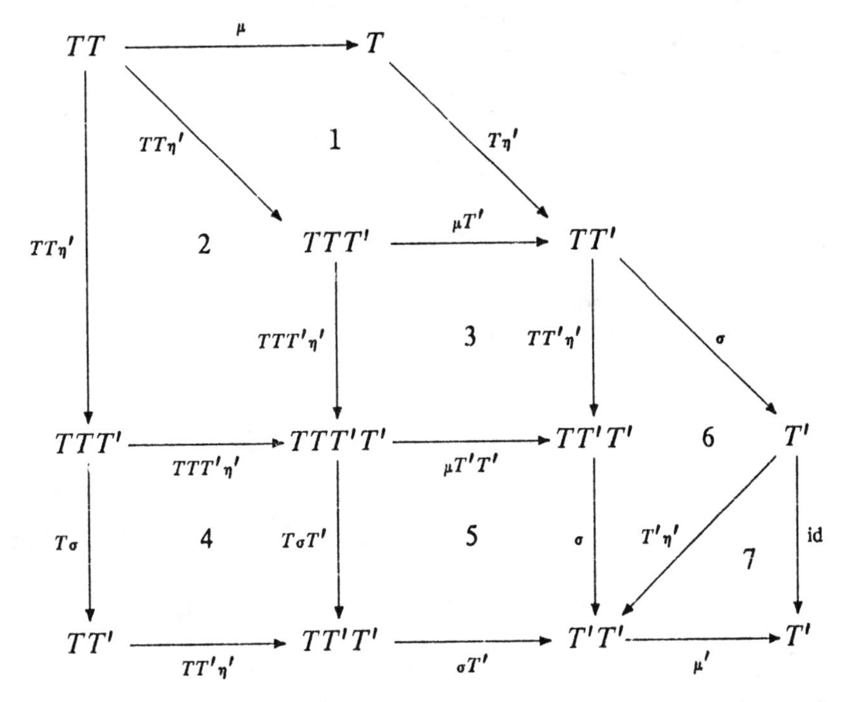

In this diagram, square 1 commutes because μ is a natural transformation, square 3 because $\mu T'$ is, square 4 because $T\sigma$ is, and square 6 because σ is.

As for the other squares in the diagram, square 2 is the identity, square 5 is diagram (4) above, and square 7 is part of diagram (1) of the definition of triple in Section 3.1.

Squares 1, 3, 4 and 6 of diagram (6) are all examples of part (a) of Exercise (GOD), Section 1.3. For example, to see how square 1 fits, take $\mathcal{B}$, $\mathcal{C}$ and $\mathcal{D}$ of the exercise to be $\mathcal{C}$, $F = \mathrm{id}$, $G = T'$, $H = T^2$ and $K = T$, and $\kappa = \eta'$, $\mu = \mu$. Thus square 1 is $\eta'\mu$.

Corollary 2. *With* T *and* T' *as in Proposition 1, suppose* $\sigma : T'T \to T'$ *is such that* $\sigma \circ \sigma T' = \sigma \circ T'\mu$ *and* $\sigma \circ T'\eta = \mathrm{id}$. *Then* $\alpha = \sigma \circ \eta T : T \to T'$ *is a morphism of triples.*

Proof. This is Proposition 1 stated in $\mathcal{C}at^{\mathrm{op}}$ (which means: reverse the functors but not the natural transformations).

Theorem 3. *There is a bijection between morphisms* $\alpha : \mathsf{T} \to \mathsf{T}'$ *and functors* $V : \mathcal{C}^{\mathsf{T}'} \to \mathcal{C}^{\mathsf{T}}$ *for which*

(7)

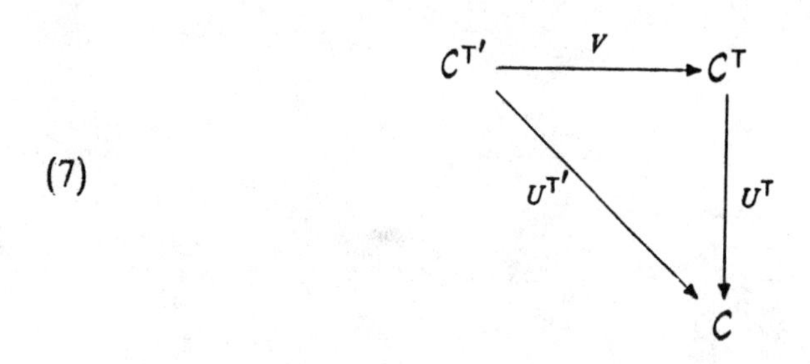

commutes. This bijection preserves composition.

Proof. Suppose $\alpha : \mathsf{T} \to \mathsf{T}'$ is a morphism of triples. Define $U^\alpha : \mathcal{C}^{\mathsf{T}'} \to \mathcal{C}^{\mathsf{T}}$ by

$$U^\alpha(A, a : T'A \to A) = (A, a \circ \alpha A),$$

and for an algebra map $f : A \to B$, $U^\alpha f = f$. $U^\alpha(A, a)$ is a T-algebra: The unitary law follows from (1) and the other law from the commutativity of this diagram:

(8)

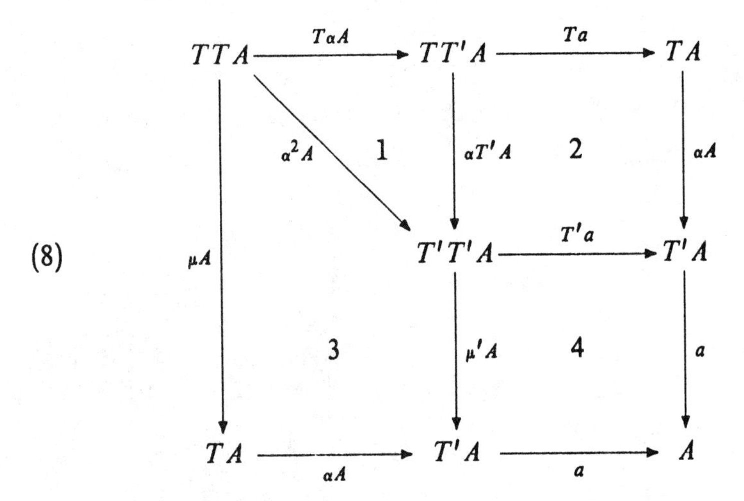

Here triangle 1 commutes by definition of of α^2, square 2 because α is a natural transformation, square 3 is diagram (2), and square 4 because a is a structure map (diagram (1), Section 3.2).

Seeing that U^α is a functor is left as an exercise, as is the functoriality of the operation which associates U^α to α.

Conversely, suppose that $V : \mathcal{C}^{\mathrm{T}'} \to \mathcal{C}^{\mathrm{T}}$ is a functor making (7) commute. If we apply V to the free algebra $(T'A, \mu' A)$, the result must be a T-algebra $(T'A, \sigma A)$ with the same underlying object $T'A$. The fact that σA is a T-algebra structure map means immediately that σ is a natural transformation and it satisfies the hypotheses of Proposition 1. Hence $\alpha = \sigma \circ T\eta' : T \to T'$ is a morphism of triples, as required.

We will conclude by showing that the association $\alpha \to U^\alpha$ is inverse to $V \to \sigma \circ T\eta'$. One direction requires showing that for $\alpha : T \to T'$ and any object A, $\alpha A = \mu' A \circ \alpha T' A \circ T\eta' A$. This follows from the commutativity of the following diagram, in which the square commutes because α is a natural transformation, the triangle commutes by definition and the bottom row is the identity by the definition of triple.

(9)

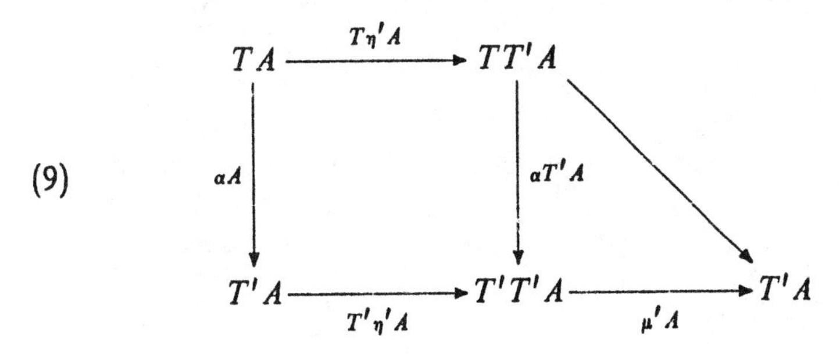

The other direction is more complicated. Suppose we are given V. We must show that for any $\mathbf{T}'$-algebra (A, a),

$$V(A,a) = (A, a \circ \sigma A \circ T\eta' A), \tag{10}$$

where by definition σA is the $\mathbf{T}$-algebra structure on $T'A$ obtained by applying V to the free algebra $(T'A, \mu' A)$.

In the first place,

$$\mu' A : (T'^2 A, \mu' T' A) \to (T', \mu' A)$$

is a morphism of $\mathbf{T}'$-algebras, so because of (7), μ' is also a morphism of the $\mathbf{T}$-algebras $(T'^2 A, \sigma T' A) \to (T' A, \sigma A)$. This says the square in the diagram below commutes. Since the triangle commutes by definition of triple, (10) is true at least for images under T of free T'-algebras.

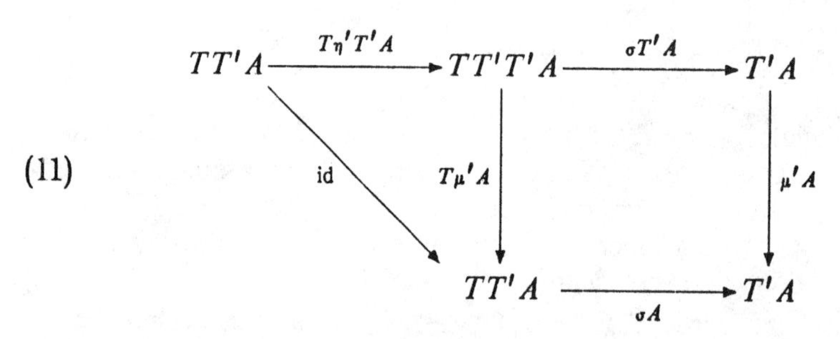

Now by Proposition 4, Section 3.3, for any $\mathbf{T}'$-algebra (A, a),

$$(T'^2 A, \mu' T' A) \underset{T'A}{\overset{\mu' A}{\rightrightarrows}} (T' A, \mu' A) \xrightarrow{\ a\ } (A, a) \tag{12}$$

is a U-contractible coequalizer diagram. Applying $U\alpha$ (where $\alpha = \sigma \circ T\eta'$) must give a U-contractible coequalizer diagram since $U\alpha$ commutes with the underlying functors. Because (10) is true of images of free algebras, that diagram is

$$(T'^2 A, \sigma T' A) \underset{T'A}{\overset{\mu' A}{\rightrightarrows}} (T' A, \sigma A) \xrightarrow{\ a\ } (A, b) \tag{13}$$

where $b = a \circ \sigma A \circ T\eta' A)$. Since V also commutes with underlying functors, applying V to (12) also gives a U-contractible coequalizer pair, with the same left and middle joints as (13) (that is how σ was defined). Its coequalizer must be $V(A, a)$ since the underlying functors create coequalizers. Thus (10) follows as required.

Exercises 3.6

(CATT). Show that for a given category $\mathcal{C}$, the triples in $\mathcal{C}$ and their morphisms form a category.

(UF). Show that U^α as defined in the proof of Theorem 3 is a functor.

(GAB). Let T be the Abelian group triple and T' the free group triple. What is the triple morphism α corresponding to the inclusion of Abelian groups into groups given by Theorem 3?

3.7. Adjoint Triples

In this section, we state and prove several theorems asserting the existence of adjoints to certain functors based in one way or another on categories of triple algebras. These are then applied to the study of the tripleability of functors which have both left and right adjoints.

Induced adjoints

Theorem 1. *In the following diagram (not supposed commutative) of categories and functors,*

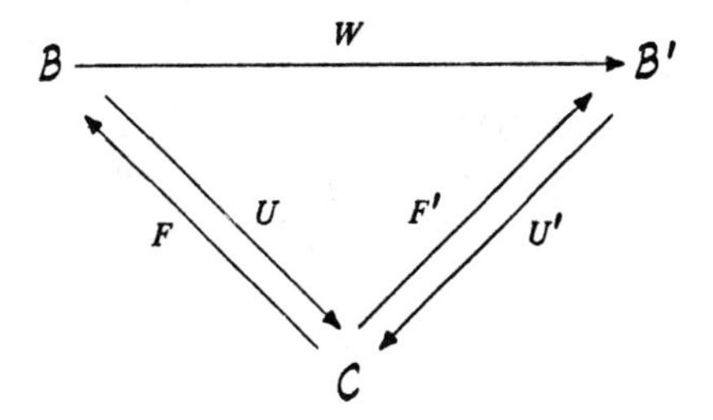

suppose that

(i) *F is left adjoint to U,*
(ii) *F' is left adjoint to U',*
(iii) *WF is naturally isomorphic to F',*
(iv) *U is tripleable, and*
(v) *W preserves coequalizers of U-contractible pairs.*

Then W has a right adjoint R for which $UR \cong U'$.

Proof. We will define R by using Corollary 2 of Section 3.6. Let the triples corresponding to the adjunctions be $\mathsf{T} = (T, \eta, \mu)$ and $\mathsf{T}' = (T', \eta', \mu')$

respectively. As usual, suppose that $\mathcal{B} = \mathcal{C}^{\mathsf{T}}$ and $U = U^{\mathsf{T}}$. Define $\sigma : T'T \to T$ so that

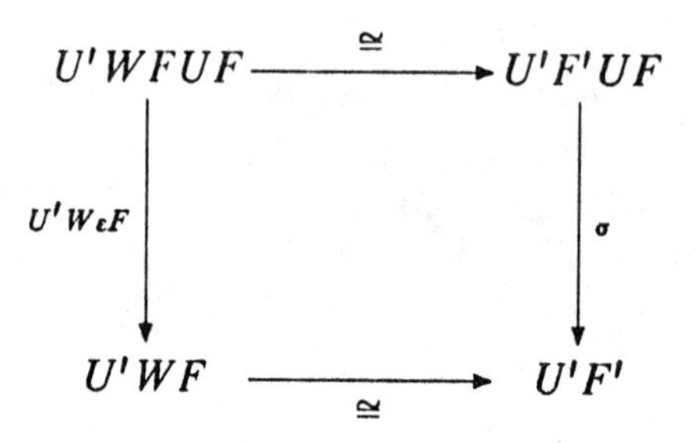

commutes.

Applying $U'W$ to diagram (4) of Section 3.1 and evaluating at F gives a diagram which, when the isomorphism of (iii) is applied, shows that $\sigma \circ \sigma T = \sigma \circ T'\mu$. An analogous (easier) proof using (2) of section 3.1 shows that $\sigma \circ T'\eta = \mathrm{id}$. Thus by Corollary 2 of Section 3.6, $\alpha = \sigma \circ \eta' T : T \to T'$ is a morphism of triples. The required functor R is $\mathcal{B}' \to \mathcal{C}^{\mathsf{T}'} \to \mathcal{C}^{\mathsf{T}}$, where the first arrow is the comparison functor and the second is the functor V induced by α.

For an object B' of $\mathcal{B}'$, we have, applying the definitions of R, the comparison functor Φ' for U', and the functor V determined by α, the following calculation:

$$\begin{aligned} URB' &= UV\Phi B' = UV(U'B', U'\varepsilon' B') \\ &= U(U'B', U'\varepsilon' B' \circ \alpha U'B') = U'B', \end{aligned}$$

so $UR = U'$ as required.

We now show that R is right adjoint to W insofar as free objects are concerned, and then appeal to the fact that algebras are coequalizers. (Compare the proof of Theorem 9 of Section 3.3). The following calculation does the first: For C an object of $\mathcal{C}$ and B' an object of $\mathcal{B}'$,

$$\begin{aligned} \mathrm{Hom}(WFC, B') &\cong \mathrm{Hom}(F'C, B') \\ &\cong \mathrm{Hom}(C, U'B') \cong \mathrm{Hom}(C, URB') \cong \mathrm{Hom}(FC, RB'). \end{aligned}$$

Now any object of B has a presentation

$$(*) \qquad FC_2 \rightrightarrows FC_1 \longrightarrow B$$

by a U-contractible coequalizer diagram. Since U is tripleable, $(*)$ is a coequalizer. Thus the bottom row of the diagram

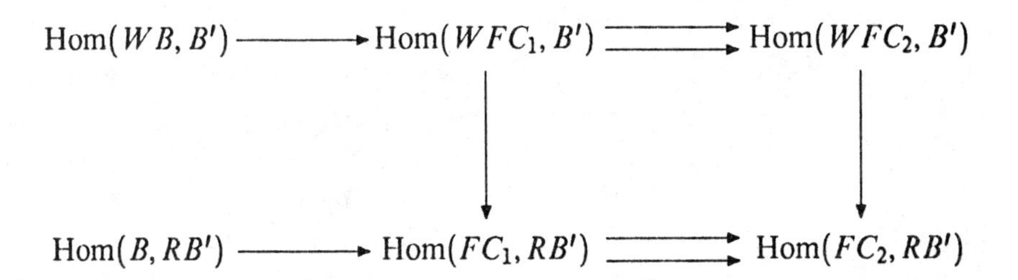

is an equalizer. By (v), W preserves coequalizers of U- contractible pairs, so the top row is an equalizer. The required isomorphism of Homsets follows.

Theorem 2. *In the following diagram (not supposed commutative) of categories and functors,*

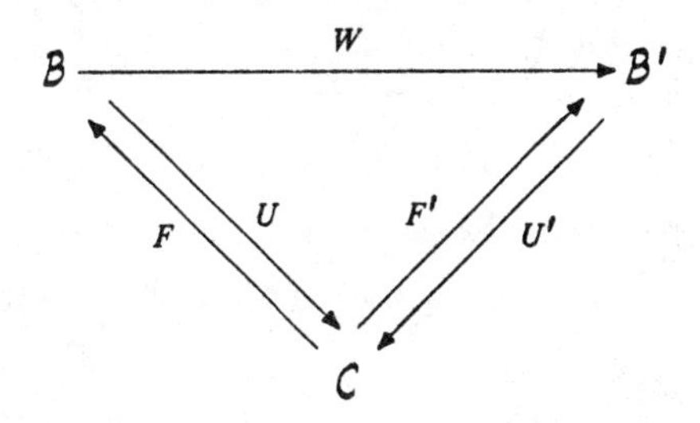

(a) *Suppose that*

- (i) F *is left adjoint to* U,
- (ii) F' *is left adjoint to* U',
- (iii) WF *is naturally isomorphic to* F',
- (iv) U *is of descent type, and*
- (v) $\mathcal{B}$ *has and* W *preserves coequalizers of* U*-contractible coequalizer pairs.*

Then W *has a right adjoint* R *for which* $UR \cong U'$.

(b) *Suppose that*

- (i) F *is left adjoint to* U,
- (ii) F' *is left adjoint to* U',
- (iii) $U'W$ *is naturally isomorphic to* U,
- (iv) U' *is of descent type, and*
- (v) $\mathcal{B}$ *has coequalizers.*

Then W *has a left adjoint* L *for which* $LF' \cong F$.

Proof. (a) We must show that the hypothesis in Theorem 1 that U is tripleable can be weakened to the assumption that it is of descent type. Consider the diagram

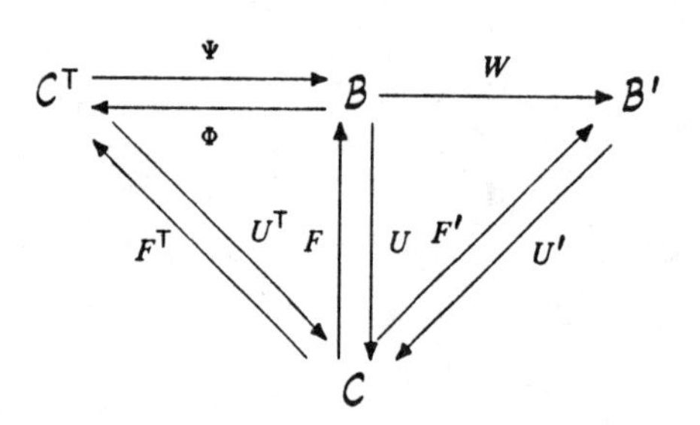

in which the left adjoint Ψ exists because $\mathcal{B}$ has coequalizers (Proposition 11, Section 3.3).

Theorem 1 implies that $L \circ \Psi$ has a right adjoint $S : \mathcal{B}' \to \mathcal{C}^{\mathsf{T}}$. To apply Theorem 1 we need to know that $L \circ \Psi \circ F^{\mathsf{T}} \cong F'$. This follows from the given fact $L \circ F \cong F'$ and the fact that the counit of the adjunction between Φ and Ψ must be an isomorphism by Exercise (EQUIII) of Section 1.9.

From this we have for objects A of $\mathcal{C}^{\mathsf{T}}$ and B' of $\mathcal{B}'$ that

$$\begin{aligned}\operatorname{Hom}(\Phi\Psi A, SB') &\cong \operatorname{Hom}(L\Psi\Phi\Psi A, B')\\ &\cong \operatorname{Hom}(L\Psi A, B') \cong \operatorname{Hom}(A, SB').\end{aligned}$$

Thus by Exercise (REFL)(c) of Section 1.9, every object of the form SB' is ΦB for some B in $\mathcal{B}$. Since Φ is full and faithful, this allows the definition of a functor $R : \mathcal{B}' \to B$ for which $S = \Phi \circ L$.

The following calculation then shows that R is right adjoint to L:

$$\begin{aligned}\operatorname{Hom}(LB, B') &\cong \operatorname{Hom}(L\Psi\Phi B, B') \cong \operatorname{Hom}(\Phi B, SB')\\ &\cong \operatorname{Hom}(\Phi B, \Phi RB') \cong \operatorname{Hom}(B, RB').\end{aligned}$$

(b) If $F'C$ is an object in the image of F', then we have

$$\begin{aligned}\operatorname{Hom}(F'C, WB) &\cong \operatorname{Hom}(C, U'WB)\\ &\cong \operatorname{Hom}(C, UB) \cong \operatorname{Hom}(FC, B)\end{aligned}$$

which shows that FC represents the functor $\operatorname{Hom}(F'C, W-)$. Moreover, the Yoneda lemma can easily be used to show that maps in $\mathcal{B}'$ between objects in the image of $\mathcal{F}'$ give rise to morphisms in $\mathcal{B}$ with the required naturality properties. Thus we get a functor L defined at least on the full subcategory whose objects are the image of F'. It is easily extended all of $\mathcal{B}'$ by letting

$$F'C_2 \rightrightarrows F'C_1 \longrightarrow B'$$

be a coequalizer and defining LB' so that

$$FC_2 \rightrightarrows FC_1 \longrightarrow LB'$$

is as well. The universal mapping property of coequalizers gives, for any object B of $\mathcal{B}$ the diagram below in which both lines are equalizers,

$$\begin{array}{ccccc}
\mathrm{Hom}(LB', B) & \longrightarrow & \mathrm{Hom}(FC_1, B) & \rightrightarrows & \mathrm{Hom}(FC_2, B) \\
 & & \downarrow \cong & & \downarrow \cong \\
\mathrm{Hom}(B', WB) & \longrightarrow & \mathrm{Hom}(F'C_1, WB) & \rightrightarrows & \mathrm{Hom}(F'C_2, WB)
\end{array}$$

from which the adjointness follows.

Theorem 3 (Butler). *In the situation*

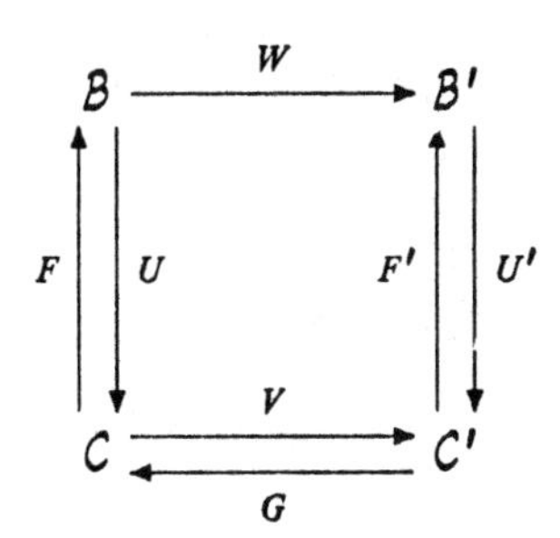

(a) *Suppose:*

(i) *F is left adjoint to U,*
(ii) *F' is left adjoint to U',*
(iii) *$W \circ F \cong F' \circ V$,*
(iv) *G is right adjoint to V,*
(v) *U is of descent type, and*
(vi) *$\mathcal{B}$ has and W preserves coequalizers of U-contractible coequalizer pairs.*

Then W has a right adjoint.

(b) *Suppose:*

(i) *F is left adjoint to U,*
(ii) *F' is left adjoint to U',*
(iii) *$V \circ U \cong U' \circ W$,*
(iv) *G is left adjoint to V,*
(v) *U' is of descent type, and*
(vi) *$\mathcal{B}$ has coequalizers.*

Then W has a left adjoint.

Proof. (a) Apply Theorem 2(a) to the diagram

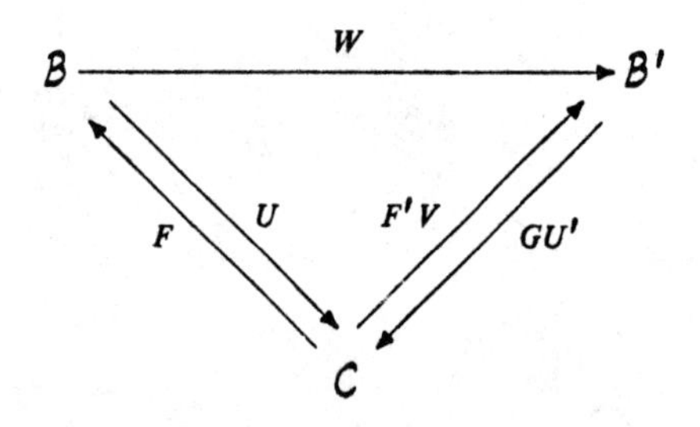

(b) Apply Theorem 2(b) to the diagram

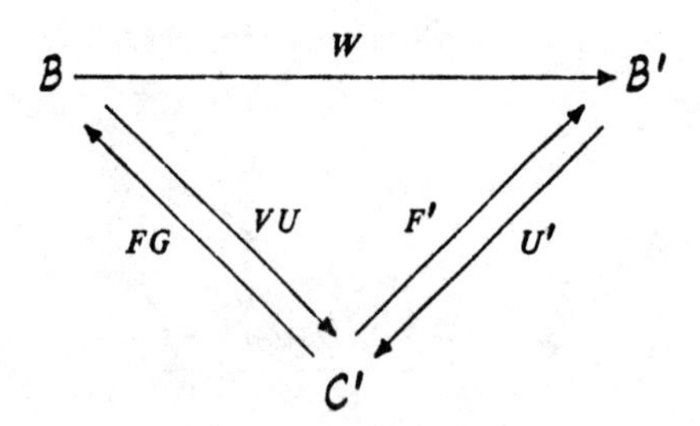

Adjoint triples

By an **adjoint triple** in a category $\mathcal{C}$, we mean

(i) A triple $\mathsf{T} = (T, \eta, \mu)$ in $\mathcal{C}$,
(ii) A cotriple $\mathsf{G} = (G, \varepsilon, \delta)$ in $\mathcal{C}$, for which
(iii) T is left adjoint to G.

We say in this case that T is left adjoint to G.

A functor $U : \mathcal{B} \to \mathcal{C}$ is **adjoint tripleable** if it is tripleable and cotripleable (the latter means that $U^{\mathrm{op}} : \mathcal{B}^{\mathrm{op}} \to \mathcal{C}^{\mathrm{op}}$ is tripleable). Theorem 6 below implies, among other things, that an adjoint tripleable functor results in an adjoint triple.

Proposition 4. *Let $\mathcal{B}$ be a category with finite limits and colimits, and $U : \mathcal{B} \to \mathcal{C}$ a functor. Then U is adjoint tripleable if and only if it has left and right adjoints and reflects isomorphisms.*

Proof. The existence of adjoints implies the preservation of the required equalizers and coequalizers.

Theorem 5. *Let T be a triple in $\mathcal{C}$ and suppose that T has a right adjoint G. Then G is the functor part of a cotriple G in $\mathcal{C}$ for which $\mathcal{C}^{\mathsf{T}}$ is equivalent to $\mathcal{C}_{\mathsf{G}}$ and the underlying functor U^{T} has left and right adjoints which induce T and G respectively.*

Conversely, let $U : \mathcal{B} \to \mathcal{C}$ be a functor with right adjoint R and left adjoint L. Let $\mathsf{T} = (T, \eta, \mu)$ be the triple induced by L and U and $\mathsf{G} = (G, \epsilon, \delta)$ the cotriple in $\mathcal{C}$ induced by U and R. Then T is left adjoint to G and the category $\mathcal{C}^{\mathsf{T}}$ of T-algebras is equivalent to the category $\mathcal{C}_{\mathsf{G}}$ of G-coalgebras.

Proof. To prove the first statement, let G be right adjoint to T and consider the diagram

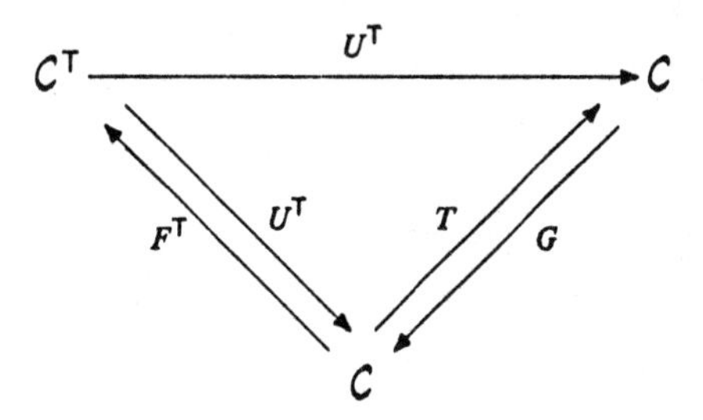

Theorem 1 implies that U^{T} has a right adjoint $R^{\mathsf{T}} : \mathcal{C} \to \mathcal{C}^{\mathsf{T}}$ for which $U^{\mathsf{T}} \circ R^{\mathsf{T}} = G$. By Proposition 4, U^{T} is tripleable and cotripleable. Hence $\mathcal{C}^{\mathsf{T}}$ is equivalent to $\mathcal{C}_{\mathsf{G}}$.

To do the converse, the adjunction between T and G is seen from the calculation

$$\mathrm{Hom}(ULC, C') \cong \mathrm{Hom}(LC, RC') \cong \mathrm{Hom}(C, URC').$$

Now the *first* half of the theorem yields a right adjoint R^{T} to U^{T} which induces G, so $\mathcal{C}^{\mathsf{T}}$ is equivalent to $\mathcal{C}_{\mathsf{G}}$.

We saw in Section 3.1 that for any monoid M, the functor $M \times (-)$ is the functor part of a triple in $\mathcal{S}et$. This functor has the right adjoint $\mathrm{Hom}(M, -)$, so is part of an adjoint triple, and the underlying functor $\mathcal{S}et^M \to \mathcal{S}et$ is adjoint tripleable. Analogously, if K is a commutative ring and R a K-algebra, then $R \otimes_K - : \mathcal{M}od K \to \mathcal{M}od K$ has a right adjoint $\mathrm{Hom}_K(R, -)$ and so gives rise to an adjoint triple. The algebras for this triple are the modules over the K-algebra R (R modules in which the action of K commutes with that of R). If $K = Z$, we just get R-modules.

In Section 6.7 we will make use of the fact that in a topos, functor categories (functors from a category object to the topos) are adjoint tripleable. Exercise (Fcs) asks you to prove this for $\mathcal{S}et$. The general situation is complicated by the problem of how to define that functor category in a topos.

Exercises 3.7

(Fcs). Let $\mathcal{C}$ be a small category and $\mathcal{E}$ the category of functors from $\mathcal{C}$ to $\mathcal{S}et$. There is an underlying functor $\mathcal{E} \to \mathcal{S}et/\mathcal{O}b(\mathcal{C})$. Show that this functor is adjoint tripleable. (Hint: One way to approach this is to use the Yoneda lemma to determine what the left or right adjoint must be on objects of the form $1 \to \mathcal{O}b(\mathcal{C})$ and then use the fact that a set is a coproduct of its elements and left and right adjoints preserve colimits and limits respectively.)

(YFcs). Deduce the Yoneda lemma from Exercise (Fcs).

3.8. Historical Notes on Triples

It is very hard to say who invented triples. Probably many scientific discoveries are like that. The first use of them was by Godement [1958] who used the flabby sheaf cotriple to resolve sheaves for computing sheaf cohomology. He called it the "standard construction" and presumably intended by that nothing more than a descriptive phrase. It seems likely that he never intended to either create or name a new concept.

Nonetheless Huber [1961] found these constructions useful in his homotopy theory and now did name them standard constructions. He also provided the proof that every adjoint pair gave rise to one, whatever it was called. He commented later that he proved that theorem because he was having so much trouble demonstrating that the associative identity was satisfied and noticed that all his standard constructions were associated with adjoints.

As remarked in Section 3.2, Kleisli [1965] and independently Eilenberg-Moore [1965] proved the converse. Although Hilton had conjectured the result, it was Kleisli [1964] who had an application. He wanted to show that resolutions using resolvent pairs (essentially pairs of adjoint functors) and those using triples give the same notion of resolution. Huber's construction gave the one direction and Kleisli's gave the other.

Eilenberg and Moore also gave them the name by which they are known here: triples. Although we do not regard this name as satisfactory we do not regard the proposed substitutes as any better. In this connection, it is worth mentioning that when asked why they hadn't found a better term, Eilenberg replied that they hadn't considered the concept very important and hadn't thought it worth investing much time in trying to find a good name. (By contrast, when Cartan-Eilenberg [1956] was composed, the authors gave so much thought to naming their most important concept that the book reached proof stage with blanks inserted before they found the exact term.)

At the same time, more or less, Applegate [1965] was discovering the connection between triples and acyclic models and Beck [1967] (but the work was substantially finished in 1964) was discovering the connection with homology. In addition, Lawvere [1963] had just found out how to do universal algebra by viewing an algebraic theory as a category and an algebra as a functor. Linton was soon to connect these categories with triples. In other words triples were beginning to pervade category theory but it is impossible to give credit to any one person. The next important step was the tripleableness theorem of Beck's which in part was a generalization of Linton's results. Variations on that theorem followed (Duskin [1969], Paré [1971] and acquired arcane names, but they all go back to Beck and Linton. They mostly arose either because of the failure of tripleableness to be transitive or because of certain special conditions.

Butler's theorems—Theorem 3 above includes somewhat special cases of two of them—are due to a former McGill University graduate student, William Butler. They consisted of a remarkable series of 64 theorems, 12 on the existence of adjoints and 52 on various technical results on tripleableness and related questions such as when a functor is of descent type. These theorems have never been published and, as a matter of fact, have remained unverified, except by Butler, since 1971. Within the past two years, they have been independently verified and subststantially generalized in his doctoral thesis: [1984], by another student, John A. Power, who found a few minor mistakes in the statements.

Chapter 4.

Theories

In this chapter, we explicate the naive concept of a mathematical theory, such as the theory of groups or the theory of fields, in such a way that a theory becomes a category and a model for the theory becomes a functor based on the category. Thus a theory and a model become instances of mathematical concepts which are widely used by mathematicians. This is in contrast to the standard treatment of the topic (see Shoenfield [1967], Chang and Kiesler [1973]) in which "theory" is explicated as a formal language with rules of deduction and axioms, and a model is a set with structure which corresponds in a specific way with the language and satisfies the axioms. Our theories should perhaps have been called "categorical theories"; however, the usage here is now standard among category theorists.

Our theories are, however, less general than the most general sort of theory in mathematical logic.

We will construct a hierarchy of types of theories, consisting of categories with various amounts of structure imposed on them. For example, we will construct the theory of groups as the category with finite products which contains the generic group object, in the sense to be defined precisely in Section 4.1. (The definition of group object using representable functors mentioned in Section 1.7 does not require that the category have finite products but we do not know how to handle that more general type of theory.) On the other hand, a theory of fields using only categories with finite products cannot be given, so one must climb further in the hierarchy to give the generic field.

In this chapter we consider the part of the hierarchy which can be developed using only basic ideas about limits. In the process we develop a version of Ehresmann's theory of sketches suitable for our purposes. This chapter may be read immediately after Chapter 1, except for Theorem 5 of Section 4.3. The theories higher in the hierarchy (in particular including the theory of fields) require the machinery of Grothendieck topologies and are described in Chapter 8.

A brief description of this hierarchy and its connections with different types of logical systems has been given by Lawvere [1975]. Makkai and Reyes [1977] provide a detailed exposition of the top of the hierarchy.

4.1. Sketches

Groups

An **FP-category** is a category with finite products, and an FP-functor between FP-categories is a functor which preserves finite products. The **FP theory of groups** should be an FP-category $\mathcal{G}$ containing a group object G which is generic in the sense that every group object in any FP-category is the image of G under a unique FP-functor from $\mathcal{G}$ to the category.

We will begin by constructing the FP-theory of groups from the ground up, so to speak, but then interrupt ourselves for a different approach which makes it obvious that the theory thus constructed is uniquely determined.

To construct the FP-theory of groups, we must have an object representing the group, powers of that object, morphisms representing the projection maps from those powers, and morphisms representing the operations (identity, inverse, multiplication). A law like the associative law is stated by requiring that two maps (representing the two ways of multiplying) in the category are equal.

Thus we need a category containing an object G and all its powers, specifically including G^0, which is the terminal object. It must have morphisms $m : G^2 \to G$, $i : G \to G$ and $u : G^0 \to G$ as operations. Each power G^k must have k projections $p_i : G^k \to G$, and for each k and n each n-tuple of maps $G^k \to G$ induces a map $G^k \to G^n$. A category containing all these maps must contain all their composites; for example the composites $m \circ m \times 1$ and $m \circ 1 \times m : G^3 \to G$. The group laws are equivalent to requiring that the following diagrams commute:

(1)

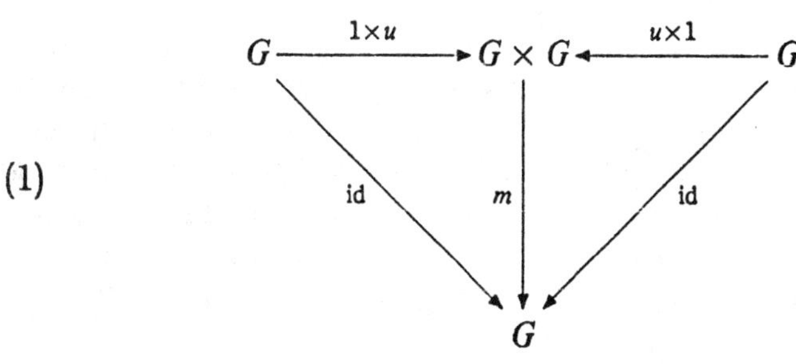

(2)

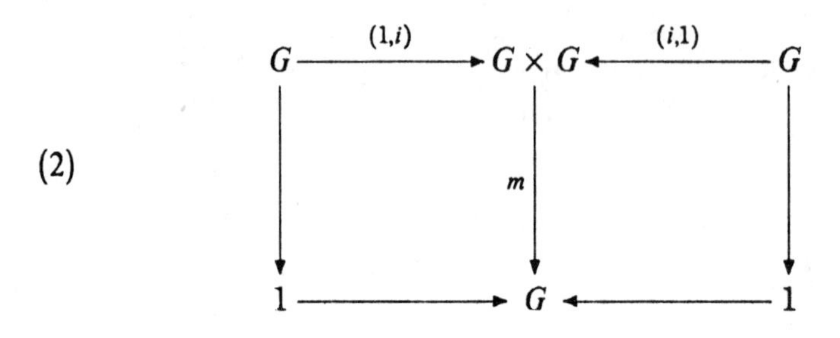

and

(3)

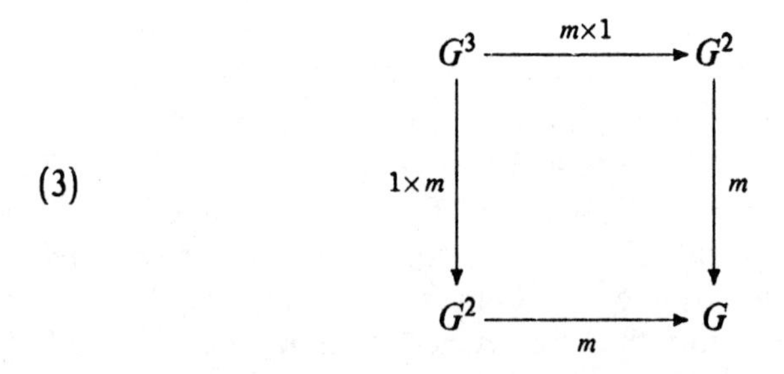

The commutativity of these diagrams, of course, forces many other pairs of arrows of the category to be equal.

At first sight, it seems reasonably clear that the data we have given determine a unique category $\mathcal{G}$. Assuming that that is the case, it is easy to see that giving a group object in a category with finite products (in other words, giving a model of the theory of groups in such a category) is the same as giving a finite-product-preserving functor from $\mathcal{G}$ to the category.

However, there are complications in carrying out the construction of $\mathcal{G}$ (even more so in the case of multi-sorted algebraic structures, about which more below). These difficulties are analogous to the difficulties in constructing the free group on a set as consisting of equivalence classes of strings: they are not insurmountable, but a construction using the adjoint functor theorem is much less fussy.

Sketches

In this section, we introduce some machinery (sketches) which will enable us to give, in Section 4.3, a formal construction of $\mathcal{G}$ and analogous categories by embedding the given data in a functor category and defining $\mathcal{G}$ to be the smallest subcategory with the required properties (having finite products in the case of groups). Our sketches are conceptually similar to, but different in detail from those of Ehresmann (see the historical notes in Section 4.5).

Recall that a graph $\mathcal{G}$ consists of a set of vertices denoted G_0 and a set of arrows denoted G_1 together with the operators $d^0, d^1 : G_1 \to G_0$ which assign to each arrow its source and target. Cones and diagrams are defined for graphs in exactly the same way as they are for categories. Note that we have carefully distinguished between cones and commutative cones and between diagrams and commutative diagrams. Commutative cones and diagrams of course make no sense for graphs.

By a **sketch** we mean a 4-tuple $\mathcal{S} = (\mathcal{G}, U, D, C)$ where $\mathcal{G}$ is a graph, $U : G_0 \to G_1$ is a function which takes each object A of G_0 to an arrow from A to A, D is a class of diagrams in $\mathcal{G}$ and C is class of cones in $\mathcal{G}$. Each cone in C goes from some vertex to some diagram; that diagram need not be in D; in fact in general it is necessary to allow diagrams which are not in D as bases of cones.

An **FP-sketch** is a sketch in which the cones are discrete; that is, there are no arrows between two distinct vertices of the base.

If $\mathcal{S}'$ is another sketch, a **morphism** of $\mathcal{S}$ into $\mathcal{S}' = (\mathcal{G}', U', D', C')$ is a graph homomorphism $h : \mathcal{G} \to \mathcal{G}'$ such that $h \circ U = U' \circ h$, every diagram in D is taken to a diagram in D', and every cone in C is taken to a cone in C'.

If $\mathcal{C}$ is a category, the **underlying sketch** $\mathcal{S} = (\mathcal{G}, U, D, C)$ of $\mathcal{C}$ has as graph the underlying graph of $\mathcal{C}$, for U the map which picks out the identity arrows of $\mathcal{C}$, takes D to be the class of all commutative diagrams of $\mathcal{C}$ and for C all the limit cones.

A **model** for a sketch $\mathcal{S}$ in a category $\mathcal{C}$ is then a sketch morphism from $\mathcal{S}$ into the underlying sketch of $\mathcal{C}$. It follows that a model forces all the diagrams of the sketch to commute and all the cones of the sketch to be limit cones (hence commutative cones).

The models in $\mathcal{C}$ form a category. The morphisms are "natural" transformations, defined in just the same way as natural transformations of functors (whose definition, after all, makes no use of composition in the domain category). The models of $\mathcal{S}$ in $\mathcal{S}et$ will be denoted $\mathcal{M}od(\mathcal{S})$. The category of *graph* morphisms from $\mathcal{S}$ to $\mathcal{S}et$ which take all the diagrams to commutative diagrams will be denoted throughout the book as $\mathcal{S}et^{\mathcal{S}}$.

The sketch of the theory of groups

We may construct a suitable FP-sketch for the theory of groups using the data given above. The objects of the graph of the sketch should be $1 = G^0$, $G = G^1$, G^2 and G^3. Note that these are just formal names for objects at this point. We will shortly introduce cones to force a model in the sense just defined to take them to the powers of one object. These powers are just those needed to state the various group laws.

The sketch for groups must have the following arrows, besides the identity arrows required by the definition of sketch:

(i) Three arrows $m : G^2 \to G$, $i : G \to G$ and $u : 1 \to G$ for the operations.

(ii) Projection arrows for cones $p_i : G^2 \to G, (i = 1, 2)$ and $q_i : G^3 \to G, (i = 1, 2, 3)$. In each case the cone is to the discrete diagram all of whose nodes are G.

(iii) An arrow $G \to 1$.

(iv) $r_i : G \to G^2, (i = 1, 2)$, which will be forced to be $\mathrm{id} \times u$ and $u \times \mathrm{id}$ respectively. To allay any confusion, we should make it perfectly clear that G^2 is in no sense yet a product. Thus the r_i have to be explicitly assumed.

(v) $s_i : G \to G^2, (i = 1, 2)$, to be (id, i) and (i, id)).

(vi) $t_i : G^3 \to G^2, (i = 1, 2)$, to be (q_1, q_2) and (q_2, q_3).

(vii) $n_i : G^3 \to G^2, (i = 1, 2)$, to be $\mathrm{id} \times m$ and $m \times \mathrm{id}$.

As its designated cones, it must have the cones given in (ii) above and the empty cone defined on 1. Its diagrams must be diagrams (1) through (3) above (with arrows renamed as just described) and forcing diagrams such as

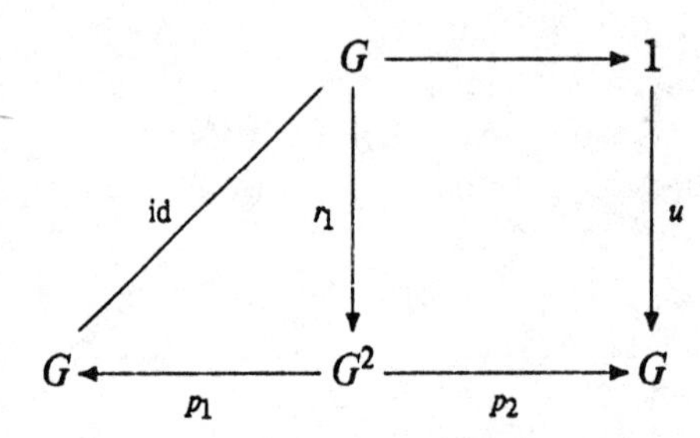

which will force the r_1 in models to be $\mathrm{id} \times u$,

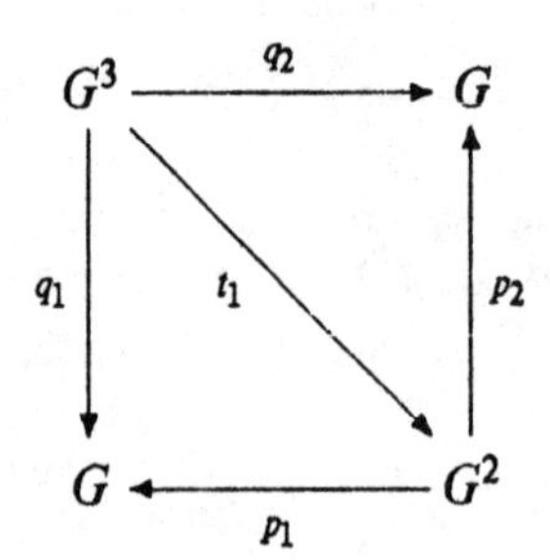

which will force t_1 to be (q_1, q_2), and

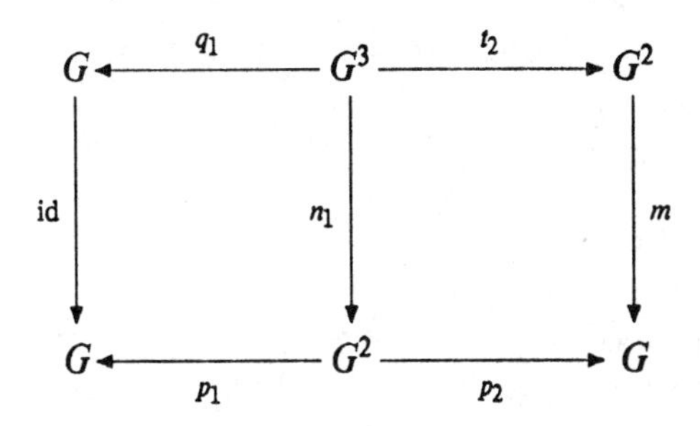

which will force n_1 to be $\mathrm{id} \times m$.

The preceding construction is involved, but the principle behind it is straightforward. You need

(i) an object to be the generic algebraic structure (group in the example above);

(ii) objects to be the powers needed to define the operations and state the laws;

(iii) arrows for identities, for the operations, and for the cones forcing the powers to be actual powers; and

(iv) diagrams to state the laws.

In the process of constructing these diagrams you may need

(v) arrows composed out of operations and projections.

The definition in (v) requires additional diagrams such as the last three in the definition of groups above.

It should be clear that algebraic structures with finitary operations satisfying universal equations (unlike fields, for example, where one needs statements like "$x = 0$ or $xx^{-1} = 1$") can all be modeled in this way.

More complicated algebraic structures may be modeled by an FP-sketch, too. As an example, consider the category of all M-actions for all monoids M. An object is a pair (M, A) where M is a monoid which acts on the set A. A morphism $(f, \varphi) : (M, A) \to (M', A')$ has f a monoid homomorphism and φ a set map for which $f(m)(\varphi(a)) = \varphi(ma)$ for $m \in M$ and $a \in A$. (One can define the category of all modules (R, M) where R is a ring and M is an R-module analogously). The sketch for monoid actions will have to contain objects M and A and objects representing M^n for $n = 0, 2, 3$ as well as products of certain powers of M with A.

The sketch for a theory for groups, for example, is essentially a direct translation of the ingredients that go into the usual definition of groups in a textbook. However, the sketch which results is not a category, and so we cannot use the machinery of category theory to study the resulting models.

In Section 4.3, using Kennison's theorem from Section 4.2, we show how to achieve our original goal of producing a theory of groups which is itself a category, in fact the category generated in a strong sense by the sketch constructed above.

4.2. Kennison's Theorem

This section is devoted to proving the following theorem, due essentially to Kennison [1968]. A generalization to base categories other than Set was proved by Freyd and Kelly [1972].

Theorem 1. *Let $\mathcal{S}$ be a small sketch. Then $\mathit{Mod}(\mathcal{S})$ is a reflective subcategory of $\mathit{Set}^{\mathcal{S}}$.*

Proof. Kennison's original proof was stated for $\mathcal{S}$ a category and for models which preserve all limits, but it works without change in the present case.

It is necessary to construct a left adjoint for the inclusion of $\mathit{Mod}(\mathcal{S})$ in $\mathit{Set}^{\mathcal{S}}$. We will do this using the adjoint functor theorem.

Limits in $\mathit{Set}^{\mathcal{S}}$ are constructed pointwise in exactly the same way as for functor categories (see Exercise (LIMFUN) of Section 1.7). It is necessary to show that a limit of a diagram of morphisms (objects in $\mathit{Set}^{\mathcal{S}}$) takes the diagrams which are part of the structure of $\mathcal{S}$ to commutative diagrams; but that is an easy exercise. To show that inclusion preserves all limits requires showing that a morphism constructed as a limit of a diagram of models takes cones to limit cones. This follows easily in exactly the same way as it does for a limit of ordinary functors (Exercise (COMLIM)).

We need the following lemma to get the solution set condition.

Lemma 2. *Let $\mathcal{S}$ be a small sketch, and $F : \mathcal{S} \to \mathit{Set}$ a function on the objects of $\mathcal{S}$. There is a cardinal $\aleph$ with the following property: If M is any model of $\mathcal{S}$ with the property that $F(A) \subseteq M(A)$ for every object A of $\mathcal{S}$, then there is a model $\hat{F}$ of $\mathcal{S}$ for which for every object A of $\mathcal{S}$,*

(i) *$F(A) \subseteq \hat{F}(A) \subseteq M(A)$, and*
(ii) *The cardinality of $\hat{F}(A)$ is less than $\aleph$.*

Given the lemma, let $E \in \mathit{Set}^{\mathcal{S}}$, let M be a model, and let $\lambda : E \to M$ be a natural transformation. Let F be the function whose value on an object A is the image of λA. The lemma provides a cardinal $\aleph(F)$ with the property that F is contained in a model $\hat{F} \subseteq M$ whose cardinality at each object of $\mathcal{S}$ is at most $\aleph(F)$. Now let $\aleph$ be the sup of the $\aleph(F)$ as F

varies over all object functions which are quotients of E. Then a solution set for E will be the set of all natural transformations $\varphi_i : E \to M_i$ for all models M_i for which for all objects A of $\mathcal{S}$, the cardinal of $M_i(A)$ is less than or equal to $\aleph$.

Proof of Lemma 2. For any such F as in the hypothesis, define a function $F^{\#}$ on objects by

$$F^{\#}(A) = \bigcup\{Mf(FB) \mid f : B \to A\}$$

(union over all arrows into A) and a function F^* by

$$F^*(A) = \bigcup\{x \in \prod FDi \mid MDdx_i = x_j \text{ for all } d : i \to j \in \mathcal{I}\},$$

the union over all those cones from A to $D : \mathcal{I} \to \mathcal{G}$ which are in the set C of cones of $\mathcal{S}$. ($\mathcal{G}$ is the underlying graph of $\mathcal{S}$ and each cone is to some diagram $D : \mathcal{I} \to \mathcal{G}$ defined on some index category $\mathcal{I}$.)

For the purpose of understanding F^*, observe that since M is a model, if $\alpha : A \to D$ is a cone, then $M\alpha : MA \to MD$ is a limit cone in $\mathbf{Set}$, whence

$$MA = \{x \in \prod MDi \mid MDdx_i = x_j \text{ for all } d : i \to j \in I\},$$

and $x_i = M\alpha_i(x)$ for $x \in MA$.

For any given F, let m denote the smallest cardinal which is greater than

(i) The cardinality of the set of arrows of $\mathcal{S}$;
(ii) the cardinality of $F(A)$ for every object A of $\mathcal{S}$; and
(iii) the cardinality of I for every cone $A \to D(I)$ of $\mathcal{S}$.

Clearly, $F^{\#}(A)$ has cardinality less than m and it is not much harder to see that $F^*(A)$ has cardinality less than m^m (one is quantifying over sets—the sets of projections α_i—of cardinality less than m).

We construct a transfinite sequence of maps F_α, beginning with $F_0 = F$. For an ordinal α, if F_α has been defined, then $f_{\alpha+1} = (F_\alpha)^{\#*}$. If α is a limit ordinal, then for each object A,

$$F_\alpha(A) = \bigcup\{F_\beta(A) \mid \beta < \alpha\}.$$

According to the observations in the preceding paragraph, at each stage of this construction the bound m on the cardinality of the sets $F(A)$ may do no more than exponentiate —and this bound is a function of $\mathcal{S}$ and the function F we started with, but independent of M.

Now let $\hat{F} = F_\gamma$, where γ is the smallest ordinal of cardinality greater than the cardinality of the set of all arrows of $\mathcal{S}$. (Note that γ is a limit ordinal). If we can show $\hat{F}$ is the object function of a model, we are done.

In the first place, if $f : B \to A$, then $Mf(\hat{F}(B) \subseteq \hat{F}(A)$. For if $x \in \hat{F}(B)$, then $x \in F_\beta(B)$ for some ordinal $\beta < \gamma$, whence $Mf(x) \in F_{\beta+1}(A) \subseteq \hat{F}(A)$. Thus one can define $\hat{F}(f)$ to be the restriction of Mf to $\hat{F}(B)$ and make $\hat{F}$ a morphism of sketches.

To show that $\hat{F}$ preserves the limits in the sketch $\mathcal{S}$, let $\alpha : A \to D$ be a cone of $\mathcal{S}$, where $D : I \to G$. If $x \in \hat{F}(A)$, then $x \in MA$, so for each $d : i \to j$ in I, $MDdx_i = x_j$, so $\hat{F}Ddx_i = x_j$ because $\hat{F}Dd$ is the restriction of MDd. Hence

$$\hat{F}(A) \subseteq \{x \in \prod \hat{F}Di \mid \hat{F}Ddx_i = x_j \text{ for all } d : i \to j \text{ in } I\} = \lim \hat{F}D.$$

Conversely, suppose $x \in \prod \hat{F}Di$ has the property that $\hat{F}Ddx_i = x_j$ for all $d : i \to j$ in I. For each i, $x_i \in \hat{F}Di$, so there is some ordinal $\beta_i < \gamma$ for which $x_i \in F_{\beta_i}Di$. Now since $x \in MA$, $x_i = M\alpha_i(x)$, and we can assume that if $\alpha_i = \alpha_j$ then $\beta_i = \beta_j$. (For it is perfectly possible that I have greater cardinality than the set of arrows of $\mathcal{S}$). Thus the set of all β_i has cardinality less than or equal to the cardinality of the set of arrows of $\mathcal{S}$. That means there is a single ordinal $\beta < \gamma$ for which $x_i \in F_\beta Di$ for all objects i of I. Thus it follows from the definition that $x \in F_{\beta+1}(A) \subseteq \hat{F}(A)$ which must therefore be $\lim \hat{F}D$.

Note: Exercise (RFLK) shows that the information in the proof of the lemma yields a more precise description of the left adjoint.

Exercises 4.2

(RFLK). Let $E \in \mathbf{Set}^{\mathcal{S}}$, let M be a model of $\mathcal{S}$, and let $\lambda : F \to S$ be a natural transformation.

(i) Show that the model $\hat{F}$ constructed above is the smallest submodel of M through which λ factors. We say λ is **dense** if $\hat{F} = M$.

(ii) Prove that if $\alpha, \beta : M \to N$ are natural transformations of models, λ is dense, and $\alpha \circ \lambda = \beta \circ \lambda$, then $\alpha = \beta$.

(iii) Let $\varphi_i : E \to M_i$ be the solution set for E constructed in the text. Then the φ_i determine a map $E \to \prod M_i$; let $\hat{E}$ be the smallest model through which this map factors. Show that the function $E \to \hat{E}$ is the object function of a left adjoint to the inclusion of $\mathbf{Mod}(\mathcal{S})$ in $\mathbf{Set}^{\mathcal{S}}$.

(COMLIM)°. Let $\mathcal{A}$ be a category and C be a class of diagrams in $\mathcal{A}$, each of which has a limit in $\mathcal{A}$. Suppose that I is index category and

$D : \mathcal{I} \to \mathit{Set}^{\mathcal{A}}$ a diagram of functors such that each value of D preserves all the limits in the class C. Show that $\lim D$ also preserves the limits in C. (Note that the functor category is complete because Set is (see Exercise (LIMFUN) of Section 1.7).)

4.3. Finite-Product Theories

Given an FP-sketch $\mathcal{S}$, we want to construct the FP-category which will be the theory generated by the sketch. Letting $\mathcal{X}$ be the unknown theory, let us discover properties it must have until enough of them emerge to characterize it.

$\mathcal{X}$ will be the generic model of $\mathcal{S}$, so there will be a sketch morphism $m : \mathcal{S} \to \mathcal{X}$. Moreover, composing with m should induce a bijection between the models of $\mathcal{X}$ considered as a sketch with all its product cones and the models of $\mathcal{S}$. Now because $\mathcal{X}$ is a category and its cones are products, by Yoneda we can get a canonical embedding $y : \mathcal{X}^{\mathrm{op}} \to \mathit{Mod}(\mathcal{X})$: For an object X of $\mathcal{X}$, $y(X)(X') = \mathrm{Hom}(X, X')$. $y(X)$ is a model because representable functors preserve limits and—like all functors —preserve commutative diagrams. With the requirement that m induce an equivalence between $\mathit{Mod}(\mathcal{X})$ and $\mathit{Mod}(\mathcal{S})$, this produces the following diagram of sketch morphisms, in which e is the equivalence and $u = e \circ y \circ m^{\mathrm{op}}$ by definition. The composite across the top is the Yoneda embedding Y.

(1)

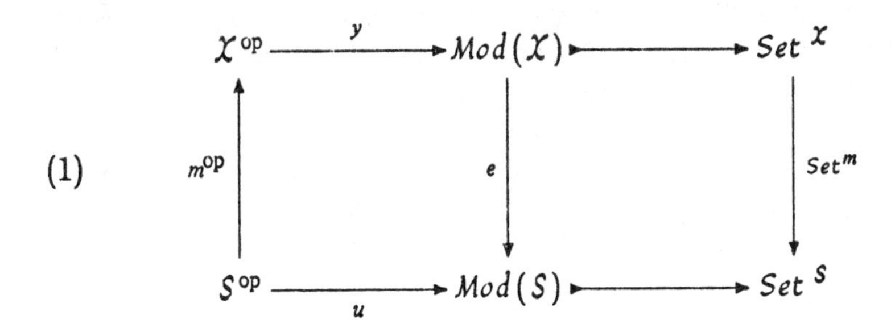

Because e is an equivalence, m has to be an epimorphism in the category of sketches. This would seem to mean that every object in $\mathcal{X}$ is a product of objects in $\mathcal{S}$. Moreover, Y, hence y, is full and faithful and $\mathcal{X}^{\mathrm{op}}$ is closed under finite sums and contains the image of $\mathcal{S}^{\mathrm{op}}$. This suggests that we try to construct $\mathcal{X}$ by *defining* u and letting $\mathcal{X}^{\mathrm{op}}$ be the full FP-subcategory generated by the image of u. We will construct u by constructing $\mathcal{X}$ in the special case that $\mathcal{S}$ has no cones, and then bootstrap up to the FP case using Kennison's theorem.

Theorem 1. *Given a sketch $\mathcal{S}$ with no cones (i.e., a graph with diagrams) there is a category $\mathcal{C}$ and a model $m : \mathcal{S} \to \mathcal{C}$ such that composition with m is an equivalence between $\mathbf{Set}^{\mathcal{S}}$ and $\mathbf{Set}^{\mathcal{C}}$.*

Proof. We constructed the free category for a graph with no diagrams in Exercise (GRADJ) of Section 1.9. Here, $\mathcal{C}$ can be obtained as a quotient category from this free category by factoring out the smallest congruence making every diagram commute (see Exercise (QUOT) of Section 1.1). Then m is the map into the free category followed by the quotient map. The required equivalence follows from the definition of "free" and "quotient".

In this case, u is the composite $\mathbf{Set}^{m} \circ j \circ m^{\mathrm{op}}$, where $j : \mathcal{C}^{\mathrm{op}} \to \mathbf{Set}^{\mathcal{C}}$ is the Yoneda embedding. u has the Yoneda-like property that for any model $F : \mathcal{S} \to \mathbf{Set}$,

$$\text{(Y)} \qquad \mathrm{Hom}(u(S), F) \cong F(S).$$

This follows since m is full and faithful and $uS = \mathrm{Hom}(mS, m(-))$; then use Yoneda. Note that, as the composite of full and faithful functors, u is full and faithful.

To get the case when $\mathcal{S}$ does have nontrivial cones, we must have a map $u : \mathcal{S} \to \mathbf{Mod}(\mathcal{S})$ rather than into $\mathbf{Set}^{\mathcal{S}}$. We have a map $u_1 : \mathcal{S}^{\mathrm{op}} \to \mathbf{Set}^{\mathcal{S}}$, namely the map constructed above called u for sketches with no cones (when $\mathcal{S}$ has no cones it is the u we want because then $\mathbf{Set}^{\mathcal{S}} = \mathbf{Mod}(\mathcal{S})$). We take u to be $k \circ u_1$, where k is the adjoint to the inclusion of $\mathbf{Mod}(\mathcal{S})$ in $\mathbf{Set}^{\mathcal{S}}$ given by Kennison's Theorem (Section 4.2). It is a trivial consequence of adjointness that property (Y) continues to hold in this case.

As suggested in our heuristic argument at the beginning of the section, we define $\mathrm{FP}(\mathcal{S})$, the **FP-theory generated by** $\mathcal{S}$, to be the full FP-subcategory of $\mathbf{Mod}(\mathcal{S})^{\mathrm{op}}$ generated by the image of u^{op}. In the rest of this section, we will use the factorization

(2)

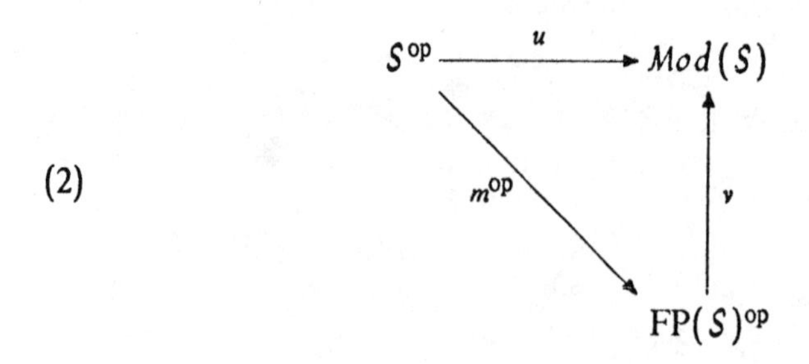

where v is the inclusion.

A theorem on adjoints

To see that FP($\mathcal{S}$) has the same models as $\mathcal{S}$, we need the following theorem.

Theorem 2. *In the diagram of categories and functors,*

(3)

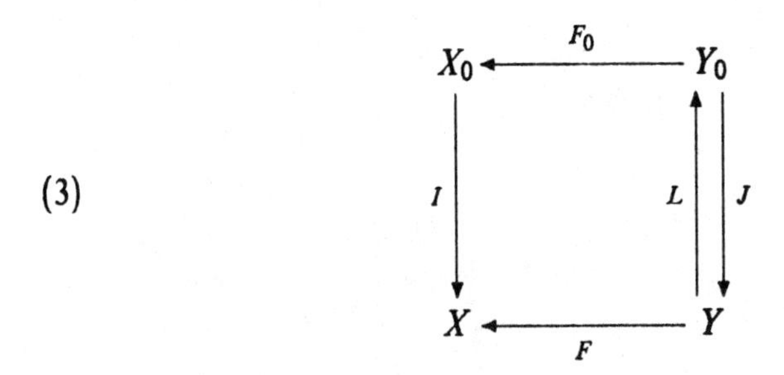

suppose that L is left adjoint to J. Then

(a) *If I is full and faithful, F has a left adjoint E and FJ naturally equivalent to IF_0, then $E_0 = LEI$ is left adjoint to F_0;*

(b) *If J is full and faithful, F has a right adjoint R, I has a left adjoint K and KF is equivalent to F_0L, then there is a unique functor $R_0 : X_0 \to Y_0$ for which $JR_0 \cong RI$; R_0 is right adjoint to F_0.*

Proof. For (a),

$$\begin{aligned}\mathrm{Hom}(LEIX_0, Y_0) &\cong \mathrm{Hom}(EIX_0, JY_0) \cong \mathrm{Hom}(IX_0, FJY_0)\\ &\cong \mathrm{Hom}(IX_0, IF_0Y_0) \cong \mathrm{Hom}(X_0, F_0Y_0).\end{aligned}$$

As for (b), first note that since J is full and faithful, $LJL \cong L$ so that we have,

$$\begin{aligned}\mathrm{Hom}(JLY, RIX_0) &\cong \mathrm{Hom}(FJLY, IX_0) \cong \mathrm{Hom}(KFJLY, X_0)\\ &\cong \mathrm{Hom}(F_0LJL, X_0) \cong \mathrm{Hom}(F_0LY, X_0)\\ &\cong \mathrm{Hom}(KFY, X_0) \cong \mathrm{Hom}(FY, IX_0)\\ &\cong \mathrm{Hom}(Y, RIX_0).\end{aligned}$$

Therefore, by Exercise (Refl)(c) of Section 1.9, there is a unique Y_0 which we will denote R_0X_0, for which $JY_0 \cong RIX_0$. Verification of the adjunction equation is easy, so that R_0 extends to a right adjoint to F_0 by the pointwise construction of adjoints.

Properties of morphisms of FP-sketches

Theorem 3. *Let $f : S_1 \to S_2$ be a morphism of FP-sketches. Then composition with f defines a map $f^* : Mod(S_2) \to Mod(S_1)$ which has a left adjoint $f_\#$. Moreover, if $u_i = v_i \circ m_i^{op}$ is the embedding of S_i^{op} into $Mod(S_i)$, then there is a map* $\mathrm{FP}(f)^{op}$ *which is the unique map making the top square in the diagram below commute, and also the unique map making the middle square commute.*

Proof. We have the following diagram:

(4)

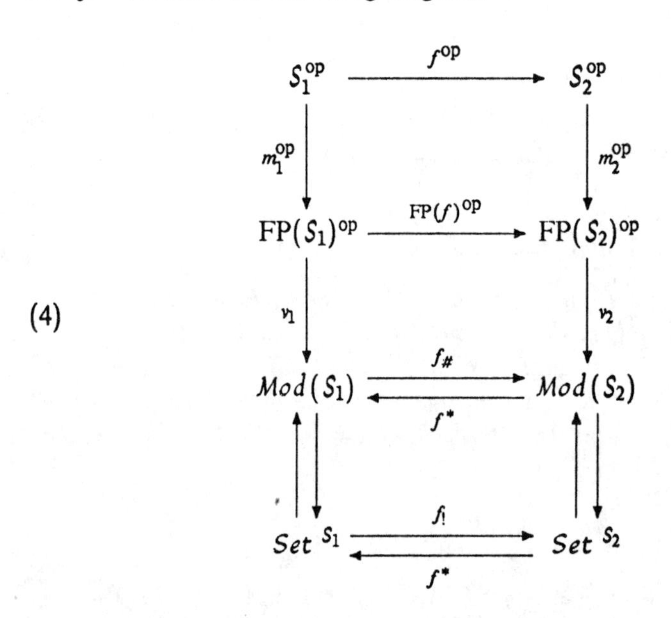

In this diagram, the lower f^* is induced by composing with f, and since f takes cones to cones, f^* restricts to a map with the same name on models. $f_!$ is the left Kan extension (Section 1.9). The inclusions of models into the functor categories have left adjoints by Kennison's theorem, and $f_\#$ exists by Theorem 2(a).

The commutativity $f_\# \circ u_1 = u_2 \circ f^{op}$ follows from this computation, where the last two isomorphisms follow from property (Y):

$$\begin{aligned}\mathrm{Hom}(f_\# u_1(X), M_2) &\cong \mathrm{Hom}(u_1(X), f^* M_2) \cong \mathrm{Hom}(u_1(X), M_2 \circ f) \\ &\cong M_2(f(X)) \cong \mathrm{Hom}(u_2(f(X)), M_2).\end{aligned}$$

Here X an object of S_1 and M_2 a model of S_2.

Since $f_\#$ is a left adjoint, it preserves colimits, whence $f_\#^{op}$ preserves limits. Moreover, $f_\#^{op}(S_1) \subseteq S_2 \subseteq \mathrm{FP}(S_2)$, $\mathrm{FP}(S_1)$ is the full closure of S_1 under finite products and $\mathrm{FP}(S_2)$ is full and closed under finite products, so it follows that $f_{\#}(\mathrm{FP}(\mathcal{S}_1)^{\mathrm{op}}) \subseteq \mathrm{FP}(\mathcal{S}_2)^{\mathrm{op}}$. We let $\mathrm{FP}(f)^{\mathrm{op}}$ be the restriction of $f_{\#}$. Clearly it makes both squares commute. Uniqueness for the middle one follows from the fact that v_2 is monic, and for the top one from the fact that $\mathrm{FP}(\mathcal{S}_1)^{\mathrm{op}}$ is the FP-subcategory generated by the image of m_1^{op}.

Theorem 4. *Let $\mathcal{S}$ be an FP-sketch. Then m is a model of $\mathcal{S}$. Moreover, any model of $\mathcal{S}$ in a category $\mathcal{C}$ with finite products has a unique extension along m to a model of* $\mathrm{FP}(\mathcal{S})$ *in $\mathcal{C}$. Furthermore, $\mathcal{M}od(\mathcal{S})$ is equivalent to $\mathcal{M}od(\mathrm{FP}(\mathcal{S}))$, where $\mathrm{FP}(\mathcal{S}))$ is considered a sketch whose cones are all the finite product cones.*

Proof. Just as in Exercise (PRES) of Chapter 1.7, if u embeds $\mathcal{S}$ into a subcategory of $\mathcal{S}et^{\mathcal{S}}$, then u^{op} preserves everything that every functor in that subcategory preserves. It follows that u^{op} and hence m is a model of $\mathcal{S}$. Now let $\mathcal{S}_1 = \mathcal{S}$ and $\mathcal{S}_2 = \mathcal{C}$ in Theorem 3. The fact that $\mathcal{C}$ is a category and has finite products and u_2 is a functor and preserves finite products implies that $\mathcal{C}$ is equivalent to $\mathrm{FP}(\mathcal{C})$. (Note that $\mathrm{FP}(\mathcal{S})$ is the closure of $\mathcal{S}$ under two operations – composition of arrows and finite products—and $\mathcal{C}$ is closed under both of them).

Fine points

Observe that models of a sketch $\mathcal{S}$ which happens to be a category with finite products and whose cones happen to be all finite product cones are the same as FP-functors from the category.

It follows from the fact that $\mathcal{M}od(\mathcal{S})$ is equivalent to $\mathcal{M}od(\mathrm{FP}(\mathcal{S}))$ that m is both mono and epi in the category of FP-sketches. Nevertheless, m need not be surjective on objects, nor reflect isomorphisms (a property more relevant than injectivity on objects). For example, the embedding of the sketch for groups into the theory of groups is not surjective, since the latter has all powers of G.

As for reflecting isomorphisms, consider the sketch obtained from the sketch for groups by adding an arrow $G \to G^2$ and a diagram forcing m to be invertible. This forces all the powers of G to be isomorphic.

Single-sorted theories

An FP-theory $\mathcal{T}h$ is **single-sorted** (or an **algebraic theory**) if there is an object G of $\mathcal{T}h$ with the property that every object of $\mathcal{T}h$ is a power of G. Most of the familiar categories in mathematics which are categories of models of FP-theories are actually models of single-sorted theories. This includes examples like groups, rings, monoids, R-modules for a fixed ring

R (each element of the ring is a unary operation), and even unlikely looking examples such as the category of all modules (but *not* the category of all monoid actions (Exercise (RMFP)).

A set of objects in an FP-theory Th is a **set of sorts** for the theory if Th is the smallest FP-subcategory of Th containing the set of objects. The examples just mentioned, especially the last, illustrate that the number of sorts for a theory is not well-defined, although single-sorted is well-defined. Even though the category of all modules *can* be presented as algebras for a single sorted theory, that is conceptually not the most reasonable way to present it.

Single-sorted FP-algebras are tripleable over sets

The proof of the following theorem requires material from Chapters 3, 8 and 9, but we include it here because of the connection it makes between theories and tripleability.

Theorem 5. *The category of models in* Set *of an FP-theory* Th *is tripleable over* Set *if and only if it is the category of models of a single-sorted FP-theory.*

Proof. We make use of Theorem 5 of Section 9.1. The category of models of an FP-theory is exact, as we will see in Section 8.4. If Th is single-sorted, then its generating object (lying in the category of models via the embedding u given above) is a regular projective generator. The converse follows immediately from Theorem 5 of Section 9.1.

Exercises 4.3

(VFF)°. Show that m^{op} in diagram (2) is full and faithful (but, as pointed out in the text, not necessarily either injective or surjective on objects).

(RMFP). Show that the category of all modules (defined analogously to the category of all monoid actions defined in Section 4.1) is given by a single-sorted theory. (Hint: The underlying set of (R, M) where R is a ring and M is an R-module is $R \times M$).

(SSFP)°. Show that if Th is a single sorted FP-theory, and $Mod(Th)$ the category of models of Th, then Th is equivalent to the full subcategory of $Mod(Th)^{op}$ consisting of the finitely generated free algebras.

Note: The following two exercises require familiarity with Theorem 5 above, parts of Section 3.4 and Theorem 5 of Section 9.1.

(MAFP). Show that the category of all monoid actions (for all monoids) defined in Section 4.1 is the category of models of an FP-theory but not of a single-sorted FP-theory. (Use Theorem 5, and show that it cannot have a single projective generator. Remember the set acted upon can be empty. Or else use the following exercise.)

(EVTT)°. (a) Prove that if A is a set of sorts for a theory $\mathcal{Th}$, then $\mathcal{Mod}(\mathcal{Th})$ is tripleable over $\mathcal{Set}^A$.

(b) Prove that the full subcategory of $\mathcal{Set}^A$ consisting of functions whose values are either always empty or never empty satisfies VTT (see Section 3.5) over $\mathcal{Set}$, with underlying functor taking a function to the product of its values. (Its left adjoint takes a set to a constant function.)

(c) Prove that the following three statements about an FP-theory $\mathcal{Th}$ are equivalent:

(i) $\mathcal{Th}$ is equivalent to a single-sorted FP-theory.

(ii) If M is a model of $\mathcal{Th}$ in $\mathcal{Set}$, then either MA is empty for every object A of $\mathcal{Th}$ or MA is nonempty for every object A of $\mathcal{Th}$.

(iii) The underlying functor $\mathcal{Mod}(\mathcal{Th}) \to \mathcal{Set}$ reflects isomorphisms. (Hint: Show that if $\mathcal{Th}$ is generated by a single object X then evaluation at X reflects isomorphisms).

(OPS)°. Show that the elements of the free algebra on n generators for a single sorted theory are in 1-1 correspondence with the set of n-ary operations. (Hint: The n-ary operations are the maps $G^n \to G$ in $\mathcal{Th}$ which are the same as arrows $G \to G^n$ in the category of models. Now use adjointness.)

4.4. Left Exact Theories

An **LE-category** is a left exact category, i.e. a category which has all finite limits. An **LE-functor** between LE-categories is one which preserves finite limits. An **LE-sketch** is a sketch, all of whose cones are over finite diagrams.

Given an LE-sketch $\mathcal{S}$, we can construct the **LE-theory** associated to $\mathcal{S}$, denoted LE($\mathcal{S}$), in exactly the same way that we constructed FP($\mathcal{S}$) in Section 4.3. As there, we also get a generic LE-model $m : \mathcal{S} \to \mathrm{LE}(\mathcal{S})$ and we can prove the following two theorems. In Theorem 1, v_i is defined as in Section 4.3.

Theorem 1. *Let $f : \mathcal{S}_1 \to \mathcal{S}_2$ be a morphism of LE-sketches. Then f induces a map $f^* : \mathcal{Mod}(\mathcal{S}_2) \to \mathcal{Mod}(\mathcal{S}_1)$ which has a left adjoint $f_\#$. Moreover, if, for $i = 1,2$, $u_i = v_i \circ m_i^{\mathrm{op}}$ is the embedding of $\mathcal{S}_i^{\mathrm{op}}$ into*

$Mod(S_i)$, then there is a map $\mathrm{LE}(f)^{\mathrm{op}}$ *which is the unique map making each of the squares in the following diagram commute.*

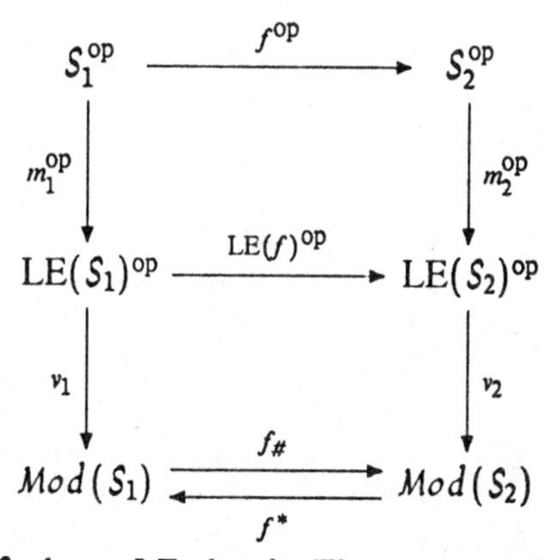

Theorem 2. *Let S be a LE-sketch. Then any model of S in an LE-category C has a unique extension along m to a model of* $\mathrm{LE}(S)$ *in C. Moreover, $Mod(S)$ is equivalent to $Mod(\mathrm{LE}(S))$ where the latter is considered a sketch whose cones are all the cones over finite diagrams.*

The following corollary is immediate from Theorem 2.

Corollary 3. *Every FP-theory has an extension to an LE-theory which has the same models in any LE-category.*

Thus for example there is an LE-theory of groups. Besides the powers of the generic group G, it contains constructions which can be made from the powers of G and the arrows in the FP-theory by forming finite limits. Since the models preserve these limits, and homomorphisms of groups are just natural transformations of the models, it follows that homomorphisms in a fixed LE-category must preserve all constructions which can be made on the groups using finite limits in the theory of groups. For example, homomorphisms must preserve the subset $\{(x,y) \mid xy = yx\}$ of the product of a group with itself, since the latter is an equalizer in the LE-theory of groups. Another example is the subset consisting of elements whose order divides 2 (or any other fixed integer). This subset is the equalizer of the homomorphism which is identically 1 and the squaring map, both of which can be expressed in the theory of groups.

The center of a group can be described as $\{x \in G \mid \text{ for all } y \in G, xy = yx\}$. This does not have the form of a limit because of the universal quantifier inside the definition. Moreover, there can be no clever way to express the center as a finite limit, because it is not preserved by homomorphisms.

In chapter 8, we are going to study in some detail the properties of categories of models. However, one important property of models of LE-theories will be used before that and we give the result here.

We must begin with a definition. A diagram $D : \mathcal{I} \to \mathcal{C}$ is called **filtered** if

(i) given any two objects i and j of $\mathcal{I}$ there is a third object k to which i and j both map, and

(ii) given two arrows

$$i \overset{f}{\underset{g}{\rightrightarrows}} j$$

of $\mathcal{I}$ there is an arrow $h : j \to k$ such that $D(h) \circ D(f) = D(h) \circ D(g)$.

The slight awkwardness of this definition is the price we must pay for using index graphs instead of index categories. We believe it is worth it for the reasons mentioned at the end of Chapter 1. A colimit taken over a filtered diagram is called a **filtered colimit**. The main significance is that filtered colimits commute with limits in $\mathcal{S}et$ and many other interesting categories (Exercise (FILT)).

Theorem 4. *The category of set-valued models of a left exact theory has arbitrary limits and all filtered colimits; moreover, these are preserved by the set-valued functors of evaluation at the objects of the theory.*

Proof. Let $\mathcal{T}h$ be an LE-theory. If $\{M_i\}$ is a diagram of models, the fact that limits commute with limits implies that the pointwise limit $\lim M_i$ is a model and is evidently the limit of the given diagram. Similarly, the fact that finite limits commute with filtered colimits implies that the filtered colimit of models is a model. To say that these limits and filtered colimits are computed "pointwise" is the same thing as saying that they are preserved by the evaluation.

Finitely presented algebras

A **finitely presented algebra** for an equational theory (that is, a single-sorted FP-theory) is an algebra which is a coequalizer of two arrows between finitely generated free algebras. Since the LE theory associated with an equational theory is the finite limit closure of the sketch in the dual of the category of models, it can also be viewed as the dual of the finite colimit closure of the category of finitely generated free algebras in the category of algebras and hence contains every finitely presented algebra. Moreover, it clearly consists of exactly that category if and only if that category is cocomplete.

Theorem 5. *Let $\mathcal{C}$ be the category of algebras for an equational theory. Then the category of finitely presented algebras is finitely cocomplete, hence is the dual of the associated LE theory.*

Proof. We let $F(n)$ denote the free algebra on n generators for a finite integer n. Since the sum of coequalizer diagrams is a coequalizer and the sum of finite free algebras is finite free, it is evident that this category has finite sums. So we must show that in any coequalizer diagram

$$A' \rightrightarrows A \longrightarrow A''$$

if A' and A are finitely presented, so is A''. Consider the picture

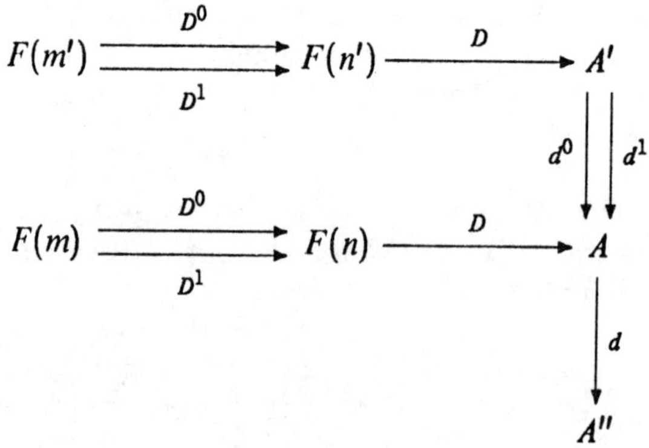

$$\begin{array}{ccccc}
F(m') & \overset{D^0}{\underset{D^1}{\rightrightarrows}} & F(n') & \xrightarrow{D} & A' \\
 & & & & d^0 \downarrow\downarrow d^1 \\
F(m) & \overset{D^0}{\underset{D^1}{\rightrightarrows}} & F(n) & \xrightarrow{D} & A \\
 & & & & \downarrow d \\
 & & & & A''
\end{array}$$

in which the two rows and the column are coequalizers. The properties of freeness, in conjunction with the fact that $D : F(n) \to A$ is surjective, allow us to find arrows $d^0, d^1 : F(n') \to F(n)$ making the diagram serially commutative. Furthermore, by replacing the top row by the sum of the top row and the second, making the d^i be the identity on the second component, everything remains as is and now both pairs of d^i have a common right inverse in such a way that the diagram remains serially commutative. By applying the same trick to the top row we can suppose that the top row is also a reflexive coequalizer. Thus the result follows from the following lemma whose proof we leave as an exercise (Exercise (DIAG)).

Lemma 6. *In any category, if the top row and right column are reflexive coequalizers and the middle column is a reflexive parallel pair, then the* *diagonal sequence is a coequalizer.*

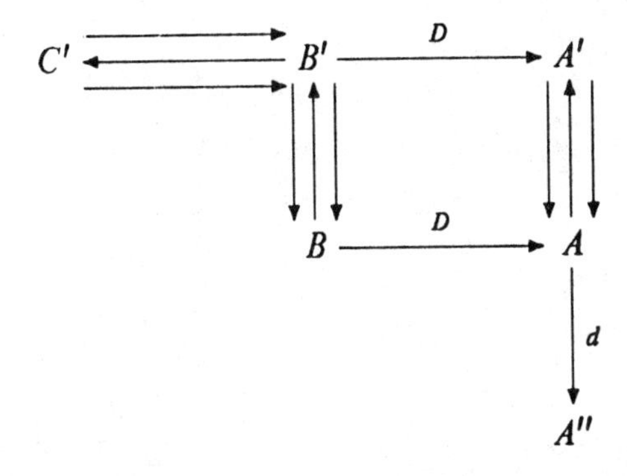

Examples of LE-theories

The main point of LE-theories is that many mathematical structures are models of LE-theories. In the remainder of this section, we will show that

(i) posets and order-preserving maps,
(ii) categories and functors,
(iii) LE-categories and LE-functors, and
(iv) toposes and logical functors

are all categories of models of LE-theories. None of these is the category of models of an FP-theory (see Exercise (NOTFP)).

Toposes are the subject of Chapter 2, and logical functors are defined in Section 5.3. We define them below to maintain the independence of this chapter from Chapter 2.

We first observe that to make an arrow $f : B \to A$ become a mono in all the LE-models of the sketch, one only need add the cone

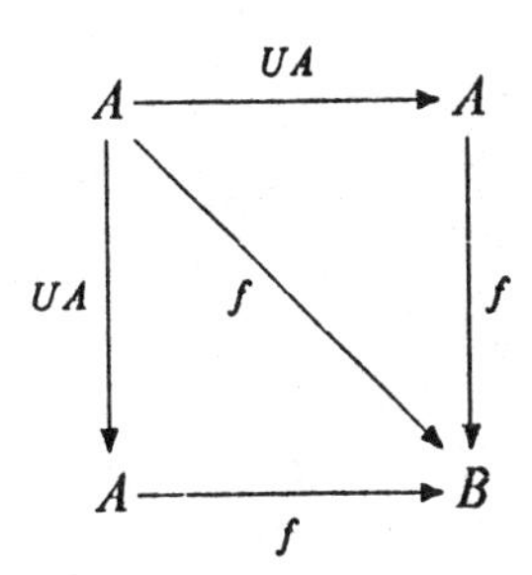

to the set of cones of the sketch. This is shorthand for saying, "Add the cone with vertex A whose base is the diagram

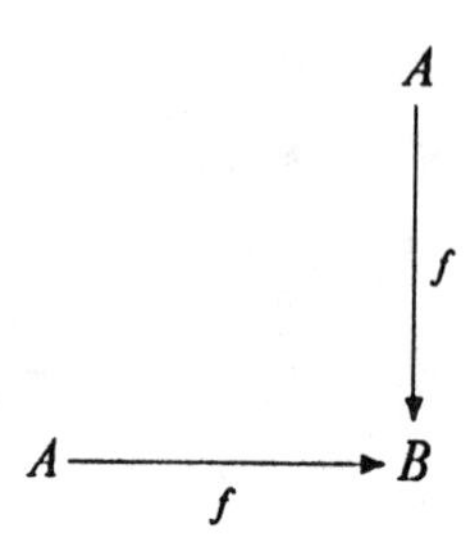

and whose transition arrows are UA, UA and f to the sketch". The diagram then must become a pullback in any model (and UA must become id_A), forcing f to be monic. (Note that this cone is made up of data already given in the sketch).

Similarly, to construct a pullback of a diagram

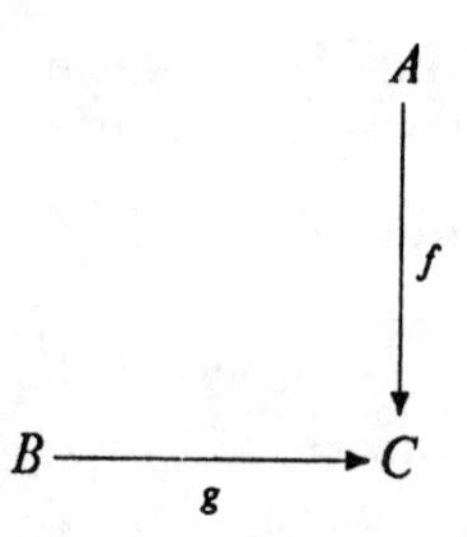

where the objects and arrows are already in the diagram, one adds an object P and arrows $p_1 : P \to A$, $p_2 : P \to B$, and $p_3 : P \to C$ and makes these data a cone of the sketch.

To force the following diagram

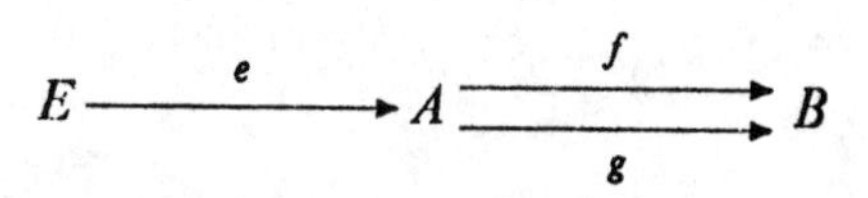

to become an equalizer in the models, one must add a cone with vertex E and arrows to the diagram

$$A \underset{g}{\overset{f}{\rightrightarrows}} B$$

The latter diagram must *not* be included as one of the diagrams in the sketch as that would obviously force $f = g$.

One can construct other limits in a similar way. In the sequel, that is what we mean when we say that a sketch must have an arrow "which is to be a monic" or an object "which is to be a limit" of some given diagram.

Thus we can describe the sketch for posets as containing the following items.

(i) An object S, which will become the underlying set of the poset.
(ii) An object S^2 to be the product of S and S.
(iii) An object R (the relation) and a monic $i : R \to S^2$.

To force R to be reflexive, you need an object Δ, a monic $\delta : \Delta \to S^2$ and a cone forcing it to be the equalizer of the projections. Then add an arrow $r : \Delta \to R$ and a diagram

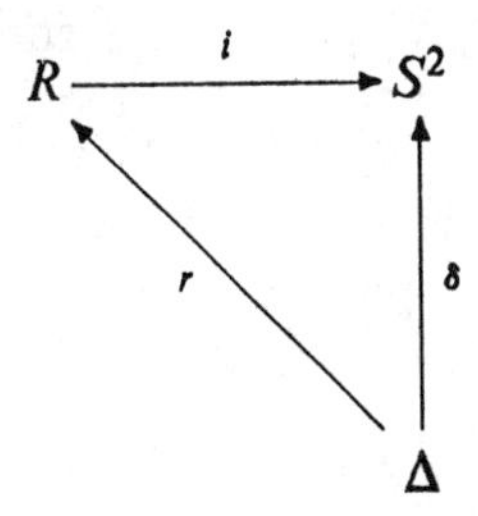

You could instead construct a common right inverse to the projections of R onto S but we need Δ anyway.

For antisymmetry, add an arrow $S^2 \to S^2$ and a diagram forcing it to be the switching map. Then add an arrow $s : R \to R$ and a diagram forcing it to be the restriction of the switching map. With this you can add an object A and a cone forcing it to be the fiber product $R \times_S R$ where the first projection is UR and the second is s. Thus A must become the set

$$\{(r, r') \mid (r, r') \in R \text{ and } (r', r) \in R\}.$$

Note that A must become a subobject of R in the model (the pullback of a monic is a monic), and so a subobject of S^2. Antisymmetry is simply the requirement that there is a monic e from A to Δ and a diagram forcing the inclusion of A in S^2 to factor through it.

Transitivity can be attained by constructing the pullback

$$P = [(r_1, r_2, r_3) \mid (r_1, r_2) \in R \text{ and } (r_2, r_3) \in R]$$

and an arrow $p : P \to R$ with a diagram forcing $p(r_1, r_2, r_3) = (r_1, r_3)$.

We will see in Chapter 8 (Theorem 1 of Section 8.4) that, since the category of posets is not regular, it cannot be expressed as models of an FP-theory.

Categories

The work of constructing a sketch whose models are categories was essentially done in Section 1.1, where categories were defined by commutative diagrams. Thus the sketch for categories must contain objects A (to be the set of arrows), O (to be the set of objects) P, and Q, along with arrows $d^i : A \to O, (i = 0, 1)$, $u : O \to A$ and $m : P \to A$, cones making $P = [(f, g) \mid d^0(f) = d^1(g)]$ and

$$Q = [(f, g, h) \mid d^0(f) = d^1(g) \text{ and } d^0(g) = d^1(h)].$$

It must also contain the diagrams (i) through (iv) of Section 1.1; to include these diagrams, we have to add the four arrows in those diagrams not already in the sketch, such as $1 \times m$, and diagrams forcing those four arrows to be what they should be, in much the same way as we added arrows to get the FP-sketch for groups in Section 4.1. We omit the details. (Note that we do not need to add arrows to be id_O or id_A because of the incorporation of the function U in the definition of sketch).

By omitting some of these arrows, you get a sketch for the category of graphs and morphisms of graphs. However, that category is actually given by an FP-theory (see Exercise (GRLE)).

Left exact categories

By adding appropriate data to the LE-theory of categories, one can force the models to be left exact categories.

To get left exactness, we force the existence of a terminal object and of pullbacks. We do pullbacks in considerable detail as an example of how to do other constructions later; the terminal object is done by similar methods (but more easily) and is omitted.

To the sketch for categories we add an object CC, which is to be the set of pairs of arrows with common codomain, i.e.,

$$CC = \{(f, g) \in A \times A \mid d^1(f) = d^1(g)\};$$

an object CD, which is to be the set of pairs of arrows with common domain; and CS, which is to be the set of commutative squares. Thus CC and CD must be the vertices of the following ones:

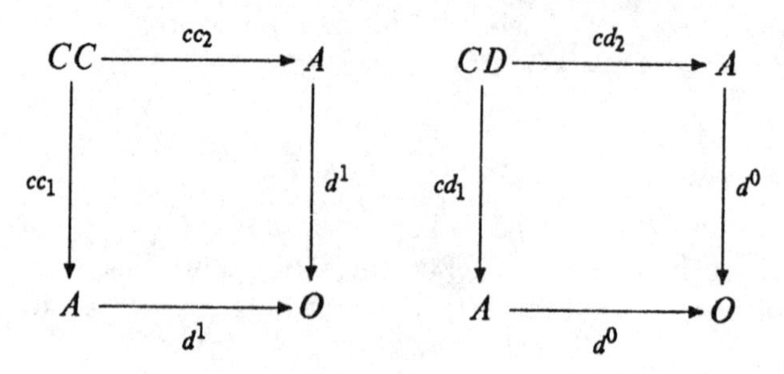

Similarly, CS must be a cone which in an interpretation becomes

$$[(f,g,h,k) \mid d^0(f) = d^0(g), d^1(h) = d^1(k), d^1(f) = d^0(h),$$
$$d^1(g) = d^0(k), \text{ and } m(h,f) = m(k,g)].$$

This comes equipped with four projections $s_i : CS \to A$. (Don't confuse these pullbacks with the pullbacks which we are trying to force the existence of *in* the models.)

We also add an arrow $t : CS \to CC$ which projects a commutative square onto its lower right half; this is forced by adding the diagram

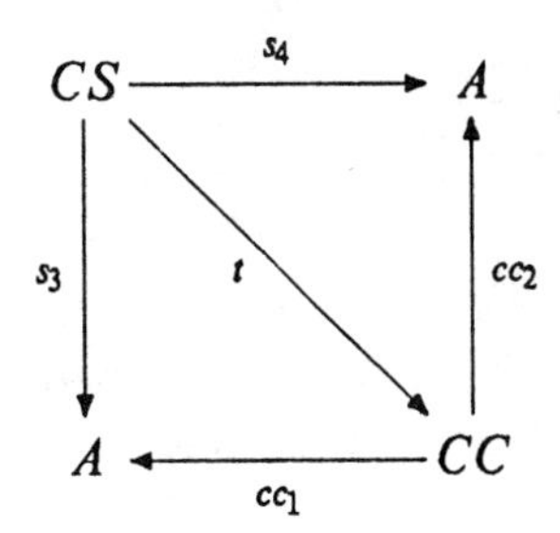

to the sketch.

Forming the pullback of something in CC must be an arrow $\lambda : CC \to CD$; on this we must impose equations forcing the codomains of the two arrows which make up $\lambda(f,g)$ to be the domains of f and g respectively. These are the diagrams

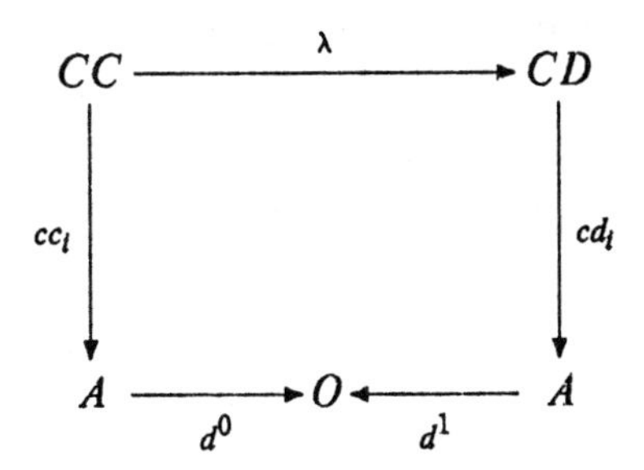

for $i = 1, 2$.

To get the pullback condition, any commutative square must have a unique arrow to the appropriate pullback square with the appropriate commutativity conditions. The existence of the arrow is assured by including an arrow $\theta : CS \to A$ in the sketch which for a given commutative square $S = (f,g,h,k)$ makes everything in the following diagram commute. Note that this diagram, unlike the ones above, is intended to be in a model, not in the sketch.

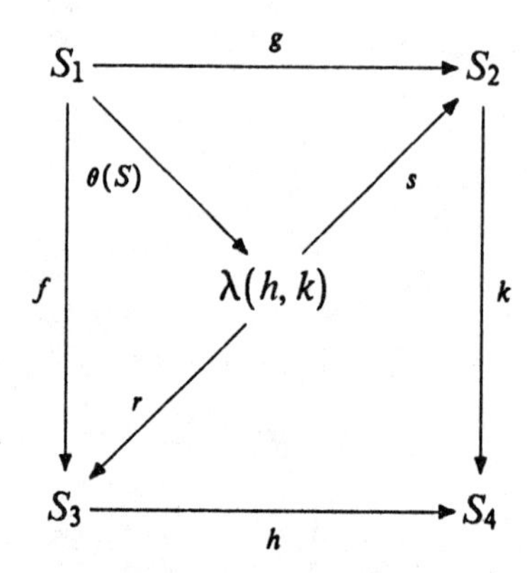

Here, $r = cd_1(\lambda(\iota(S)))$ and $s = cd_2(\lambda(\iota(S)))$.

To make everything in the preceding diagram commute requires several diagrams to be added to the sketch. These diagrams must

(i) force the domain of $\theta(S)$ to be the upper left corner of S;

(ii) force the codomain of $\theta(S)$ to be the upper left corner of $\lambda(\iota(S))$;

(iii) and make the two triangles in the preceding diagram commute.

In some cases, arrows and other diagrams defining them must be added to the sketch before these diagrams are included. For example, the diagram which makes the upper triangle commute is the right diagram below, where the left diagram defines u.

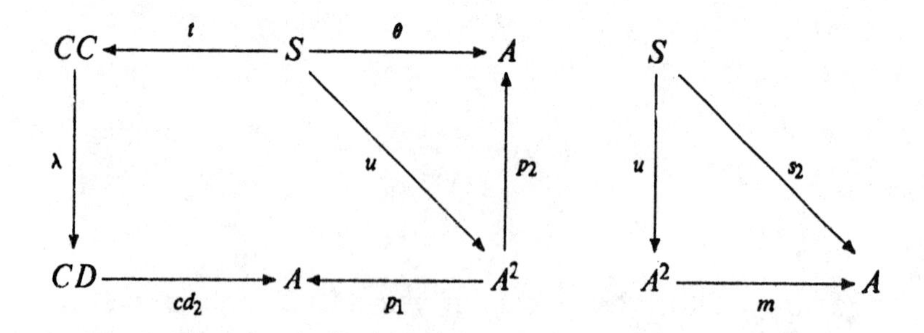

Finally, to get the uniqueness θ, we must define an object SS in the sketch which is to represent all sextuples of arrows in the following

configuration

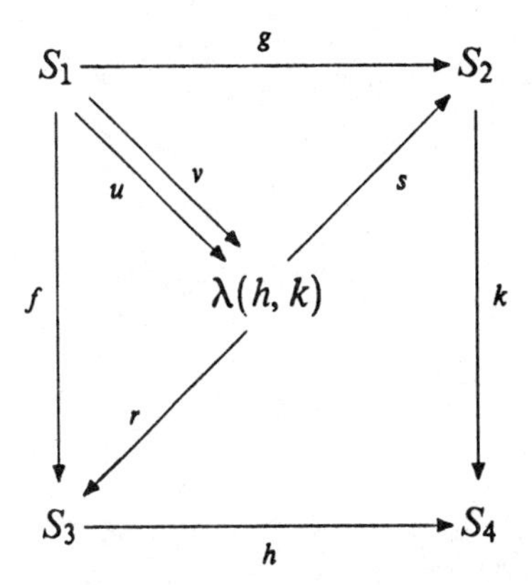

for which $r \circ u = r \circ v = f$ and $s \circ u = s \circ v = g$ (which requires adding a certain cone and some diagrams to the sketch) and an arrow $\alpha : SS \to A$, along with diagrams forcing $\alpha(a)$ to equal both u and v.

Toposes

A **topos** is an LE-category with the property that for each object X there is an object $\mathrm{P}X$ (the **power object** of X) and a subobject $e : \in X \rightarrowtail X \times \mathrm{P}X$ with the property that if $u : U \rightarrowtail X \times B$ is any subobject of $X \times B$ there are arrows $\Phi u : B \to \mathrm{P}X$ and $\Psi u : U \to \in$ for which

(i) this diagram

(1)

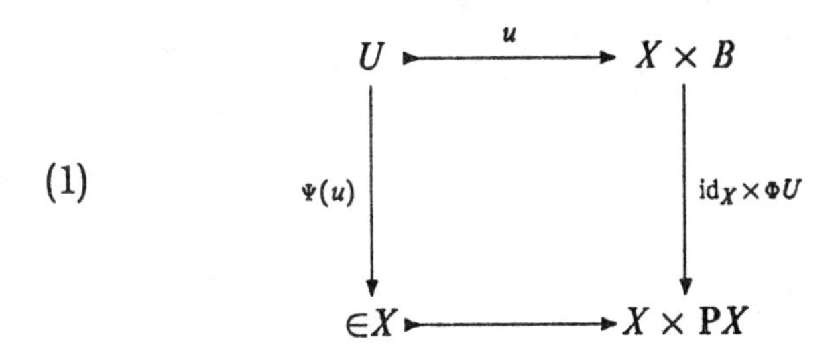

is a pullback, and

(ii) if $u' : U' \rightarrowtail X \times B$ determines the same subobject as u then $\Psi(u) = \Psi(u')$ and $\Phi(u) = \Phi(u')$. (This is equivalent to the definition in Chapter 2: the subobject $\in X$ is in fact a universal element for the functor $\mathrm{Sub}(X \times (-))$, which is therefore representable.)

Toposes are also models of a LE-theory. One obtains them by adding some data determining the power object and $\in$ to the sketch for LE-categories just given.

The first thing we need is an object M of the sketch which is to consist of all the monic arrows in A. This can be constructed as an equalizer of two arrows from A to CD, one which takes $f : X \to Y$ to the pullback of

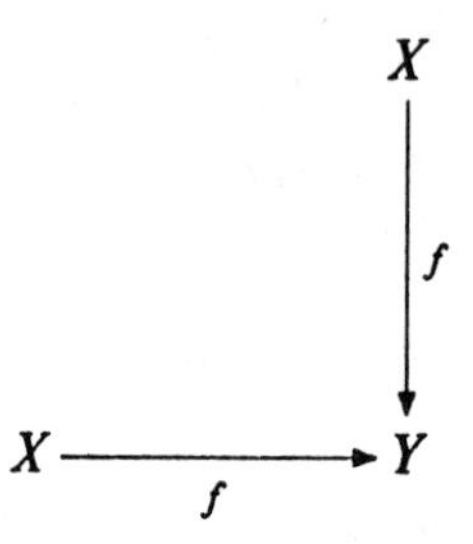

and another which takes f to

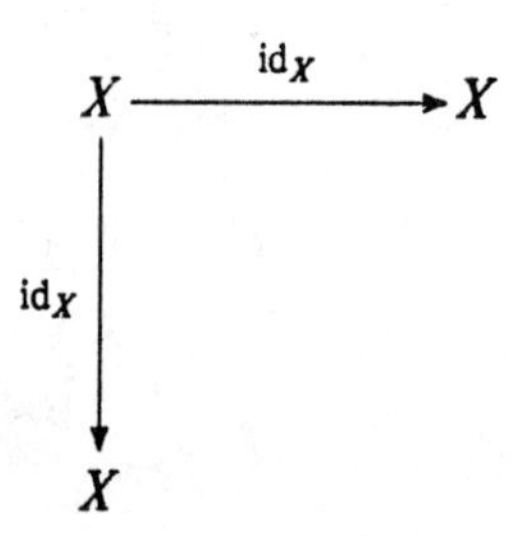

Both these arrows can be constructed by techniques used above.

We also need an arrow $P : O \to O$ which should take an object (element of O) to its power object, and an arrow $E : O \to M$ along with diagrams forcing the domain of $E(X)$ to be X and the codomain to be $X \times PX$.

Another equalizer construction will produce an object S of monos along with a specific representation of the codomain of the mono as a product of two objects; in other words,

$$S = \{(u, X, B) \mid u \text{ is monic, codomain of } u \text{ is } X \times B\}.$$

The universal property of $\in$ then requires an arrow $\varphi : S \to A$ along with diagrams forcing the domain of $\varphi(u, X, B)$ to be B and the codomain to be PX. Further constructions will give an arrow from A to A taking $f : B \to PX$ to (id_X, f). From these ingredients it is straightforward to construct an arrow $\beta : S \to CD$ which takes $u : U \to X \times B$ to

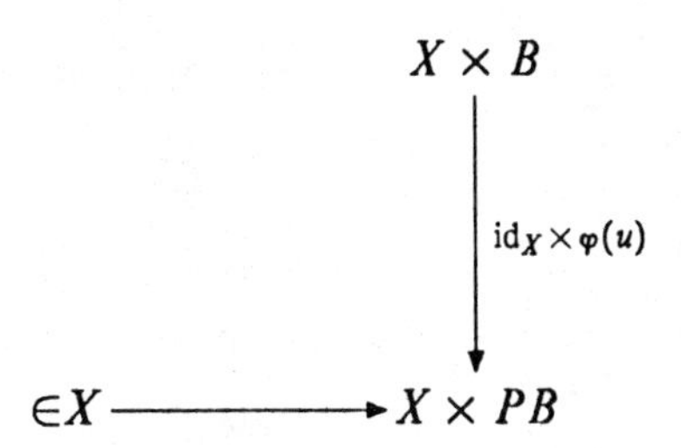

The appropriate diagram then forces the pullback of this to be (1). The uniqueness of $\varphi(u)$ can be obtained by a construction similar to that which gave the uniqueness of the arrow to a pullback.

Perhaps the most efficient way to make φ and ψ be invariant on subobjects is to construct an object T consisting of all diagrams of the form

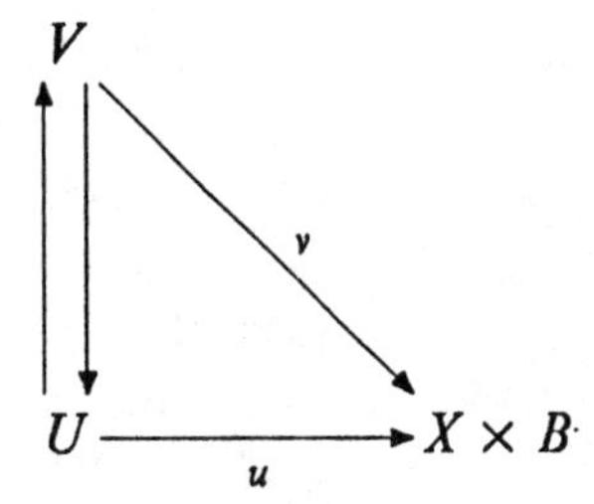

with all four arrows monic. (In a model the arrows between U and V must become inverse to each other). Then add to the sketch arrows $\varphi' : T \to A$ and $\psi' : T \to A$ and diagrams forcing $\varphi'(u,v) = \varphi(u) = \varphi(v)$ and similarly for ψ.

Exercises 4.4

(FILT)°. Suppose $D : \mathcal{I} \to \mathit{Set}$ is a filtered diagram. On the set $U = \bigcup DI$ define a relation R by letting xRy for $x \in DI$ and $y \in DJ$ if and only if there is an object K and maps $f : I \to K$ and $g : J \to K$ such that $Df(x) = Dg(y)$.

(i) Show that R is an equivalence relation.
(ii) Show that the set of equivalence classes together with the evident maps $DI \to U \to U/R$ is the colimit of D.
(iii) Show that if $\mathcal{J}$ is a finite graph and $E : \mathcal{J} \to \mathrm{Hom}(\mathcal{I}, \mathit{Set})$ is a diagram, then colim : $\mathrm{Hom}(\mathcal{I}, \mathit{Set}) \to \mathit{Set}$ preserves the limit of E.

(GRLE). One can construct an LE-theory for graphs by omitting some of the arrows, cones and diagrams in the theory for categories, as suggested in the text, or by constructing an FP-theory along the lines suggested in Exercise (GRMN) of section 3.4 and then constructing the LE-completion of that theory. Are the resulting theories equivalent (or even isomorphic) as categories?

(REC). Show that the category of right-exact small categories and right-exact functors is the category of models of an LE-theory.

(SUBS). Show that neither of these subcategories of *Set* is the category of models of an LE-theory:

(a) The full subcategory of finite sets;
(b) The full subcategory of infinite sets.

(See Exercise (INF) of Section 8.3.)

(DIAG)°. Prove Lemma 6. (Hint: Use representable functors to reduce it to the dual theorem for equalizers in *Set*. It is still a fairly delicate diagram chase.)

4.5. Notes on Theories

The motivating principle in our study of theories is: turn the mathematician's informal description of a type of structure into a mathematical object which can then be studied with mathematical techniques. The fruitfulness of the subject comes from the interplay between properties of the description (the theory, or syntax) and properties of the objects described (the models, or semantics). That LE theories are closed under filtered colimits is an example of this (Theorem 4 of Section 4.4). Many other properties are given in Theorem 1 of Section 8.4.

In classical model theory, of which a very good presentation is found in [Chang and Kiesler, 1973], the theory consists of a language, rules of inference and axioms; thus the theory is an object of formal logic. The models are sets with structure. The natural notion of morphism is that of elementary embedding. The reason for this is that inequality is always a stateable predicate.

In our treatment, the theory is a category with certain properties (FP, LE, etc. as in Chapter 4) and extra structure (a topology, as in Chapter 8), the models are functors to other categories with appropriate structure, and morphisms of models are natural transformations of these functors. This almost always gives the correct class of morphisms.

Categorical theories were developed in two contexts and from two different directions. One was by Grothendieck and his school in the context of classifying toposes (Grothendieck [1964]). These are, essentially, our geometric theories as in Chapter 8. The other source was the notion of (finitary) equational theories due to Lawvere [1963]; they are our single-sorted FP theories. Thus these two sources provided the top and bottom of our hierarchy.

The Grothendieck school developed the idea of the classifying topos for a type of structure in the late 1950's. Because of the name, we assume that they were developed by analogy with the concept of classifying space in topology.

In the 1960's, Lawvere invented algebraic theories (our single-sorted FP theories) quite explicitly as a way of describing algebraic structures using categories for theories and functors for models. His work is based on the concept of G. Birkhoff's equational classes. Of course, Birkhoff did not describe these in terms of categories, nor was his concept of lattice useful in this connection. The latter was primarily useful for describing the classes of subobjects and quotient objects. Models were described in semantic terms and it seems never to have occurred to anyone before Lawvere that the theory of groups could be thought of as a generic group. Lawvere's seminal observation that the theory of groups, for example, is a category with a group object, that a group in *Set* is a product preserving functor and that a morphism of groups is a natural transformation of functors is an idea of a different sort, rather than just an extension of existing ones.

Lawvere's work was limited to finitary equational theories, and it was Linton [1966], [1969a] who extended it to infinitary theories (not covered in this book) and made precise the relation with triples. It became clear very early that the study of infinitary theories becomes much more tractable via the Lawvere-Linton approach.

Lawvere alluded to multisorted FP theories in his thesis and even asserted—incorrectly—that the category of algebras for a multisorted theory could be realized as algebras for a single-sorted theory. Multisorted algebraic theories have recently found use in explicating the notion of data structure in computer science. See Goguen and Meseguer [1983a].

An early attempt at extending theories beyond FP was Freyd's "essentially algebraic theories" [1972] which are subsumed by our LE theories. The idea of defining algebraic structures in arbitrary categories predates Lawvere's work; for example Eckmann-Hilton [1962]. See also Bénabou [1968], [1972].

Somewhere along the line it became clear that algebraic theories had classifying toposes and that Lawvere's program of replacing theories by categories, models by functors and morphisms by natural transformations of those functors could be extended well beyond the domain of equational theories.

Ehresmann introduced sketches in the late 1960's as a way of bringing the formal system closer to the mathematician's naive description. (Our notion of sketch is even more naive than his. However, the kinship is clear and we are only too happy to acknowledge the debt.) The development of the categorical approach to general theories constituted the major work of the last part of his career. The presentation in Bastiani-Ehresmann [1973] (which contains references to his earlier work) is probably the best starting place for the interested reader. Our description of sketches and induced theories is very different from Ehresmann's. In particular, he constructs the theory generated by a sketch by a direct transfinite induction rather than embedding in models and using Kennison's Theorem. (See Kelly [1982] for a general report on the use of transfinite induction in this area.)

Ehresmann's sketches are categories with extra structure rather than graphs, but as is clear from our treatment in Section 4.1, the transition is straightforward.

The connection between logical theories and categorical theories was explored systematically by Makkai and Reyes (two logicians!) [1977] who showed that when restricted to geometric theories the two were entirely interchangeable. See also Lambek and Scott [forthcoming] and Bunge [1983].

With the advent of toposes and geometric morphisms from Lawvere-Tierney, the theory of geometric theories reached its full fruition. What was left was only to fill in the holes—special cases such as regular and finite sum theories. This was more or less clear to everyone (see Lawvere [1975]) but we have the first systematic treatment of it.

Chapter 5.

Properties of Toposes

In this chapter we will develop various fundamental properties of toposes. Some of these properties are familiar from $\mathcal{S}et$; thus, every topos has finite colimits and has internal homsets (is Cartesian closed). Others are less familiar, but are technically important; an outstanding example of this sort of property is the fact that if $\mathcal{E}$ is a topos then so is the category $\mathcal{E}/A$ of objects over an object A of $\mathcal{E}$. Our treatment makes substantial use of triple theory as developed in Chapter 3.

5.1. Tripleability of $\mathbf{P}$

Theorem 1 (Paré). $\mathbf{P} : \mathcal{E}^{\mathrm{op}} \to \mathcal{E}$ *is tripleable.*

Proof. We will use the Crude Tripleability Theorem (Section 3.5). $\mathbf{P}$ has a left adjoint, namely $\mathbf{P}^{\mathrm{op}}$ (Proposition 3 of Section 2.3). $\mathcal{E}^{\mathrm{op}}$ has coequalizers of reflexive pairs, indeed all coequalizers, because $\mathcal{E}$ has all finite limits. The other properties, that $\mathbf{P}$ reflects isomorphisms and preserves coequalizers of reflexive pairs, follow from Lemmas 4 and 5 below.

Before proving these lemmas, we illustrate the power of this theorem with:

Corollary 2. *Any topos has finite colimits.*

Proof. By Theorem 1 of 3.4, a category of algebras $\mathcal{E}^{\mathsf{T}}$ for a triple T on a category $\mathcal{E}$ has finite limits if $\mathcal{E}$ does, but $\mathcal{E}^{\mathrm{op}}$ having limits is equivalent to $\mathcal{E}$ having colimits.

The same argument implies that a topos has colimits of any class of diagrams that it has limits of. The converse of this is also true (Exercise (Lim)).

Observe that for any topos $\mathcal{E}$, the functor $T = \mathbf{P} \circ \mathbf{P}$ is a covariant endofunctor of $\mathcal{E}$. Since $\mathbf{P}$ has a left adjoint, so does T (Exercise (CADJ) of Section 1.8). Thus T is the functor part of a triple $(T, \eta : 1 \to T, \mu : T^2 \to T)$.

Lemma 3. *For any object A of $\mathcal{E}$, $\eta A : A \to \mathbf{PP}A$ is monic.*

Proof. The singleton map $\{\} : A \to \mathbf{P}A$ is monic by Proposition 1 of Section 2.3, so $\{\} \circ \{\} : A \to TA$ is also monic. The Lemma then follows from Exercise (ETAMON) of section 3.1.

Lemma 4. $\mathbf{P}$ *reflects isomorphisms.*

Proof. Suppose $f : B \to A$ in $\mathcal{E}$ and suppose $\mathbf{P}f$ is an isomorphism. Then so is $\mathbf{PP}f$. In the commutative diagram

(1)

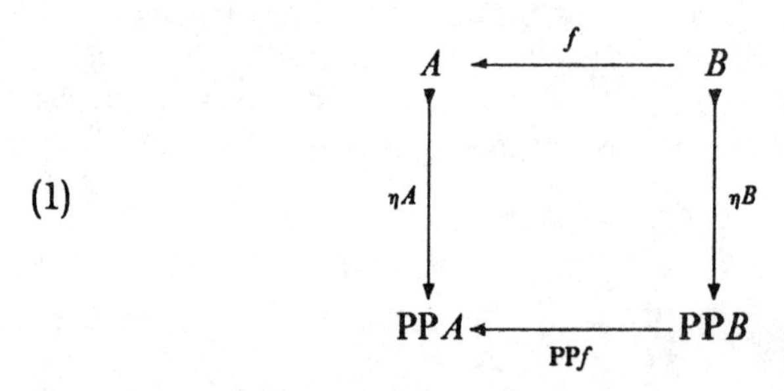

the vertical arrows are monic by Lemma 3, so f is monic. Since for any object C of $\mathcal{E}$, $\mathrm{Hom}_{\mathcal{E}}(C, \mathbf{P}f)$ is essentially the same (via the natural isomorphism φ of Section 2.1) as $\mathrm{Sub}(C \times f)$, the fact that $\mathbf{P}f$ is an isomorphism means that $\mathrm{Sub} f : \mathrm{Sub}\, A \to \mathrm{Sub}\, B$ is an isomorphism in $\mathcal{S}et$, i.e., a bijection.

Now A is a subobject of A via the identity and B is a subobject of A via f, which we now know is monic. It is easy to see that $\mathrm{Sub} f(A) = \mathrm{Sub} f(B)$ is the subobject $\mathrm{id}_B : B \to B$, so since $\mathrm{Sub} f$ is a bijection, $\mathrm{id}_A : A \to A$ and $f : B \to A$ determine the same subobject of A. It follows easily that f is an isomorphism.

Lemma 5. $\mathbf{P}$ *preserves the coequalizers of reflexive pairs. In fact, it takes them to contractible coequalizers.*

Proof. This elegant proof is due to Paré. A coequalizer diagram of a reflexive pair in $\mathcal{E}^{\mathrm{op}}$ is an equalizer diagram

$$A \xrightarrow{\;f\;} B \underset{h}{\overset{g}{\rightrightarrows}} \begin{matrix} C \\ D \end{matrix} \qquad (2)$$

in $\mathcal{E}$ in which g and h are split monos. All these diagrams are pullbacks:

(3)

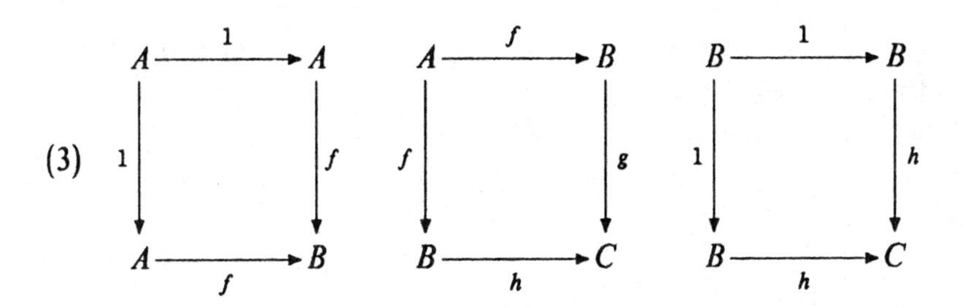

Hence, by the Beck condition and the fact that $\exists \mathrm{id}_A = \mathrm{id}_{\mathbf{P}A}$, these diagrams commute:

(4)

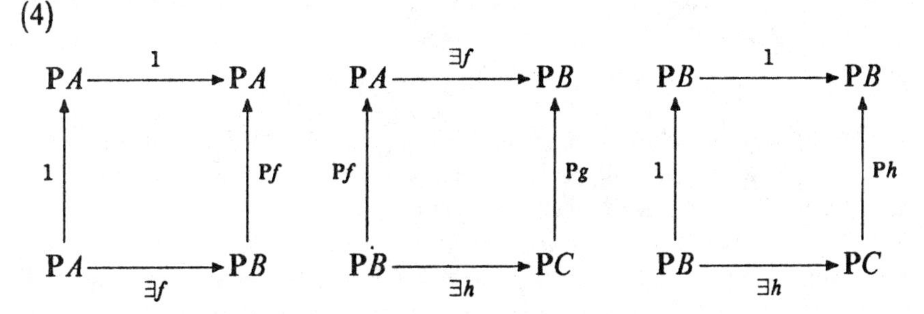

It follows from this that

$$\text{(5)} \qquad \mathbf{P}C \underset{\mathbf{P}h}{\overset{\mathbf{P}g}{\rightrightarrows}} \mathbf{P}B \xrightarrow{\mathbf{P}f} \mathbf{P}A$$

is a contractible coequalizer, with its contraction given by $\exists f$ and $\exists h$.

Exercises 5.1

(Lim)$^\circ$. Show that a topos has colimits corresponding to whatever class of limits it has and conversely. (Hint: To have a limit (resp. colimit) for all diagrams based on a graph $\mathcal{I}$ is to have a left (resp. right) adjoint to the diagonal—or constant functor—functor $\mathcal{E} \to \mathrm{Func}(\mathcal{I}, \mathcal{E})$. Use the Butler Theorem 3 of 3.7 and the tripleability of $\mathcal{E}^{\mathrm{op}} \to \mathcal{E}$ to derive the two directions.)

(Spco)$^\circ$. Prove that (5) is a contractible coequalizer diagram.

5.2. Slices of Toposes

Recall from Section 1.1 that if $\mathcal{C}$ is a category and A an object of $\mathcal{C}$, the category $\mathcal{C}/A$, called the slice of $\mathcal{C}$ by A, has as objects arrows $C \to A$ and morphisms commutative triangles.

The following theorem is heavily used in proving the embedding theorems of Chapter 7.

Theorem 1. *If $\mathcal{E}$ is a topos and A an object of $\mathcal{E}$, then $\mathcal{E}/A$ is a topos.*

Proof. $\mathcal{E}/A$ has a terminal object, namely the identity on A, and the map $\mathcal{E}/A \to A$ creates pullbacks (Exercise (PIX) of Section 1.7), so $\mathcal{E}/A$ has finite limits.

We must construct a power object for each object of $\mathcal{E}/A$.

The product of objects $B \to A$ and $C \to A$ in $\mathcal{E}/A$ is the pullback $B \times_A C$, which is an object over A, and for any object $X \to A$ of $\mathcal{E}/A$, $\mathrm{Sub}(X \to A)$ is the same as $\mathrm{Sub}(X)$ in $\mathcal{E}$. Thus given an object $f : B \to A$ of $\mathcal{E}/A$, we must construct an object $\mathrm{P}(B \to A)$ which represents $\mathrm{Sub}(B \times_A C)$ regarded as a functor of objects $g : C \to A$ of $\mathcal{E}/A$. The key to the proof lies in representing the pullback $B \times_A C$ as the equalizer $[(b, c) \mid (b, fb, c) = (b, gc, c)]$, thus given as the equalizer:

$$B \times_A C \xrightarrow{\ d\ } B \times C \overset{d_0}{\underset{d_1}{\rightrightarrows}} B \times A \times C$$

Now suppose we are given the arrow

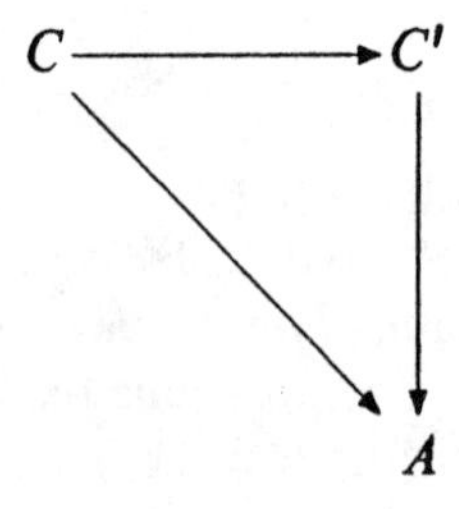

in $\mathcal{E}/A$. We have a serially commutative diagram in which both rows are equalizers of reflexive pairs:

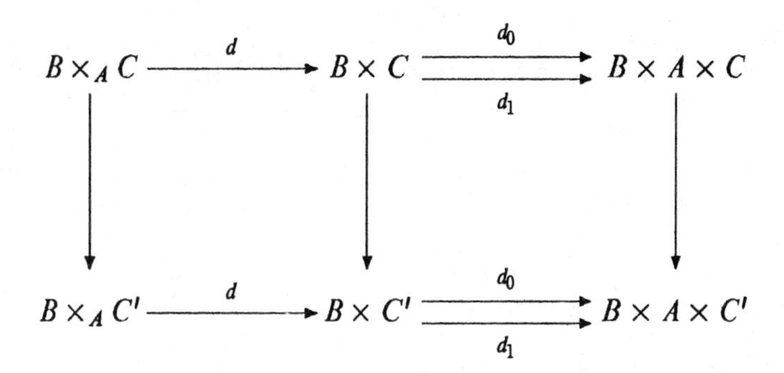

Applying **P** to this diagram gives rise to the diagram in which by Lemma 5 of the preceding section the rows are contractible coequalizers. After $\mathrm{Hom}(1, -)$ is applied, we get

(*)

$$\begin{array}{ccccc}
\mathrm{Sub}(B \times A \times C') & \overset{d_0}{\underset{d_1}{\rightleftarrows}} & \mathrm{Sub}(B \times C') & \longrightarrow & \mathrm{Sub}(B \times_A C') \\
\downarrow & & \downarrow & & \downarrow \\
\mathrm{Sub}(B \times A \times C) & \overset{d_0}{\underset{d_1}{\rightleftarrows}} & \mathrm{Sub}(B \times C) & \longrightarrow & \mathrm{Sub}(B \times_A C)
\end{array}$$

in which by the external Beck condition the rows are contractible coequalizers with contractions $d^0 \circ -$, so this diagram commutes serially too. Thus

$$\begin{array}{ccc}
\mathrm{Hom}(C', \mathbf{P}(B \times A)) & \overset{d_0}{\underset{d_1}{\rightleftarrows}} & \mathrm{Hom}(C', \mathbf{P}B) \\
\downarrow & & \downarrow \\
\mathrm{Hom}(C, \mathbf{P}(B \times A)) & \overset{d_0}{\underset{d_1}{\rightleftarrows}} & \mathrm{Hom}(C, \mathbf{P}B)
\end{array}$$

is serially commutative and has contractible pairs as rows. By the adjunction between $\mathcal{E}$ and $\mathcal{E}/A$ (see Exercise (SLADJ) of Section 1.9), the following is a serially commutative diagram of contractible pairs:

$$\begin{array}{ccc}
\mathrm{Hom}(C' \to A, \mathrm{P}(B \times A) \times A \to A) & \overset{d_0}{\underset{d_1}{\rightleftarrows}} & \mathrm{Hom}(C' \to A, \mathrm{P}B \times A \to A) \\
\downarrow & & \downarrow \\
\mathrm{Hom}(C \to A, \mathrm{P}(B \times A) \times A \to A) & \overset{d_0}{\underset{d_1}{\rightleftarrows}} & \mathrm{Hom}(C \to A, \mathrm{P}B \times A \to A)
\end{array}$$

It follows from the Yoneda lemma that we have a contractible pair

$$\mathrm{P}(B \times A) \times A \to A \rightleftarrows \mathrm{P}B \times A \to A$$

in $\mathcal{E}/A$. Coequalizers exist in $\mathcal{E}/A$; they are created by the underlying functor to $\mathcal{E}$. Using the facts that Hom functors, like all functors, preserve the coequalizers of contractible pairs, and subobjects in $\mathcal{E}/A$ are identical to subobjects in $\mathcal{E}$, it is easy to see from diagram (*) above that the coequalizer of the above parallel pair represents $\mathrm{Sub}(B \times_A -)$.

Exercise 5.2

(SF)°. Assuming, as we establish in the next section, that pullbacks of regular epis are regular epis, show that $\mathcal{E} \to \mathcal{E}/A$ is faithful if and only if $A \to 1$ is epi. (Hint: consider, for

$$B \rightrightarrows C$$

the diagram

$$\begin{array}{ccc}
A \times B & \rightrightarrows & A \times C \\
\downarrow & & \downarrow \\
B & \rightrightarrows & C
\end{array}$$

and use regularity.)

5.3. Logical Functors

Two sorts of functors between toposes have proved to be important. Logical functors are those which preserve the structure given in the definition of a topos. They will be discussed in this section. The other kind is geometric functors, which arise as an abstraction of the map induced by a continuous map between topological spaces on the corresponding categories of sheaves of sets. They will be discussed in Section 6.5.

A functor L which preserves the structure of a topos must be left exact (preserve finite limits) and preserve power objects in the sense that L applied to each power object must represent in a strong sense specified below the corresponding subobject functor in the codomain category. To make sense of this, note that any functor $L : \mathcal{E} \to \mathcal{E}'$ induces by restriction a function (also denoted L) from $\mathrm{Hom}_{\mathcal{E}}(A, B)$ to $\mathrm{Hom}_{\mathcal{E}'}(LA, LB)$ for any objects A and B of $\mathcal{E}$. Furthermore, if L is left exact (hence preserves monos and products in particular), it takes any subobject $U \rightarrowtail A \times B$ to a subobject $LU \rightarrowtail LA \times LB$.

In the following definition, we put a prime on $\mathbf{P}$ or φ to indicate that it is part of the structure of $\mathcal{E}'$.

A functor $L : \mathcal{E} \to \mathcal{E}'$ between toposes is called **logical** if it preserves finite limits and if for each object B, there is an isomorphism

$$\beta B : \mathrm{Hom}(-, L\mathbf{P}B) \to \mathrm{Sub}(- \times LB)$$

such that the following diagram commutes (φ is the natural transformation in diagram (1) of Section 2.1).

(1)

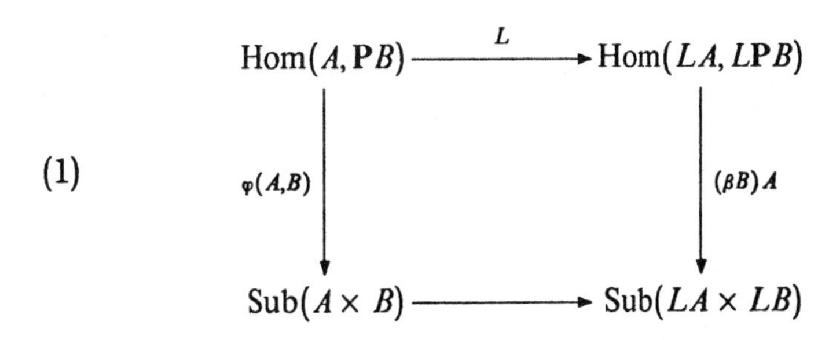

Of course, the definition implies that for every object B of E, $L\mathbf{P}B$ is isomorphic to $\mathbf{P}'LB$. We will see below that in fact the induced isomorphism is natural in B.

Proposition 1. *A functor $L : \mathcal{E} \to \mathcal{E}'$ is logical if and only if for each object B of $\mathcal{E}$ there is an isomorphism $\alpha B : L\mathrm{P}B \to \mathrm{P}'LB$ for which the induced map $\gamma B : \mathrm{Hom}(A, \mathrm{P}B) \to \mathrm{Hom}(LA, \mathrm{P}'LB)$ defined by $\gamma B(f) = \alpha B \circ Lf$ for $f : A \to \mathrm{P}B$ preserves φ in the sense that*

(2)

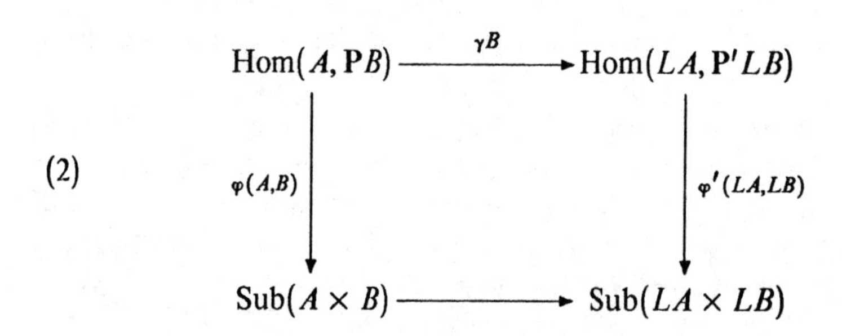

commutes.

Proof. If L is logical, so that (1) commutes, define αB to be the unique isomorphism making the triangle in (3) below commute. There is one, since $L\mathrm{P}B$ and $\mathrm{P}'LB$ represent the same functor.

(3)

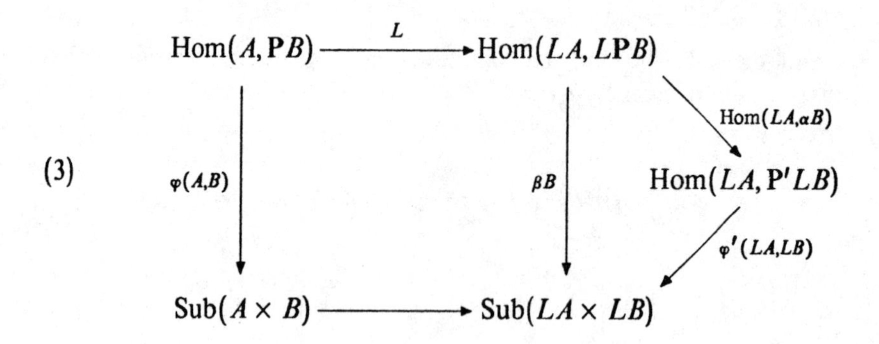

The arrow along the top of (3) is γB, so (2) commutes.

Conversely, given the arrows αB, *define* β by requiring that the triangle in (3) commute; then (1) commutes as required. β is natural in X because both small squares in (4) below commute for any $f : X \to Y$:

(4)

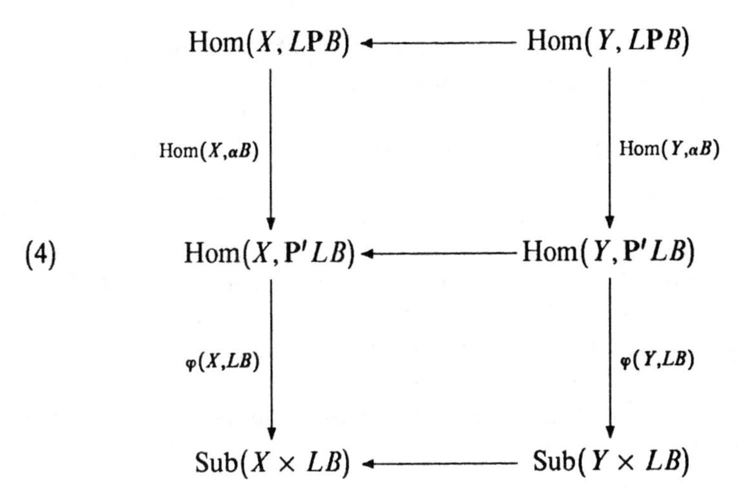

Proposition 2. *If $L : \mathcal{E} \to \mathcal{E}'$ is logical, then $\mathbf{P}' \circ L^{\mathrm{op}}$ is naturally isomorphic to $L \circ \mathbf{P}$.*

Proof. In fact, the α of Proposition 1 is a natural isomorphism. In view of the way that α is defined in terms of γ, the naturality follows from the fact that the top face in diagram (5) below commutes for any $g : B \to C$. This in turn follows from the fact that all the other faces commute and φ and φ' are isomorphisms. The left and right faces commute by definition of $\mathbf{P}$ and $\mathbf{P}'$. The bottom face commutes because L preserves monos and pullbacks.

(5)

$$\begin{array}{ccc}
\mathrm{Hom}(A, \mathbf{P}B) & \xrightarrow{\gamma B} & \mathrm{Hom}(LA, \mathbf{P}'LB) \\
\mathrm{Hom}(A,\mathbf{P}C) \uparrow & & \uparrow \mathrm{Hom}(LA,\mathbf{P}'Lg) \\
\mathrm{Hom}(A, \mathbf{P}C) & \xrightarrow{\gamma C} & \mathrm{Hom}(LA, \mathbf{P}'LC)
\end{array}$$

$$\begin{array}{ccc}
\mathrm{Sub}(A \times B) & \longrightarrow & \mathrm{Sub}(LA \times LB) \\
\mathrm{Sub}(A\times g) \uparrow & & \uparrow \mathrm{Sub}(LA\times Lg) \\
\mathrm{Sub}(A \times C) & \longrightarrow & \mathrm{Sub}(LA \times LC)
\end{array}$$

$$\varphi(A,B) : \mathrm{Hom}(A, \mathbf{P}B) \to \mathrm{Sub}(A \times B), \quad \varphi'(LA,LB) : \mathrm{Hom}(LA, \mathbf{P}'LB) \to \mathrm{Sub}(LA \times LB),$$
$$\varphi(A,C) : \mathrm{Hom}(A, \mathbf{P}C) \to \mathrm{Sub}(A \times C), \quad \varphi'(LA,LC) : \mathrm{Hom}(LA, \mathbf{P}'LC) \to \mathrm{Sub}(LA \times LC)$$

Proposition 3. *Logical functors preserve the subobject classifier.*

Proof. A logical functor preserves the terminal object since it preserves finite limits, and the subobject classifier is $\mathbf{P}1$.

Proposition 4. *Logical functors preserve finite colimits.*

Proof. Let $L : \mathcal{E} \to \mathcal{E}'$ be a logical functor. Since $\mathbf{P}$ has a left adjoint, it preserves finite limits, so $L \circ \mathbf{P}$ preserves finite limits. Hence by Proposition 2, $\mathbf{P}' \circ L^{\mathrm{op}}$ preserves finite limits. Since $\mathbf{P}'$ is tripleable, it reflects limits (Proposition 1 and Exercise (URFL) of Section 3.3); hence $L^{\mathrm{op}} : \mathcal{E}^{\mathrm{op}} \to \mathcal{E}'^{\mathrm{op}}$ must preserve finite limits. Thus L preserves finite colimits.

Proposition 5. *A logical functor L has a right adjoint if and only if it has a left adjoint.*

Proof. If L has a right adjoint, then apply Butler's Theorem 3(a) of Section 3.7 to this diagram to conclude that L^{op} has a right adjoint, whence L has a left adjoint.

(6) 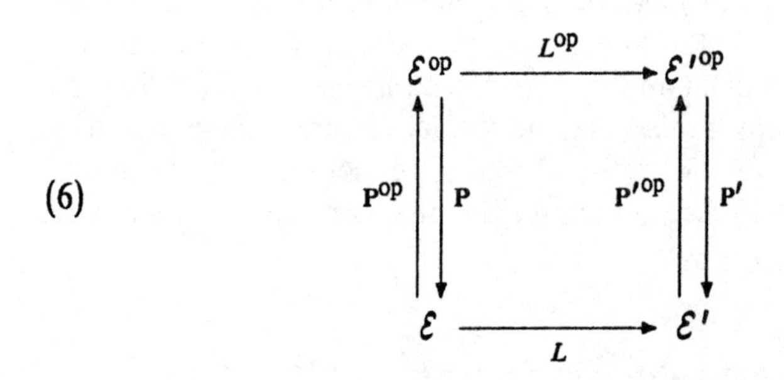

A similar argument using Butler's Theorem 3(b) of section 3.7 yields the other implication.

Theorem 6. *Let A^* be the functor which takes an object B of $\mathcal{E}$ to the object $B \times A \to A$ (the arrow is projection) of $\mathcal{E}/A$, and an arrow $f : B \to C$ to $f \times \mathrm{id}_A : B \times A \to C \times A$. Then A^**

(i) *is right adjoint to the forgetful functor,*
(ii) *is logical,*
(iii) *has a right adjoint A_*, and*
(iv) *is faithful if and only if $A \to 1$ is epi.*

Proof. (i) is Exercise (SLADJ) of Section 1.9 and (iii) follows from (ii) by Proposition 5. To prove (ii), observe first that A^* clearly preserves limits.

For a given object $C \to A$ of $\mathcal{E}/A$ and an object B of $\mathcal{E}$, diagram (1) becomes the diagram below.

$$
(7)\qquad
\begin{array}{ccc}
\mathrm{Hom}(C, \mathrm{P}B) & \xrightarrow{A^*} & \mathrm{Hom}(C \times A \to A, \mathrm{P}B \times A \to A) \\
\Big\downarrow \varphi(C,B) & & \Big\downarrow \beta(C,B) \\
\mathrm{Sub}(C \times B) & \longrightarrow & \mathrm{Sub}(C \times B \times A \to A)
\end{array}
$$

The lower right corner really is the set of subobjects of the product of $C \times A \to A$ and $B \times A \to A$ in $\mathcal{E}/A$, since that product is the pullback $[(c, a, b, a) \mid a = a] = C \times B \times A$. The bottom arrow takes a subobject u to (u, p_2).

An arrow in the upper right corner must be of the form (u, p_2) for some $u : C \times A \to \mathrm{P}B)$. We define β by requiring that $\beta(C, B)(u, p_2)$ to be the subobject $(\varphi(C \times A, B)(u))$ of $\mathrm{Sub}(C \times B \times A)$. Then β is natural in B: given $g : B' \to B$, the commutativity condition requires that

$$\mathrm{P}(\mathrm{id}_C, g, p_2)(\varphi(u), p_2) = (\varphi(\mathrm{P}g(u)), p_2),$$

which follows because φ is a natural isomorphism and so commutes with the functor $C \times - \times A$.

We must show that (7) commutes. If $f : C \to \mathrm{P}B$, the northern route around the diagram takes f to $\varphi(C \times A, B)(f, p_2)$, whereas the southern route takes it to $(\varphi(C, B)f, p_2)$. These are the same since φ commutes with the functor $- \times A$. This completes the proof of (ii). (iii) is immediate from (i), (ii) and Proposition 5.

Finally, we must prove (iv). If $A \to 1$ is epi, then since $B \times -$ has a right adjoint, it preserves epis and we conclude that $A \times B \to B$ is also epi. Now we must show that if $f \neq g$ then $A \times f \neq A \times g$. This follows from the fact just noted and the fact that this diagram commutes:

(8)

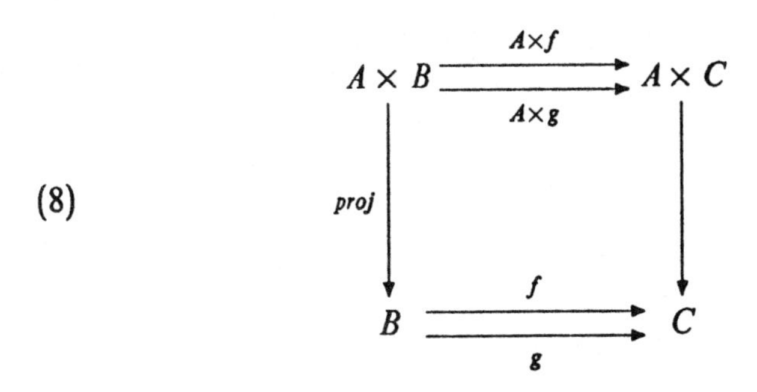

Conversely, if $f : A \to 1$ is not epi, consider $g, h : 1 \to B$ with $g \circ f \neq h \circ f$. Then

$$A \times g = A \times h : A \to A \times 1 \to A \times B$$

which contradicts faithfulness.

The right adjoint A_* is often called $\prod_A$, and the forgetful functor, which is left adjoint to A^*, is called Σ_A. This is because an object $B \to A$ of $\mathcal{E}/A$ can be thought of as an indexed family $\{B_a : a \in A\}$ of sets. In $\mathcal{S}et$, $\Sigma_A(B \to A)$, which of course is B, is the union of the family and $\prod_A(B \to A)$ is the product of the family. The two notations are both useful and suggest different, but equally correct, aspects of the story.

An object A for which $A \to 1$ is epi is said to have **global support.** (Think of a sheaf of continuous functions to see why).

Corollary 7. *If $f : A \to B$ in $\mathcal{E}$, then the pullback functor $f^* : \mathcal{E}/B \to \mathcal{E}/A$ which takes $g : X \to B$ to the pullback*

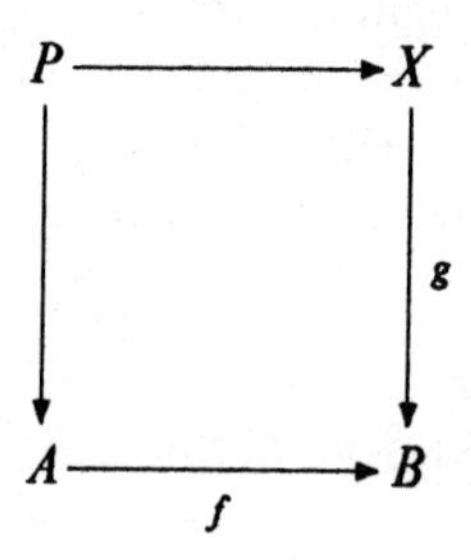

has left and right adjoints.

Proof. This follows from Theorem 6 by observing that for an object $A \to B$ of $\mathcal{E}/B$, $(\mathcal{E}/B)/(A \to B) = \mathcal{E}/A$.

Corollary 8. *In a topos, pullbacks commute with colimits. In particular, the pullback of an epimorphism is an epimorphism.*

Proof. The first sentence follows from Corollary 7. Given $f : A \to B$ epi,

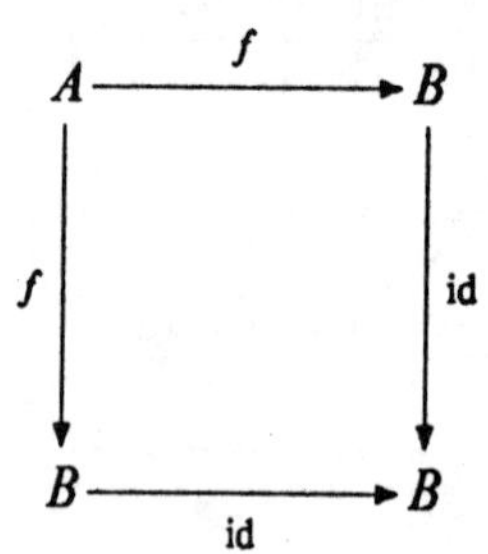

must be a pushout. Applying the functor $B' \times_B -$, which preserves pushouts, to that diagram gives

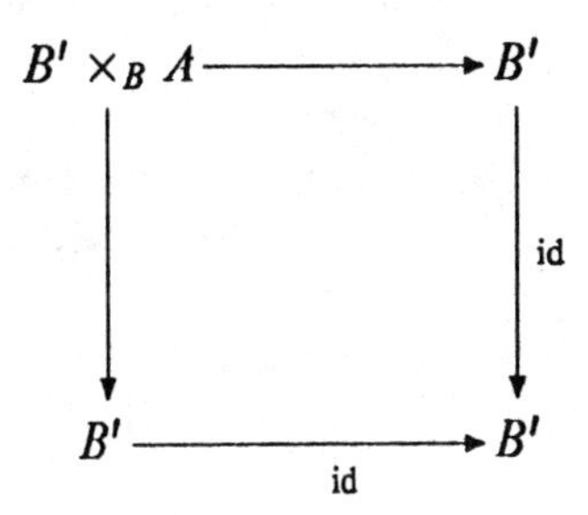

which must therefore be a pushout. Hence the map $B' \times A \to B'$ must be epi.

Exercises 5.3

(ASTAR)°. Show that in *Set*, $A_*(g : C \to A)$ can be taken to be the set of all functions from A to C which split q, or alternatively as the product of the fibers of g.

5.4. Toposes are Cartesian Closed

A category is **cartesian closed** if the functor $\mathrm{Hom}(- \times A, B)$ is representable. The representing object is denoted B^A, so that

$$\mathrm{Hom}(C \times A, B) \simeq \mathrm{Hom}(C, B^A)$$

for all objects C of $\mathcal{E}$. The object B^A is called the **exponential** of B by A. The notation B^A is used because the global elements of B^A, by the adjunction, are just the elements of $\mathrm{Hom}(A, B)$.

The general elements of B^A can also be thought of as functions. If $f \in^T B^A$, $y \in^T A$, define $f(y) = \mathrm{ev}\, A(f, y)$, where $\mathrm{ev}\, A : B^A \times A \to B$ is the counit of the adjunction (see Exercise (Cccc)). This notation has been developed extensively, for example in Kock [1982]. It can conflict with our notation $f(y) = f \circ y$ if T happens to be the same as A; Kock's treatment shows how to handle this conflict.

Theorem 1. *A topos is cartesian closed.*

We will give two constructions of the exponential. One is an easy consequence of the existence of a right adjoint to the functor $A^* (= A \times -)$

of Section 5.3; the other constructs B^A as an equalizer, from which it follows that logical functors preserve the construction.

The first construction follows from the observation that in $\mathcal{S}et$, in the notation of Section 5.3, $A_*(B \times A \to A)$ is the set of functions from A to $B \times A$ which split the structure map of $B \times A \to A$ and this is just the set of all maps A to B. This suggests trying $A_*(B \times A \to A)$ as a candidate for B^A, which works because of the following sequence of calculations:

$$\begin{aligned}\mathrm{Hom}(C \times A, B) &\simeq \mathrm{Hom}(C \times A \to A, B \times A \to A)\\ &= \mathrm{Hom}(A^*C, B \times A \to A) \simeq \mathrm{Hom}(C, A_*(B \times A \to A)).\end{aligned}$$

For the second construction, consider the canonical presentation

$$\text{(1)} \qquad \mathbf{PPPP}B \underset{\mathbf{PP}(\varepsilon B)}{\overset{\varepsilon(\mathbf{PP}B)}{\rightrightarrows}} \mathbf{PP}B \xrightarrow{\varepsilon B} B$$

of the object B in $\mathcal{E}^{op}$ (Section 3.3). Writing f, g and h for the three maps and $B' = \mathbf{P}B$, $B'' = \mathbf{PPP}B$ this becomes an equalizer

$$\text{(2)} \qquad B \xrightarrow{f} \mathbf{P}B' \underset{h}{\overset{g}{\rightrightarrows}} \mathbf{P}B''$$

in $\mathcal{E}$.

For any objects B and C, let $\varphi(B, C) : \mathrm{Hom}(C, \mathbf{P}B) \to \mathrm{Sub}(B \times C)$ be the natural isomorphism. Because it is an isomorphism, the middle and lower horizontal arrows in the following diagram are uniquely defined:

$$\text{(3)} \qquad \begin{array}{ccc}
\mathrm{Hom}(A \times C, \mathbf{P}B') & \xrightarrow{\mathrm{Hom}(A\times C, g)} & \mathrm{Hom}(A \times C, \mathbf{P}B'') \\
\uparrow{\scriptstyle \varphi(B', A\times C)^{-1}} & & \uparrow{\scriptstyle \varphi(B'', A\times C)^{-1}} \\
\mathrm{Sub}(B' \times A \times C) & \longrightarrow & \mathrm{Sub}(B'' \times A \times C) \\
\uparrow{\scriptstyle \varphi(B'\times A, C)} & & \uparrow{\scriptstyle \varphi(B''\times A, C)} \\
\mathrm{Hom}(C, \mathbf{P}(B' \times A)) & \longrightarrow & \mathrm{Hom}(C, \mathbf{P}(B'' \times A))
\end{array}$$

Let

$$g^A : \mathrm{P}(B' \times A) \to \mathrm{P}(B'' \times A)$$

be the map induced by the bottom arrow of (3) via the Yoneda lemma. In the same way, define $h^A : \mathrm{P}(B' \times A) \to \mathrm{P}(B'' \times A)$, and let the exponential B^A be defined by requiring that

$$(4) \qquad B^A \longrightarrow \mathrm{P}(B' \times A) \underset{h^A}{\overset{g^A}{\rightrightarrows}} \mathrm{P}(B'' \times A)$$

be an equalizer.

It remains to prove that $\mathrm{Hom}(C, B^A)$ is naturally isomorphic, as a functor of C, to $\mathrm{Hom}(A \times C, B)$. This follows from applying $\mathrm{Hom}(A \times C, -)$ to diagram (2) and $\mathrm{Hom}(C, -)$ to diagram (4). Both give equalizer diagrams since $\mathrm{Hom}(C, -)$ preserves equalizers. But the parallel pairs of arrows being equalized in the diagrams thus obtained are naturally isomorphic by (3) (for g) and the analog of (3) for h. Thus the left sides must be naturally isomorphic, as required.

Corollary 2. *Logical functors preserve exponentials.*

The most common definition of topos in the literature is that it is a cartesian closed category with finite limits and a subobject classifier. Exercises (ATO) and (OLDEF) of this section show that our definition is equivalent to the usual one.

Exercises 5.4

(PIO)°. Show that for any object X of any topos, $\mathrm{P}X = \Omega^X$.

(CCCC). Identify the counit $A^B \times B \to A$ in the category of sets. (Hint: this counit in any cartesian closed category is called ev.)

(ATO). Let $f : B \to C$ be an arrow in a topos. Show that $\mathrm{P}f : \mathrm{P}C \to \mathrm{P}B$ corresponds by the adjunction defining Ω^B to the arrow

$$\mathrm{ev} \circ (1 \times f) : \mathrm{P}C \times B \to \Omega,$$

where ev is the arrow of Exercise (CCCC).

(OLDEF). Use Exercise (ATO) to prove that a cartesian closed category with a subobject classifier and all finite left limits is a topos.

(EL). Construct the following isomorphisms in any Cartesian closed category in such a way that they are natural in both variables.

(a) $(B \times C)^A \cong B^A \times C^A$.
(b) $(C^A)^B \cong C^{(A \times B)}$.

(FPCC). Show that in a cartesian closed category, if $f, g \in^T B^A$ and $y \in^T A$, then $(f, g)(y) = (f(y), g(y))$. (This notation assumes that the isomorphism in Exercise (EL)(a) is the identity map.)

(CCLE). Show that Cartesian closed categories are models of an LE-theory.

5.5. Exactness Properties of Toposes

In this section we deduce a number of facts about maps in a topos, most of which have to do in some way with colimits.

Epi-mono factorizations

Lemma 1. *If $f : A \to B$ is an epimorphism in a topos, and $i : B \to C$ is any map, then the induced map $A \times_C A \to B \times_C B$ is epi.*

Proof. The functor $A \times_C -$ commutes with colimits, hence preserves epimorphisms. Thus the induced map $f \times 1 : A \times_C A \to B \times_C A$ is epi. Similarly, the induced map $1 \times f : B \times_C A \to B \times_C B$ is epi, so their composite is too.

An **epi-mono factorization** of an arrow $f : A \to B$ in a category is a representation $f = m \circ e$ where m is mono and e is epi. It is easy to see that in a topos, the representation is essentially unique (Exercises (FAC) and (EPIU)). The codomain of e is called the **image** of f (see Exercise (IMGT)).

Theorem 2. *If $f : A \to B$ is any arrow in a topos, then f has an epi-mono factorization.*

Proof. Given $f : A \to B$, construct its kernel pair (h, k) and the coequalizer $q : A \to C$ of its kernel pair. Then because f coequalizes its kernel pair, there is an arrow i as in Figure 1 below, and i has a kernel pair (u, v). Since q coequalizes the kernel pair of f, there is an arrow r

as in the figure.

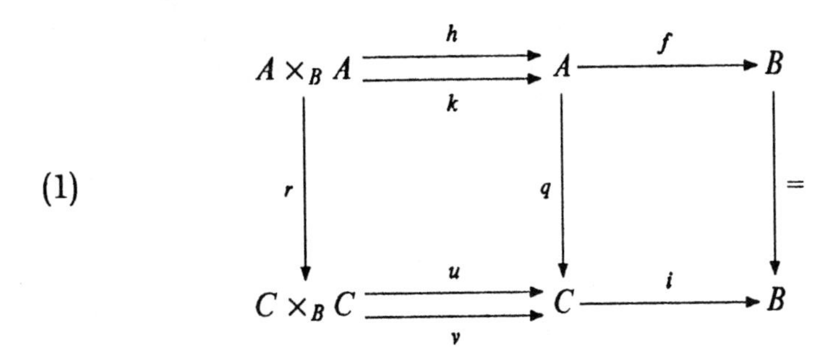

By Lemma 1, r is an epimorphism. Since q coequalizes h and k, $u \circ r = v \circ r$, so $u = v$. Thus, because the two arrows in its kernel pair are the same, i is a monomorphism. The required factorization is then $f = i \circ q$.

Exactness properties

Proposition 3. *Every epimorphism in a topos is regular.*

Proof. Assume f in Figure 1 is epi. Then the map i must also be epi. It is mono, as we proved, so is regular mono (Corollary 4 of Section 2.3). But a map which is both regular mono and epi is an isomorphism. Hence f is a coequalizer, namely of its kernel pair, as required.

A **regular category** is a category with finite limits and coequalizers in which the pullback of a regular epi is a regular epi. Proposition 3 and preceding results imply that a complete topos is a regular category.

The set of subobjects of an object forms a partially ordered set, and so one may ask when a pair of subobjects has an intersection (greatest lower bound) or union (least upper bound). In a category with finite limits, intersections always exist: the intersection of subobjects B and C of D is the pullback

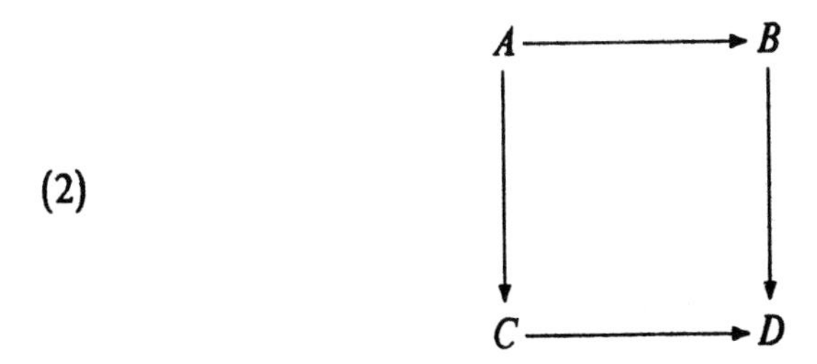

in which all the arrows are monomorphisms because the pullback of a mono is a mono.

Unions, however, are harder. In general categories, they are not colimits in a simple way (see Exercise (AMS)). In a topos, however, the situation is quite simple.

Proposition 4. *In a topos, the union of any two subobjects B and C of an object D is the image of the arrow from $B + C$ to D induced by the inclusions of B and C in D:*

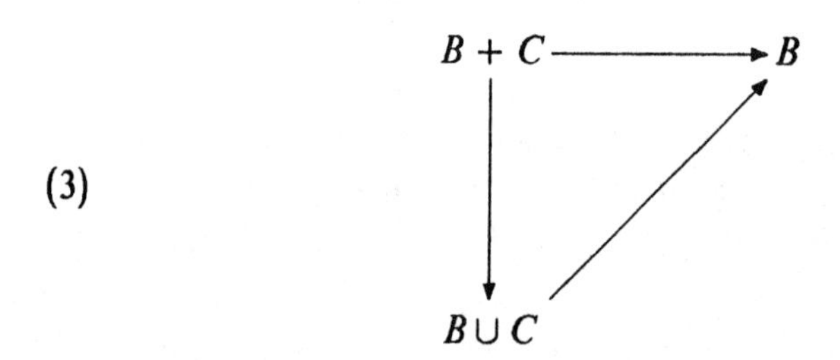

Proof. Trivial.

Proposition 5. *Suppose that (2) is a pullback diagram in which all arrows are mono. Then the induced arrow $B +_A C \to D$ is monic.*

Proof. The idea of the proof is to show that $B +_A C$ is isomorphic to $B \cup C$, so that the induced arrow is the inclusion. Construct the kernel pair of the arrow from $B + C$ to D in diagram (3):

$$(B + C) \times_D (B + C) \rightrightarrows B + C \longrightarrow B \cup C \tag{4}$$

The resulting diagram is a coequalizer, by Theorem 2 and Proposition 4. Since the pullback commutes with colimits,

$$\begin{aligned}(B + C) \times_D (B + C) &\simeq B \times_D B + B \times_D C + C \times_D B + C \times_D C \\ &= B + B \cap C + C \cap B + C.\end{aligned}$$

However, a map coequalizes the two arrows $B + B \cap C + C \cap B + C \to B + C$ if and only if it coequalizes the two arrows $B \cap C \to B + C$ (because the two arrows agree on the first and last components and interchange the middle two). So

$$B \cap C \rightrightarrows B + C \longrightarrow B \cup C$$

is a coequalizer. This means that

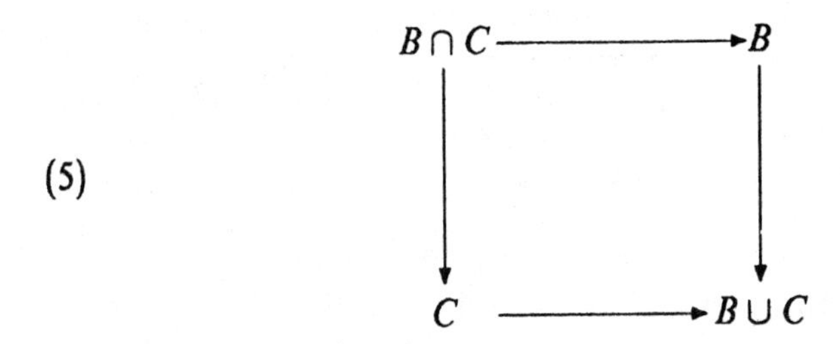

is a pushout (as well as a pullback) diagram, hence $B \cup C = B +_A C$, as required.

As an application of Proposition 5, we have

Theorem 6. *A functor between toposes which preserves monomorphisms, finite products and cokernel pairs is left exact.*

Proof. Let $F : \mathcal{E} \to \mathcal{E}'$ satisfy the conditions of the theorem. It is sufficient to show that F preserves equalizers. So let

$$A \longrightarrow B \rightrightarrows C$$

be an equalizer. An easy argument using elements shows that this is equivalent to the following diagram being an equalizer.

$$A \longrightarrow B \rightrightarrows B \times C$$

Then the two maps to $B \times C$ are mono (split by the projection onto B). It is easily seen that we have a pullback

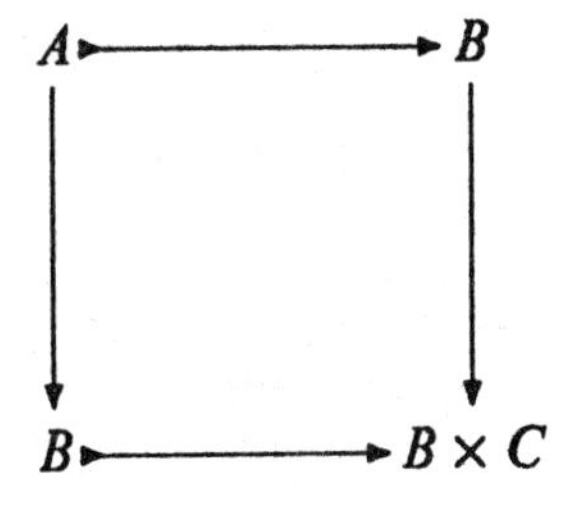

It follows from Proposition 5 that

$$B +_A B \longrightarrow B \times C$$

is monic. Applying F, we have

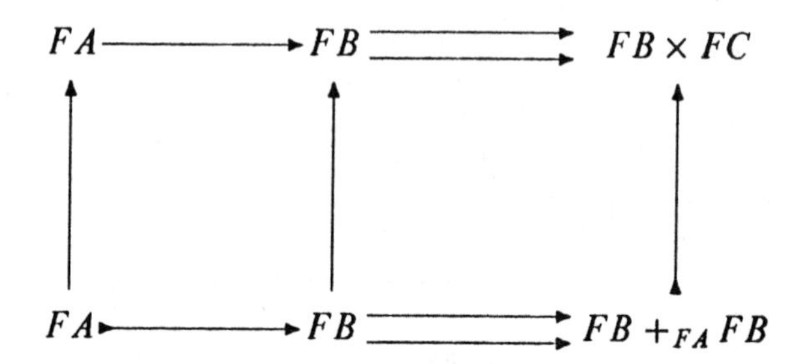

The lower row is an equalizer because in a topos every mono is the equalizer of its cokernel pair. An easy diagram chase shows that the upper row must be an equalizer, too.

Since a topos has colimits, it has an initial object 0. As in $\mathcal{S}et$, where 0 is the empty set, the initial object is the codomain of just one arrow, its own identity arrow. An initial object with this property up to equivalence is said to be **strict.**

Proposition 7. *If $f : A \to 0$, then $A \cong 0$.*

Proof. Given $f : A \to 0$, $(f, \mathrm{id}_A) : A \to 0 \times A$ is split monic. Since $A \times -$ commutes with colimits, $A \times 0 \simeq 0$, so $A \to 0$ is a split monic. But any map to 0 is epic, so A is isomorphic to 0.

Corollary 8. *Any map $0 \to A$ is monic.*

Proof. Its kernel pair is $0 \times_A 0 \to 0$.

Exercises 5.5

(FAC)°. A **factorization system** in a category consists of two classes M and E of arrows with the properties that

(i) Both M and E contain all identity arrows and are closed under composition with isomorphisms on both sides.

(ii) Every arrow can be factored as an arrow of E followed by an arrow of M.

(iii) ("Diagonal fill-in property"). In any diagram

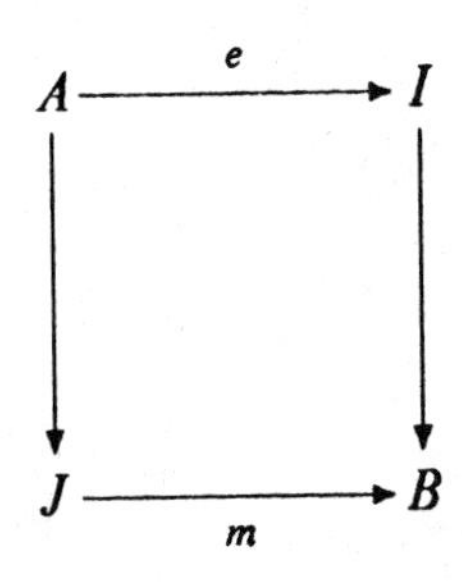

with $m \in M$ and $e \in E$, there is a unique arrow from I to J making both triangles commute.

Formulate the way in which the factorization in (ii) is unique and prove it.

(FAC2)°. (a) Show that in any factorization system any map which satisfies the diagonal fill-in property with respect to every map of M belongs to E. That is, if f has the property that whenever there is a commutative square

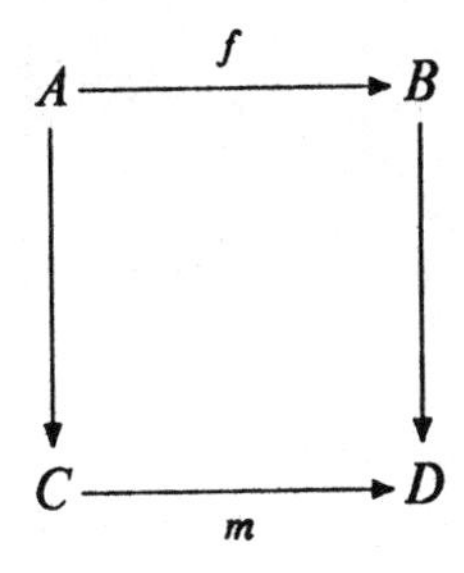

with $m \in M$, then there is an arrow from B to C making both triangles commute, then f belongs to E. (Hint: Let $f = m \circ e$ with $e \in E$ and $m \in M$. Consider the square

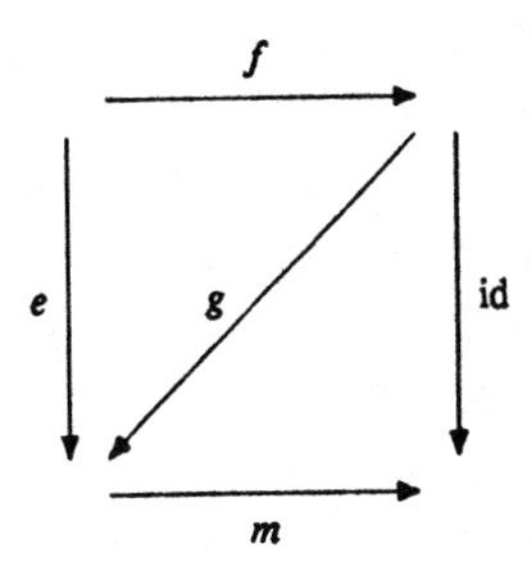

and then the square

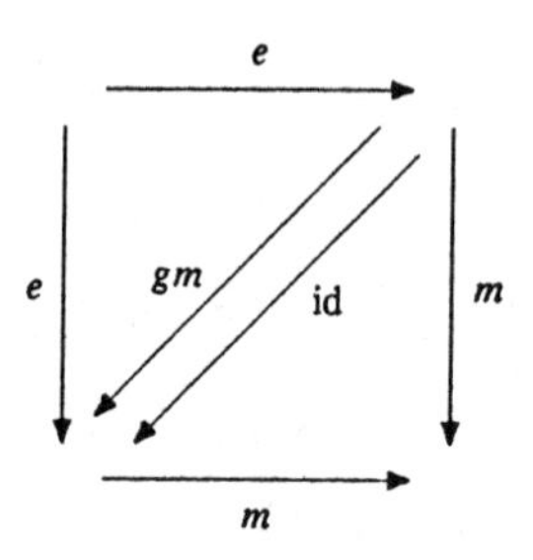

and use the uniqeness property of the diagonal fill-in.)

Conclude

(b) Every map in $M \cap E$ is an isomorphism and every isomorphism belongs to $M \cap E$.

(c) E is closed under composition.

(d) In a pushout square,

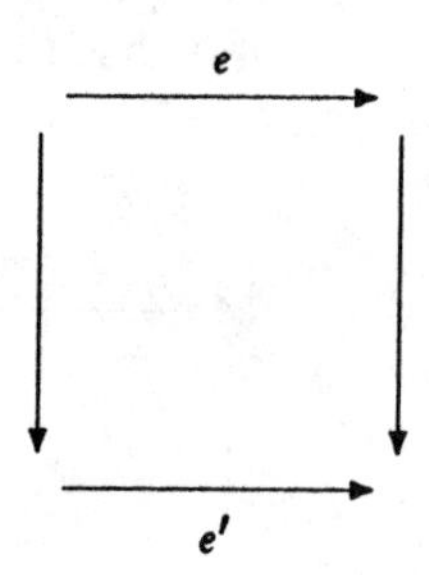

if $e \in E$, then so is e'.

(e) If M consists of monos, then every split epi belongs to E.

Of course, the duals of all these properties also hold.

(EPIU). Show that in a factorization system in which either the class E consists of all the epis or the class M consists of all the monos, the uniqueness of the diagonal fill-in does not have to be assumed, only its existence.

(RGFAC)°. Show that in a regular category the monos and regular epis form a factorization system. (Hint: The hypothesis is too strong; all that is needed is that a pullback of a regular epi be an epi.)

(IMGT)°. The **image** of an arrow $f : C \to B$ in a category, if it exists, is a subobject I of B through which f factors, with the property that if f factors through any subobject J of B, then I is a subobject of J. Prove that a topos has images.

(DIAG3)°. Prove that the construction in diagram (3) makes $B \cup C$ the least upper bound of the subobjects B and C.

(FTOP). Show that there is a factorization system in the category of topological spaces in which E consists of quotient maps with dense images and M consists of injective maps to closed subspaces. Show that there are maps in E which are not epimorphisms.

(TOPF). Find at least three other factorization systems in the category of topological spaces and continuous maps.

(SOI)°. Show that in a topos, $f + g$ is an isomorphism if and only if f and g are.

(INITU)°. Prove that an object in a topos is the initial object if and only if it has exactly one subobject.

(AMS). The group $Z_2 \times Z_2$ has three subgroups of order two.

(a) Show that the union of any two of them in the subobject lattice of $Z_2 \times Z_2$ is the whole group.
(b) Show that the pushout in the category of all groups of any two subgroups of order 2 over their intersection is an infinite group generated by two generators of order 2. (Hint: The group of isometries of the metric space of integers is infinite and is generated by two elements of order 2, namely rotation around 0 and rotation around 1/2.)

(SOO). Let $\mathcal{E}$ be a topos and $\mathcal{O}$ the full subcategory of subobjects of 1.

(a) Show that $\mathcal{O}$ is a reflective subcategory of $\mathcal{E}$. (Hint: Take an object X to the image of $X \to 1$.)
(b) Show that the left adjoint L of the inclusion I in (a) preserves products. (Hint: Use the fact that the pullback of an epi is an epi.)

(SASO)°. Use the previous exercise and the results of Section 5.3 to show that for any object A of a topos $\mathcal{E}$, the canonical functor $\mathrm{Sub}(A) \to \mathcal{E}/A$ has a left adjoint which preserves products.

(FAEX)°. Let $f : A \to B$ be an arrow in a topos. Let $\sum f$ and $\prod f$ be the left and right adjoints of the pullback functor f^* of Corollary 7.

(a) Show that f induces an arrow $f^{-1} : \mathrm{Sub}\, B \to \mathrm{Sub}\, A$ by pulling back which is the restriction of f^*.
(b) Show that the restriction of $\prod f$ is a right adjoint $\forall f : \mathrm{Sub}\, A \to \mathrm{Sub}\, B$ to f^{-1}.
(c) Show that f^{-1} has a left adjoint $\exists f$. (Hint: Define $\exists f$ to be $\exists i$ where $\exists$ is defined as in Section 2.4 and i is the inclusion of $\mathrm{Im} f$ in B.)

(DD). (The doctrinal diagram). Show that in the following diagram in which the arrows are all defined above,

(a) $\exists f \circ L \cong L \circ \sum f$,
(b) $I \circ \forall f \cong \prod f \circ I$,
(c) $f^* \circ I \cong I \circ f^{-1}$, and
(d) $f^{-1} \circ L \cong L \circ f^*$.

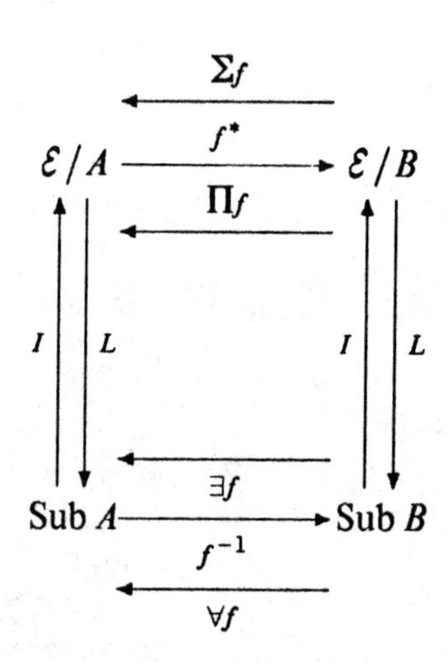

5.6. The Heyting Algebra Structure on Ω

Heyting algebras

Intuitionistic logic is a weakening of classical logic intended to allow only "positive" proofs; in other words, proof by contradiction is ruled out. Just as the concept of Boolean algebra results from abstracting the properties of "and", "or" and "not" in classical logic, the concept of Heyting algebra arises by abstracting the properties of "and", "or" and "implies" (the latter must be taken as a primitive in intuitionistic logic) in the logic developed by Brouwer, Heyting and others (see A. S. Troelstra [1977]). Intuitionistic logic arises naturally in toposes; it is not the result of a philosophical position on the part of those who do topos theory, although many people interested in constructive mathematics have been attracted to the subject.

A Heyting algebra is a lattice with some extra structure. We denote the infimum of a and b in a lattice by $a \wedge b$ and the supremum by $a \vee b$. The ordering is denoted by $\leq$. The maximum and minimum elements of the lattice, if they exist, are T and F respectively. Then a **Heyting algebra** is a lattice with a minimum and an additional binary operation "$\rightarrow$" satisfying the requirement that for all elements a, b and c, $c \le a \to b$ if and only if $a \wedge c \le b$.

This definition has a number of consequences. A Heyting algebra has a maximum element, namely $F \to F$. The operation $\to$ has the properties that if $a \le b$ then $b \to c \le a \to c$ and $c \to a \le c \to b$; this last fact means that for fixed a, $a \to -$ is a functor (regarding the lattice as a category) which is right adjoint to $a \wedge -$. Thus a Heyting algebra is a lattice with minimum which is Cartesian closed as a category.

For any element a in a Heyting algebra, one defines $\neg a$ as $a \to F$. The operation $\neg$ has only some of the properties of the classical "not". For example, $a \le \neg\neg a$, but one cannot prove that $\neg\neg a \le a$, that $a \vee \neg a = T$, or the DeMorgan laws. In the exercises, you are asked to verify these and other facts about Heyting algebras.

There are two important types of examples of Heyting algebras.

(i) Any Boolean algebra is naturally a Heyting algebra, defining $a \to b$ to be $\neg a \vee b$.
(ii) The lattice of open sets of any topological space X is a Heyting algebra. Here, meet and join are intersection and union, and $A \to B$ is the interior of $(X - A) \cup B$. A special case of this, useful for constructing counterexamples, is the **Sierpinski space**: the set $S = \{0,1\}$ with the empty set, S and $\{1\}$ as the only open sets.

The lattice of open dense subsets of a topological space, together with the empty set, also forms a Heyting algebra in the same way.

More information on Heyting algebras may be found in Rasiowa and Sikorski [1963], where they are called "pseudocomplemented lattices".

The Heyting algebra structure on Ω

For an object A of a topos, intersection and union of subobjects make $\mathrm{Sub}(A)$ a lattice. Both operations induce functions $\mathrm{Sub}(A) \times \mathrm{Sub}(A) \to \mathrm{Sub}(A)$, thus functions from

$$\mathrm{Hom}(A,\Omega) \times \mathrm{Hom}(A,\Omega) \cong \mathrm{Hom}(A,\Omega \times \Omega)$$

to $\mathrm{Hom}(A,\Omega)$. These two functions are natural in A: Suppose X and Y are subobjects of B and $f : A \to B$. We must show that $\mathrm{Sub}(f)(X \cup Y) = \mathrm{Sub}(f)(X) \cup \mathrm{Sub}(f)(Y)$ and similarly for intersection. Now $\mathrm{Sub}(f)(X)$ is computed by pulling back along f. The intersection and union are respectively a pullback and a pushout in $\mathcal{E}$ (diagram 5, Section 5.5) and the pullback functor has both a left and a right adjoint by Corollary 7, Section 5.3. Thus it takes limits and colimits in $\mathcal{E}/B$ to limits and colimits in $\mathcal{E}/A$. Colimits in $\mathcal{E}/A$ are the same as colimits in $\mathcal{E}$. As for limits, if D is a diagram in $\mathcal{E}/A$, a limit of D in $\mathcal{E}/A$ is the same as the limit of $D \to A$ in $\mathcal{E}$, which is the sort of limit we have here. Thus $\mathrm{Sub}(f)$ preserves intersection and union, as required.

Since the induced maps $\mathrm{Hom}(A, \Omega \times \Omega) \to \mathrm{Hom}(A, \Omega)$ are therefore natural, Yoneda gives us binary operations $\wedge : \Omega \times \Omega \to \Omega$ and $\vee : \Omega \times \Omega \to \Omega$.

The order relation and the arrow operation are obtained this way: The equalizer

$$[(B, C) \mid B \cap C = B] = [(B, C) \mid B \leq C]$$

of the two maps $\wedge$ and the first projection from $\Omega \times \Omega$ to Ω is a subobject of $\Omega \times \Omega$ and so corresponds to a pullback

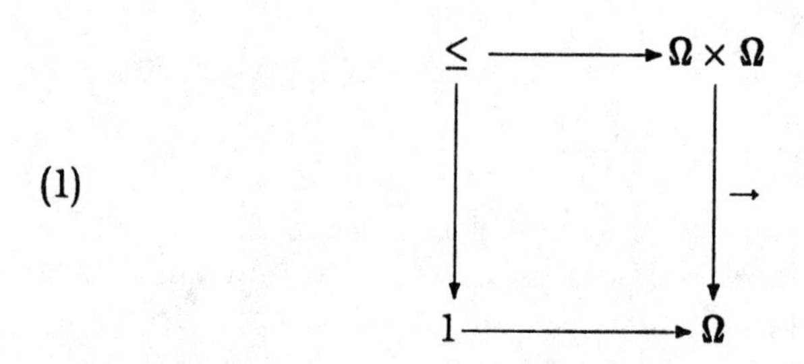

Because $\mathrm{Hom}(A, -)$ preserves pullbacks and $\mathrm{Sub}(A) \cong \mathrm{Hom}(A, \Omega)$, the order relation

$$\leq_A = \{(B, C) \mid B \subseteq C \subseteq A\}$$

on $\mathrm{Sub}(A)$ for any object A is then obtained by pulling back:

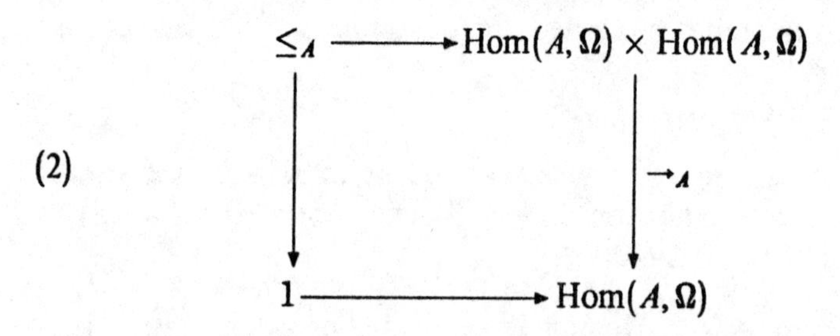

The following lemma follows immediately from the definitions.

Lemma 1. *Let x and y be elements of Ω defined on A. Let the corresponding subobjects of A be B and C respectively. Then:*

(a) *$x \leq y$ if and only if $B \leq_A C$.*
(b) *The subobject corresponding to $x \wedge y$ is $B \wedge C$.*
(c) *The subobject corresponding to $x \vee y$ is $B \vee C$.*
(d) *The subobject corresponding to $x \to y$ is $B \to C$.*

Lemma 2. *If B and C are subobjects of A, then the following are equivalent:*

(a) $B \leq_A C$
(b) $B \to_A C = A$.

Proof. $B \leq_A C$ is equivalent to $x \leq y$ by Lemma 1(a). That is equivalent to $x \to y =$ true (diagram (1)), which is equivalent to $B \to C = A$ by Lemma 1(d).

From now on, we will drop the subscript A on the relation $\leq_A$ and the operation $\to_A$ on $\mathrm{Sub}(A)$.

Lemma 3. *If B, C and D are subobjects of A, then*

(a) $D \wedge (B \to C) = D \wedge B \to D \wedge C$, *and*
(b) $D \leq B \to C$ *if and only if* $D \wedge B \leq D \wedge C$.

Proof. Intersecting by D is pulling back along the inclusion of D in A, and a pullback of a pullback is a pullback; that proves (a). By Lemma 1, $D \leq B \to C$ if and only if $D \wedge (B \to C) = D$. By (a) and Lemma 1 applied to $\mathrm{Sub}(D)$, that is true if and only if $D \wedge B \leq D \wedge C$.

Theorem 4. *For any object A in a topos $\mathcal{E}$, $\mathrm{Sub}(A)$ is a Heyting algebra with the operations defined above.*

Proof. The minimum is evidently the subobject $0 \to A$. $\mathrm{Sub}(A)$ is a lattice, so all that is necessary is to prove that $D \leq B \to C$ if and only if $D \wedge B \leq C$. If $D \leq B \to C$, then $D \wedge B \leq D \wedge C \leq C$ by Lemma 2(b). Conversely, if $D \wedge B \leq C$, then clearly $D \wedge B \leq D \wedge C$ because $\wedge$ is the greatest lower bound operation; then the result follows from Lemma 2(b) again.

Corollary 5. *Ω is a Heyting algebra, with minimum the unique element $0 \to \Omega$ and $\wedge$, $\vee$ and $\to$ as defined above.*

Proof. This follows immediately from Lemma 1.

When the Heyting algebra in Ω is that of a Boolean algebra, we call the topos **Boolean**. This is equivalent to saying that every subobject of an object has a complement.

The category of sheaves over a topological space is a topos whose subobject classifier is the sheaf whose value at an open set U is the set of open subsets of U, with restriction given by intersection. (Exercise (OMT) of Section 2.3.) The natural Heyting algebra structure on that sheaf is the Heyting algebra structure of Corollary 5.

Exercises 5.6

(HEYT)°. Prove the following facts about Heyting algebras:

(a) $F \to F$ is the maximum of the lattice.
(b) If $a \leq b$ then $b \to c \leq a \to c$ and $c \to a \leq c \to b$.
(c) $a \to b = T$ if and only if $a \leq b$.
(d) $b \leq a \to b$.
(e) $a \wedge (a \to b) = a \wedge b$. Hence $a \wedge (a \to b) \leq b$.
(f) $a \wedge \neg a = F$.
(g) $a \leq \neg\neg a$.
(h) $\neg a = \neg\neg\neg a$
(i) $a \leq b$ implies $\neg b \leq \neg a$

(EXH). Show that a finite lattice can be made into a Heyting algebra by a suitable choice of "$\to$" if and only if it is distributive. Show also that every chain is a Heyting algebra. What is the double negation of an element in the latter case?

(MAL). (Freyd) Show that the category of Heyting algebras is a Mal'cev category: that means that there is a ternary operation $\mu(a, b, c)$ with the properties that $\mu(a, a, c) = c$ and $\mu(a, b, b) = a$. To define μ, first define

$$a \leftrightarrow b = (a \to b) \wedge (b \to a),$$

and then let

$$\mu(a, b, c) = ((a \leftrightarrow b) \leftrightarrow c) \wedge (a \leftrightarrow (b \leftrightarrow c)).$$

Chapter 6.

Permanence Properties of Toposes

This chapter is concerned with certain constructions on a topos which yield a topos. We have already seen one such construction: a slice $\mathcal{E}/A$ of a topos $\mathcal{E}$ is a topos. The most important construction in this chapter is that of the category of "sheaves" in a topos relative to a "topology". When the topos is a category of presheaves on a space and the topology is the "canonical" one, the "sheaves" are ordinary sheaves. The category of sheaves in a topos (relative to any topology, canonical or not) turns out to be a topos.

The concept of topology is an abstraction of the concept of all coverings, which at one level of abstraction is a "Grothendieck topology" and at a higher level is a "topology on a topos". An important connection with logic is signalled by the fact that the double negation operator on a topos is a topology in this sense.

We find it convenient here to start with the more abstract (but easier to understand) idea of a topology on a topos first. Later in the chapter we talk about Grothendieck topologies and prove Giraud's Theorem (Theorem 1 of Section 6.8) which characterizes categories of sheaves for a Grothendieck topology.

We will also consider categories of coalgebras for a left exact cotriple in a topos, and of algebras for an idempotent left exact triple. Both these categories are also toposes (the latter are actually sheaves for a topology) and the constructions yield an important factorization theorem (Section 6.5) for geometric morphisms.

6.1. Topologies

A **topology** on a category with pullbacks is a natural endomorphism $\mathbf{j}$ of the contravariant subobject functor which is

(i) **idempotent:** $\mathbf{j} \circ \mathbf{j} = \mathbf{j}$,

(ii) **inflationary**: $A_0 \subseteq \mathbf{j}A_0$ for any subobject A_0 of an object A (where we write $\mathbf{j}A_0$ for $\mathbf{j}A(A_0)$ as we will frequently in the sequel), and
(iii) **order-preserving**: if A_0 and A_1 are subobjects of A and $A_0 \subseteq A_1$, then $\mathbf{j}A_0 \subseteq \mathbf{j}A_1$.

See Exercise (IND) for the independence of (ii) and (iii).

When $\mathbf{j}$ is a topology on a category in which subobjects are representable by an object Ω, then using the Yoneda Lemma, $\mathbf{j}$ induces an endomorphism of Ω which is idempotent, inflationary and order-preserving, and conversely such an endomorphism induces a topology on the category. A topology in this sense on Ω can be given an equational definition in terms of intersection and truth (Exercise (TOP)).

A subobject A_0 of an object A is **$\mathbf{j}$-closed** in A if $\mathbf{j}A(A_0) = A_0$ and **$\mathbf{j}$-dense** in A if $\mathbf{j}A(A_0) = A$. Observe that $\mathbf{j}A(A_0)$ is $\mathbf{j}$-closed by idempotence, and A is $\mathbf{j}$-dense in A because $\mathbf{j}$ is inflationary. When $\mathbf{j}$ is understood, we often write "dense" and "closed".

A topology is superficially like a closure operator on a topological space. However, it does not preserve finite unions (in fact we will see later that it does preserve finite intersections) and to this extent the terminology "dense" and "closed" is misleading. However, it is standard in the literature, so we retain it.

Let's start with some examples.

(a) This example shows how the pasting property of a sheaf motivated the definition of topology. Let X be a topological space and $\mathcal{E}$ the category of presheaves (functors from the opposite of the open set lattice to $\mathcal{S}et$) on X. Let F be a presheaf. Define an endofunction $\mathbf{j}F$ of the set of subfunctors of F by requiring that for an open set U of X and a subfunctor G of F, $\mathbf{j}F(G)(U)$ is the set of elements $x \in FU$ for which there is a cover $\{U_i \to U\}$ such that for all i, $x|U_i \in G(U_i)$.

It is easy to see that if $U \subseteq V$ and $y \in \mathbf{j}F(G)(V)$ then the restriction of y in FU is in $\mathbf{j}F(G)(U)$, so that $\mathbf{j}F(G)$ is really a subfunctor of F. Then the maps $\mathbf{j}F$ are the components of a natural endomorphism of the subobject functor which is a topology on $\mathcal{E}$.

To verify this requires proving that $\mathbf{j}$ is a natural transformation and that it satisfies (i)-(iii) of the definition of topology. We prove the hardest, naturality, at the end of this section and leave the rest to you.

Observe that any sheaf is $\mathbf{j}$-closed with respect to the topology $\mathbf{j}$.

(b) In any topos, $\neg\neg$ is a topology (Exercise (DN)). The proof is implicit in the results and exercises to Section 5.6. We will see that when a topos is regarded as a theory, then the sheaves for the double-negation topology force a Booleanization of the theory. Those familiar with logic should note that the word "force" is used advisedly (see Tierney [1976]).

(c) In any topos, if $U \rightarrowtail 1$, there is a "least destructive" topology $\mathbf{j}$ for which $\mathbf{j}0 = U$, namely that which for a subobject $A_0 \rightarrowtail A$ has $\mathbf{j}A(A_0) = A_0 \cup A \times U$ (note $A \times U \rightarrowtail A \times 1 = A$). This has the property that if $U \leq V \leq 1$ then V is closed in 1.

(d) Topologies exist in categories which are not toposes, too. There is a topology on the category of Abelian groups which assigns to each subgroup B of an Abelian group A the subgroup

$$\{a \in A \mid \text{there is a positive integer } n \text{ for which } na \in B\}.$$

This subgroup is the kernel of the composite

$$A \longrightarrow A/B \longrightarrow \frac{A/B}{t(A/B)}$$

where t denotes the torsion subgroup.

We should think of $\mathbf{j}A(B)$ as the set of all elements of A which are "almost in" B. Equivalently, it may be thought of as elements which are "almost zero" mod B. Topologies on additive categories are often called **torsion theories**.

Properties of topologies

We state here some technical lemmas which will be used many times later.

Lemma 1. *If*

(1)

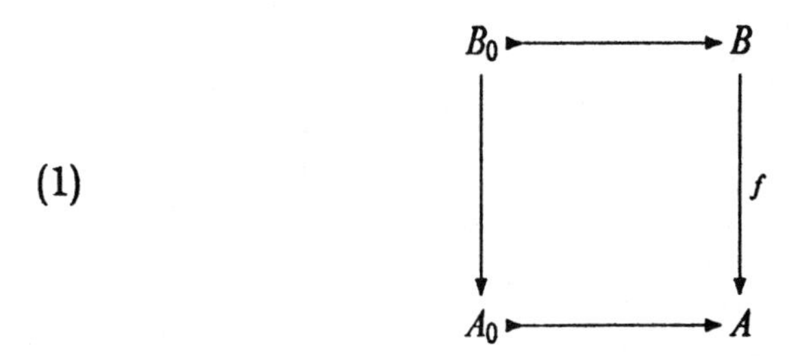

is a pullback, then there is a (necessarily unique) arrow $\mathbf{j}B(B_0) \to \mathbf{j}A(A_0)$

for which

(2)

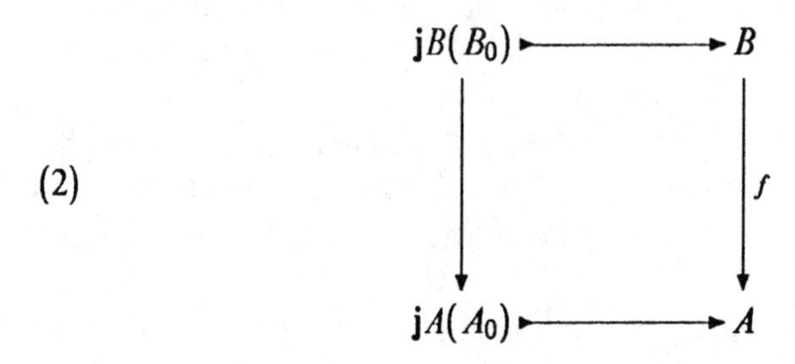

is also a pullback.

Proof. Follow A_0 around the two paths of the following diagram.

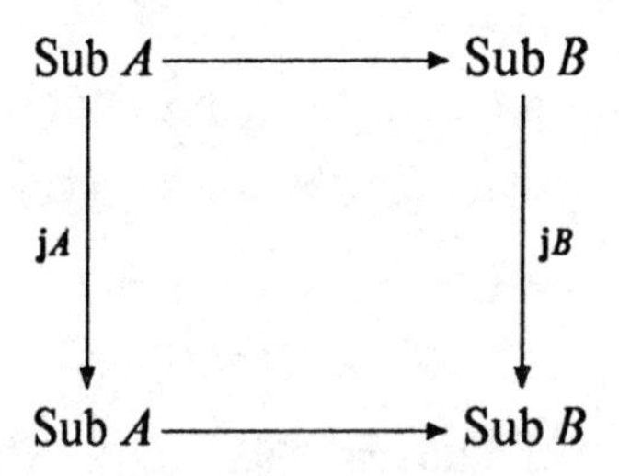

Exercise (TPPB) gives a converse to Lemma 1.

Lemma 2. *Let $\mathbf{C}$ be a category with a topology $\mathbf{j}$, A be an object of $\mathbf{C}$ and B, C, D be subobjects of A. Then*

(a) *If $C \subseteq B$, then $\mathbf{j}B(C) = B \cap \mathbf{j}A(C)$.*
(b) *If $C \subseteq B$, then $\mathbf{j}B(C) \subseteq \mathbf{j}A(C)$.*
(c) *$B \subseteq \mathbf{j}A(B)$ is dense and $\mathbf{j}A(B) \subseteq A$ is closed.*
(d) *The "diagonal fill-in property" of a factorization system is satisfied: if*

(3)

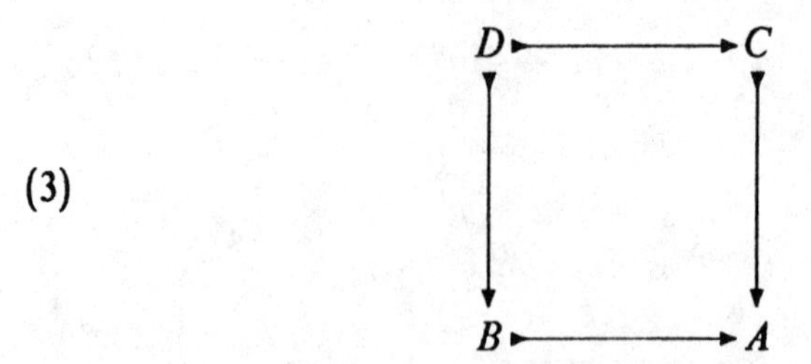

is a commutative square of monos with the top arrow dense and the bottom arrow closed, then $C \subseteq B$.

(e) *If $f : A' \to A$ is any map in $\mathcal{C}$ and B is dense (resp. closed) in A, then $f^{-1}(B)$ is dense (resp. closed) in A'.*

(f) *If B and C are both dense (resp. closed) in A then $B \cap C$ is dense (resp. closed) in A.*

(g) *If $C \subseteq B \subseteq A$ and both inclusions are dense (resp. closed), then C is dense (resp. closed) in A.*

(h) *$\mathbf{j}A(B)$ is characterized uniquely by the facts that B is dense in $\mathbf{j}A(B)$ and $\mathbf{j}A(B)$ is closed in A.*

Proof. (a) is a special case of Lemma 1 (if you ever get stuck trying to prove something about a topology, try using the fact that a topology is a natural transformation) and (b) is immediate from (a). (c) is immediate from (a) applied to $B \subseteq jA(B)$. For (d), apply $\mathbf{j}$ in the diagram to get

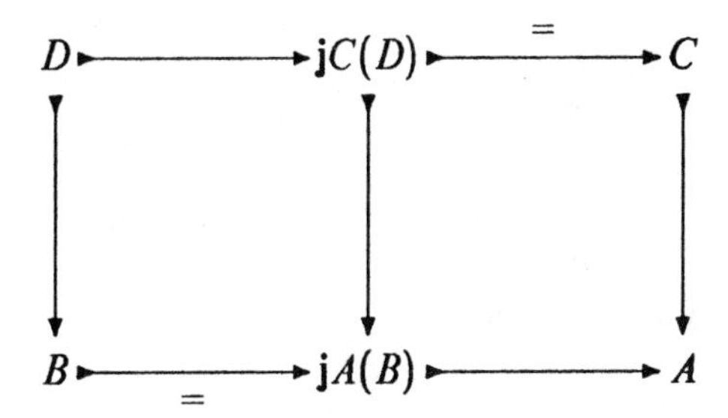

from which the conclusion is immediate. The "dense" half of (e) is a special case of Lemma 1, and the other half is true in any factorization system (Exercise (FAC2) of Section 5.5). Exactly the same is the case for (f), while both parts of (g) are true in any factorization system. Finally (h) follows from the uniqueness of image in a factorization system. The factorization system is on the category with the same objects as $\mathcal{C}$ and the monos as maps.

Proposition 3. *Let B and C be subobjects of A. Then $\mathbf{j}A(B \cap C) = \mathbf{j}A(B) \cap \mathbf{j}A(C)$.*

Proof. It follows from Lemma 2(f) that $B \cap C$ is dense in $\mathbf{j}A(B) \cap \mathbf{j}A(C)$ and that $\mathbf{j}A(B) \cap \mathbf{j}A(C)$ is closed in A. By Lemma 2(h), this characterizes $\mathbf{j}A(B \cap C)$.

Proposition 4. *In a category with pullbacks which has a topology $\mathbf{j}$, suppose the left vertical arrow in the following commutative square is a dense mono,* *and the right vertical arrow is a closed mono.*

(4)

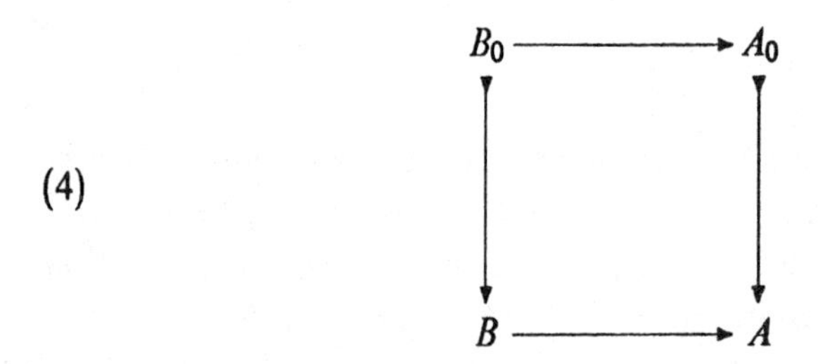

Then there is a map from B to A_0 making both triangles commute.

Proof. The inverse image (pullback) of A_0 is closed in B by Lemma 2(e) and dense because it contains B_0, so that the inverse image is B. The conclusion now follows easily.

Naturality of j for spatial sheaves

Here we outline the proof that the map $\mathbf{j}$ of Example (a) is natural. We must prove that if F and F' are presheaves and $\lambda : F \to F'$ is a natural transformation, then

(5)

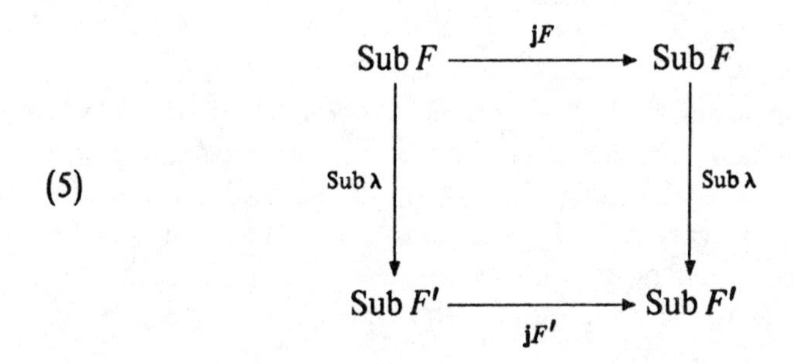

commutes.

If G' is a subfunctor of F', then $G = \text{Sub}\,\lambda(G')$ if for each open U, GU is the inverse image of $G'U$ along λU. This is because limits are constructed pointwise in a functor category like $\mathcal{E}$.

Using this notation, it is necessary to show that for every open U, the inverse image of $\mathbf{j}F'(G')(U)$ along λU is $\mathbf{j}F(G)(U)$. To see this, suppose $y \in \mathbf{j}F'(G')(U)$ and $\lambda U(x) = y$ for some $x \in FU$. Then on some cover $\{U_i\}$ of U, $y|U_i \in G'U_i$ for every i. Then by definition, $x|U_i \in GU_i$ and so $x \in \mathbf{j}F(G)(U)$. Conversely, it is clear that if $x \in \mathbf{j}F(G)(U)$ then $\lambda(x) \in \mathbf{j}F'(G')(U)$

Exercises 6.1

(IND). Find an idempotent endomorphism of the three element chain which is inflationary but not order-preserving and one which is order-preserving but not inflationary.

(TPPB)°. Suppose that for each object A of a topos there is an idempotent, inflationary, order-preserving map $\mathbf{j}A : \operatorname{Sub} A \to \operatorname{Sub} A$ with the property that whenever

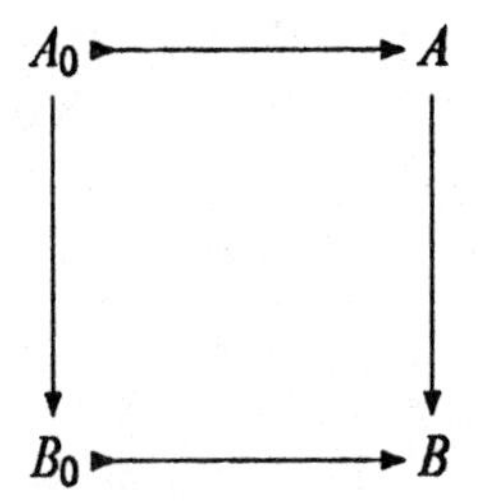

is a pullback, then there is an arrow $\mathbf{j}A(A_0) \to \mathbf{j}B(B_0)$ for which

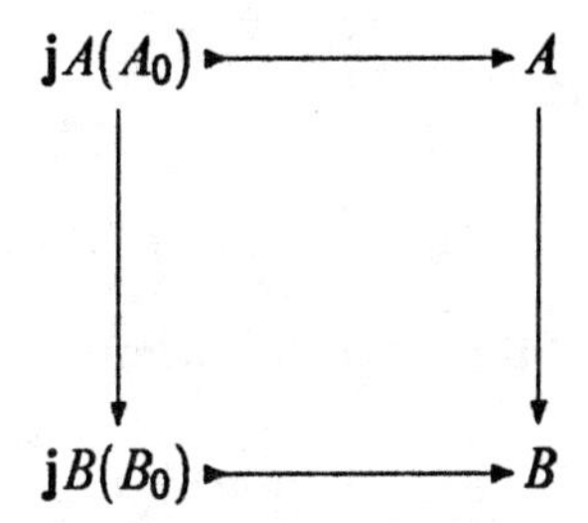

is also a pullback.

Show that these functions constitute a natural endomorphism of $\mathbf{P}$ and so induce a topology on the topos.

(NATJ)°. Prove that the natural transformation $\mathbf{j}$ of Example (a) is a topology.

(TOP)°. Prove that a topology on a category in which subobjects are representable can be given as an endomorphism $\mathbf{j}$ of the representing object which is idempotent, takes true to true, and commutes with intersection.

(DN)°. Use the results and exercises of Section 5.6 to show that there is a topology $\mathbf{j}$ on any topos such that for any subobject $A_0 \subseteq A$, $\mathbf{j}A_0 = \neg\neg A_0$.

6.2. Sheaves for a Topology

In this section we define what it means for an object in a topos to be a sheaf for a topology on the topos, and construct an "associated sheaf functor".

Separated objects

Let $\mathbf{j}$ be a topology on a topos. An object A is **j-separated** (or simply "separated" if $\mathbf{j}$ is understood) if A is a closed subobject of $A \times A$ via the diagonal. We will form the separated quotient of an object A.

Proposition 1. *Let $R(A)$ be the closure of A in $A \times A$. Then for any object B, an element $(f, g) \in^B RA$ if and only if the equalizer of f and g is dense in B.*

Proof. If B_0 is that equalizer then the outer square and hence by Lemma 1 of Section 6.1 the right hand square of

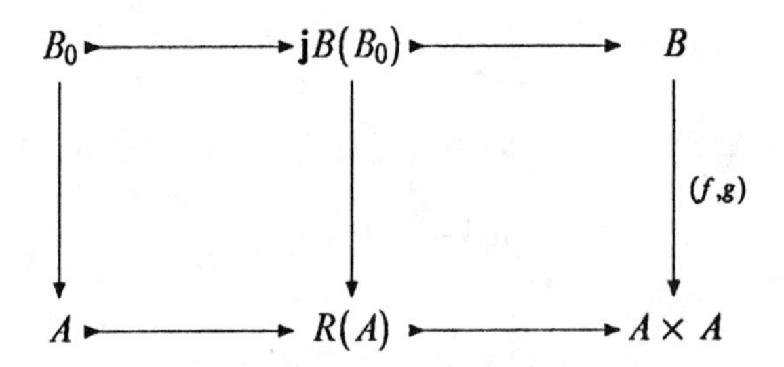

are pullbacks. The conclusion is now evident.

Corollary 2. *$R(A)$ is an equivalence relation on A.*

Proof. Reflexivity and symmetry are clear. If (f, g) and (g, h) are elements of $R(A)$ defined on B then the equalizer of f and h contains the intersection of those of f and g and g and h, each of which is dense. But by Lemma 2(f) of 6.1 the intersection of two dense subobjects is dense.

In Abelian groups we form the torsion-free quotient by factoring out the elements which are "almost zero". In analogy with this construction, we will form the separated quotient of A by identifying pairs of elements which are "almost equal". Thus we form the quotient

$$R(A) \rightrightarrows A \twoheadrightarrow S(A)$$

which we can do because equivalence relations in a topos are effective. Note that A is separated if and only if $A = SA$.

Proposition 3. *$S(A)$ is $\mathbf{j}$-separated. If $A \rightarrowtail B$, then $S(A) \rightarrowtail S(B)$.*

Proof. The diagram

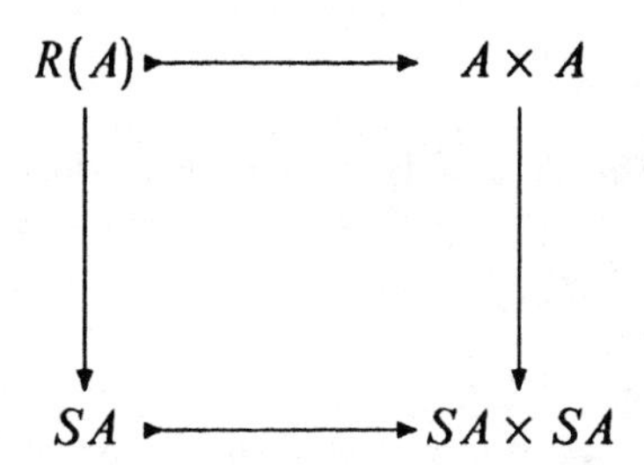

is a pullback (standard because RA is the kernel pair of $A \to SA$). It follows from the fact that the pullback of an epi is an epi that the image of RA in $SA \times SA$ is the diagonal SA. If we apply $\mathbf{j}$ we get the pullback

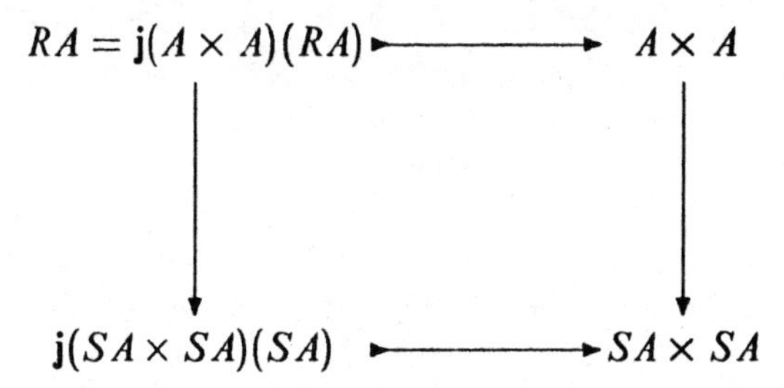

and the vertical arrows are epic so that $\mathbf{j}(SA \times SA)(SA) = SA$. Thus SA is separated. As for the second assertion, when $A \rightarrowtail B$ is monic, the diagram

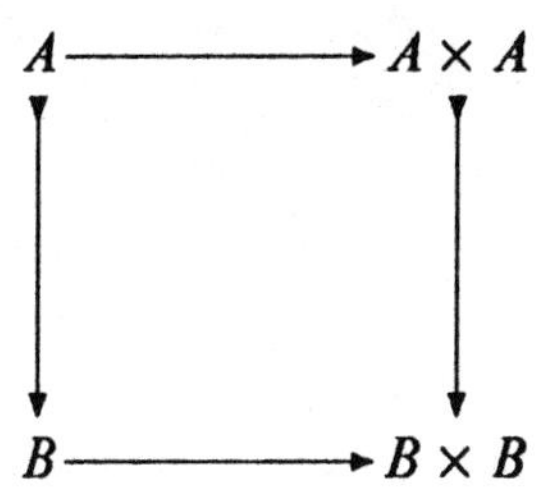

is a pullback. Then apply $\mathbf{j}$ to get a pullback

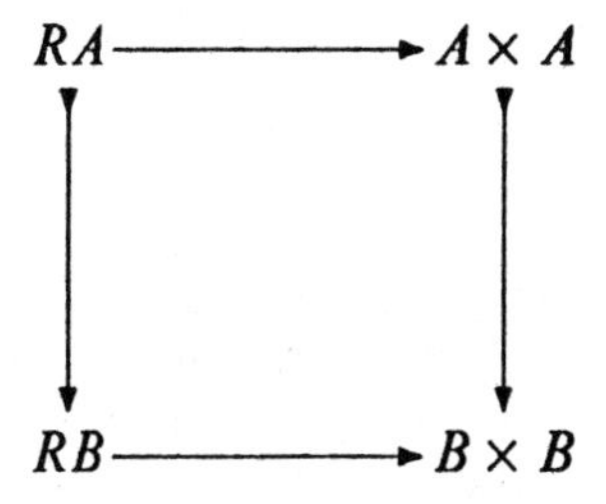

Since RA and RB are equivalence relations on A and B, respectively, it follows from Exercise (EQCLS) that $SA \to SB$ is mono.

Sheaves for a topology

An object in a topos is **absolutely closed** for a topology $\mathbf{j}$ if it is $\mathbf{j}$-closed as a subobject of any separated object. An object A in a topos is a **sheaf** for a topology $\mathbf{j}$ if it is $\mathbf{j}$-separated and absolutely closed.

For any object A, let FA denote the object $S(\mathbf{j}(\mathbf{P}SA)(SA))$ (using the singleton map to include SA in $\mathbf{P}SA$). We will show that FA is a sheaf and that F is the left adjoint of the inclusion of the full subcategory of sheaves. F is the **associated sheaf functor** (or **sheafification).**

Proposition 4. *If A is separated, the map $A \to FA$ is a $\mathbf{j}$-dense mono. If A is a sheaf, then $A \to FA$ is an isomorphism.*

Proof. Let A be separated. A is included in $\mathbf{j}(\mathbf{P}A)(A)$, so by Proposition 3, $SA = A$ is included in $S(\mathbf{j}(\mathbf{P}A)(A))$ which is FA because A is separated.

To show that the inclusion is dense, let $B = \mathbf{j}(\mathbf{P}A)(A)$ and let C be the inverse image of A along the map $B \to FA$, as in the diagram

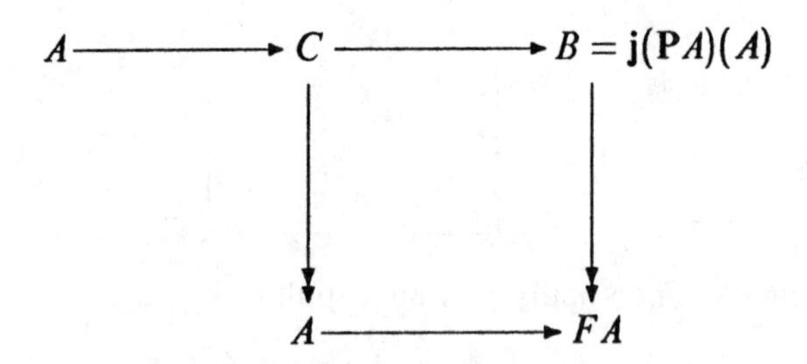

Apply $\mathbf{j}$ to this diagram using Lemma 1 of Section 6.1 and the top row becomes the identity on B so the bottom row must also become the identity because the vertical arrows are epic.

Lemma 5. *Let $B_0 \to B$ be a $\mathbf{j}$-dense inclusion. Then any map $B_0 \to A$ can be extended to a map $B \to FA$.*

Proof. In the diagram

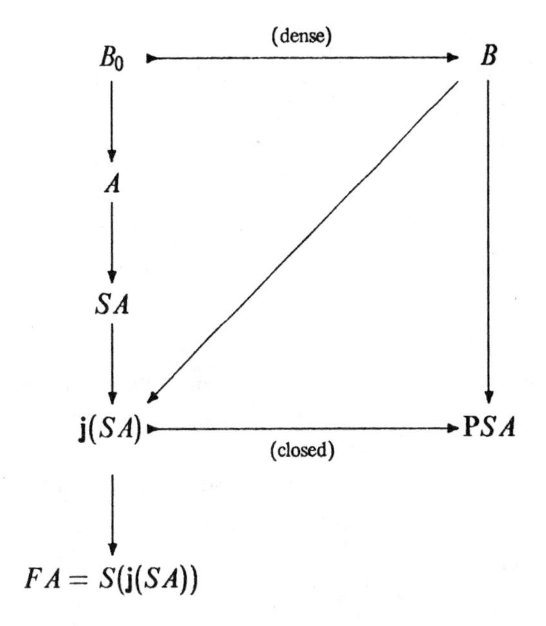

(in which $\mathbf{j}$ means $\mathbf{j}(PSA)$), the rightmost vertical arrow exists because power objects are injective (Exercise (Inj) of Section 2.1) so the diagonal arrow exists by Lemma 2(d) of Section 6.1.

Proposition 6. *Two maps to a separated object which agree on a dense subobject are equal.*

Proof. Consider the diagram

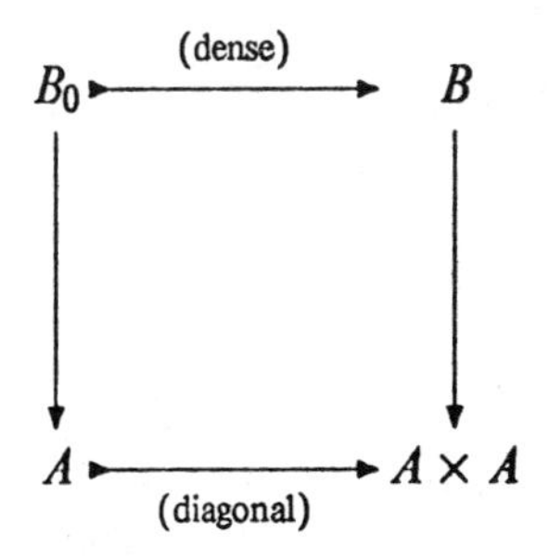

where the right arrow is induced by the given arrows. By Lemma 2(d) of Section 6.1 that arrow factors through the diagonal, as required.

Proposition 7. *Let A be separated and $B_0 \to B$ be a $\mathbf{j}$-dense inclusion. Then any map $B_0 \to FA$ can be extended to a unique map $B \to FA$.*

Proof. Consider the diagram

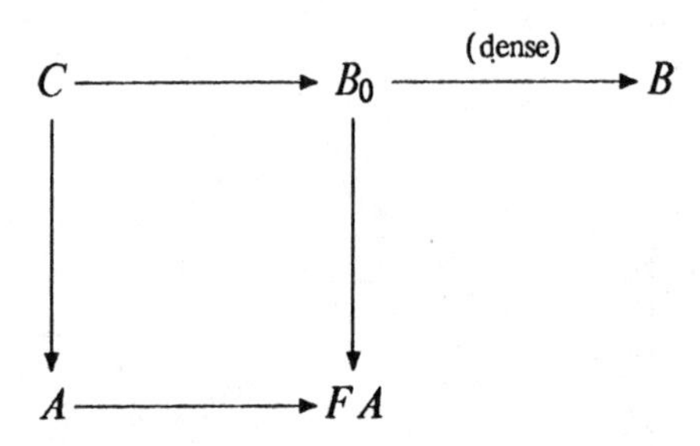

where the square is a pullback. The composite along the top is a dense inclusion by Proposition 4 and the fact that the composite of dense maps is dense. The requisite map from B to FA exists by Lemma 5. That map and the map from B_0 to FA agree on C and so are equal by Proposition 6. The uniqueness follows similarly.

The following proposition shows that the essence of being a sheaf has survived our process of abstraction.

Proposition 8. (a) *A separated object A is a sheaf if and only if whenever $B_0 \to B$ is dense then any map $B_0 \to A$ has an extension to a map $B \to A$.*

(b) *An arbitrary object A is a sheaf if and only if whenever $B_0 \to B$ is dense then any map $B_0 \to A$ has a unique extension to a map $B \to A$.*

Remark: It follows readily from this proposition that if $\mathbf{j}$ is the topology of Example (a) of section 6.1, then the category of sheaves on X is the same as the category of $\mathbf{j}$-sheaves in the presheaf category.

Proof. If A is a sheaf, these follow from Propositions 4 and 7.

Now suppose that the map extension condition holds. let $d^0, d^1 : RA \to A$ be the kernel pair of the map $A \to SA$. The equalizer of d^0 and d^1 is the diagonal of $A \times A$, which is dense in RA. Then by the version of the map extension condition in (b), $d^0 = d^1$; hence A is separated.

Suppose $m : A \to B$ is monic with B separated. Since any subobject of a separated object is separated (Exercise (Ass sh)), we may replace B by the $\mathbf{j}$-closure of A and suppose without loss of generality that $A \rightarrowtail B$

is dense. The diagram

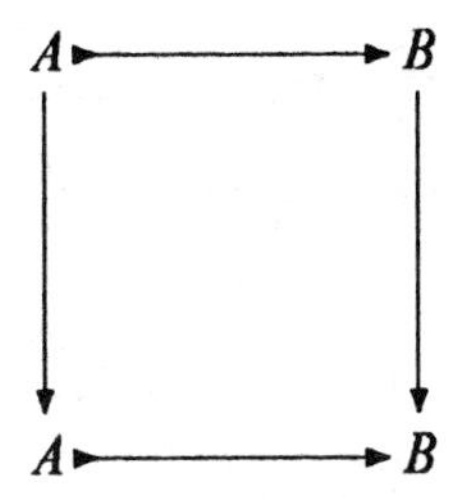

in which the vertical arrows are identities has a diagonal fill-in making the upper triangle commute given by the map extension condition. As for the lower triangle, it commutes when restricted to the dense subobject A. With B separated, this implies that the lower triangle commutes, whence $A = B$, as required.

Theorem 9. *For any object A in a topos, FA is a sheaf, and F is a functor which is left adjoint to the inclusion of the full subcategory of sheaves in the topos.*

Proof. FA is clearly separated; that it is a sheaf then follows from Propositions 7 and 8.

Any map $A \to B$ to a sheaf gives a unique map $SA \to B$, which by Proposition 7 extends to a unique map $FA \to FB = B$; this gives the adjunction.

To show that F is a functor, it is sufficient to use pointwise construction of adjoints (see Section 1.9).

Exercises 6.2

(CLSH). Give an example of a presheaf which is $\mathbf{j}$-closed in the sense of Example (a) which is nevertheless not a sheaf.

(PRDC)$^\circ$. Show that the product of dense monos is dense and the product of closed monos is closed.

(MONOTEST)$^\circ$. Suppose $B_0 \to B$ is a $\mathbf{j}$-dense inclusion with B $\mathbf{j}$-separated. Prove that a map $B \to A$ whose restriction to B_0 is monic is itself monic.

(EQCLS)°. Show that in a serially commutative diagram

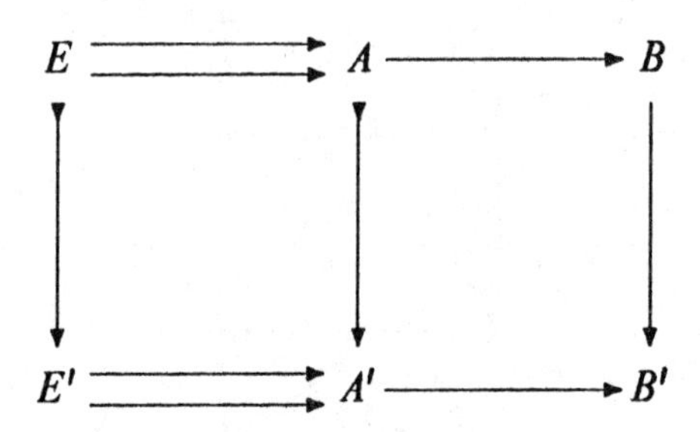

with both rows kernel pair/coequalizers and $E = E' \times_{A'} (A \times A)$, then $B \to B'$ is monic.

(ASS SH)°. (i) Show that if $A \to B$ is monic and B is separated then A is.

(ii) Show that if $A \to B$ is a dense mono where B is a sheaf then $B = FA$.

(UEZ). Show that if j is the topology of Example (iii) of Section 6.1 and E is the category of sheaves on a topological space, then $Sh_{\mathrm{j}}(\mathcal{E})$ is the category of sheaves on the complement of the open set U.

(SSR)°. Show that S is the object map of a functor which is left adjoint to the inclusion of separated objects.

6.3. Sheaves Form a Topos

The full subcategory of sheaves for a topology j in a topos $\mathcal{E}$ is denoted $\mathcal{E}_{\mathrm{j}}$. In this section we will prove that $\mathcal{E}_{\mathrm{j}}$ is a topos. We will also prove that F, which we now know is left adjoint to inclusion, is left exact, so that the inclusion is a geometric morphism.

The power object for sheaves

Using Yoneda, let $\tau A : \mathrm{P}A \to \mathrm{P}A$ denote the map induced by the natural transformation $\mathrm{j}(A \times -) : \mathrm{Sub}(A \times -) \to \mathrm{Sub}(A \times -)$. Then evidently τA is an idempotent endomorphism of $\mathrm{P}A$. Let $\mathrm{P}_{\mathrm{j}}A$ denote the splitting object—the equalizer of τA and the identity. Then for any object B, $\mathrm{Hom}(B, \mathrm{P}_{\mathrm{j}}A)$ consists of the j-closed subobjects of $A \times B$.

Proposition 1. *Let $B_0 \rightarrowtail B$ be a j-dense mono. Then for any object A, pulling back along $A \times B_0 \to A \times B$ gives a one to one correspondence between j-closed subobjects of $A \times B$ and j-closed subobjects of $A \times B_0$.*

Proof. Since $A \times B_0$ is dense in $A \times B$, it is sufficient to prove this when $A = 1$. If $B_1 \rightarrowtail B_0$ is $\mathbf{j}$-closed, then $B_2 = \mathbf{j}B(B_1) \rightarrowtail B$ is a $\mathbf{j}$-closed subobject of B. Lemma 2(a) of Section 6.1 says that $B_0 \cap \mathbf{j}B(B_1) = \mathbf{j}B_0(B_1)$, which is B_1 since B_1 is closed in B_0. If $B_3 \cap B_0 = B_1$ also, we would have two different factorizations of $B_1 \rightarrowtail B$ as dense followed by closed, which is impossible by Exercise (FAC2) of Section 5.5.

Proposition 2. $\mathbf{P_j}A$ *is a* $\mathbf{j}$*-sheaf.*

Proof. We use the characterization of sheaves of Proposition 8 of Section 5.2, which requires us to show that when B_0 is a dense subobject of B, the map

$$\mathrm{Hom}(B, \mathbf{P_j}A) \rightarrowtail \mathrm{Hom}(B_0, \mathbf{P_j}A)$$

is an isomorphism. The left side (respectively the right side) represent the set of $\mathbf{j}$-closed subobjects of $A \times B$ (respectively $A \times B_0$). By Proposition 1 those two sets are in bijective correspondence via pullback.

Theorem 3. $\mathcal{E}_{\mathbf{j}}$ *is a topos in which for a sheaf* A, $P_{\mathbf{j}}A$ *represents the subobjects.*

Proof. It follows from the diagonal fill-in property (Lemma 2(d) of Section 6.1) that a subobject of a sheaf is a sheaf if and only if it is $\mathbf{j}$-closed. Thus for a sheaf A, $P_{\mathbf{j}}A$ represents the subsheaves of A. Finite limits exist because the inclusion of the subcategory has a left adjoint.

Exactness of F

We show here that the functor F is exact, which is equivalent to showing that the inclusion of $E_{\mathbf{j}}$ into E is a geometric morphism. We begin with:

Proposition 4. *The separated reflector* S *preserves products and monos.*

Proof. That S preserves monos is Proposition 3 of Section 6.2.

The product of dense monos is dense (Exercise (PRDC) of 6.2). It follows that for an object A, $A \times A \to RA \times RA$ is dense so that $RA \times RA \subseteq R(A \times A)$. Since products commute with coequalizers, we have the following commutative diagram, in which both rows are coequalizers:

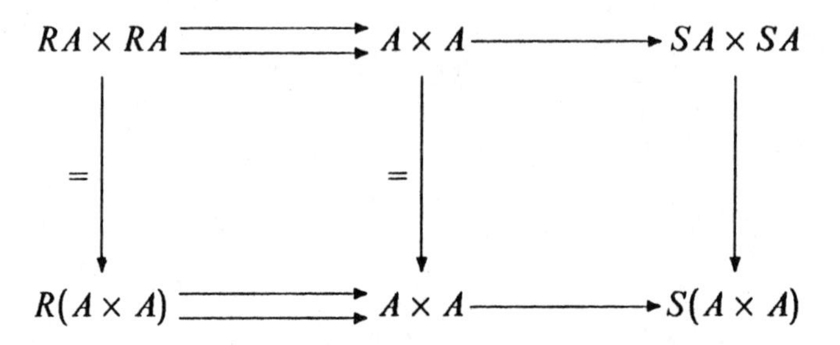

Since $A \times A$ is dense in $RA \times RA$ and $RA \times RA$ is closed in $A \times A \times A \times A$ (Exercise (PRDC) of Section 6.2), it follows from Lemma 2(h) of section 6.1 that $RA \times RA = R(A \times A)$. The required isomorphism follows immediately from the uniqueness of coequalizers.

Proposition 5. *F preserves products and monos.*

Proof. By the preceding proposition it is enough to show that F restricted to the category of separated objects preserves products and monos. That it preserves monos is obvious. Now let A and B be separated. It is clear from Exercise (PRDC) that $A \times B$ is dense in $FA \times FB$ and that the latter is a sheaf. It follows that and from Exercise (ASS SH) of Section 6.2 that $FA \times FB$ is $F(A \times B)$.

Theorem 6. *F is left exact.*

Proof. F is a left adjoint so preserves cokernel pairs. The theorem then follows from Proposition 5 above and Theorem 6 of Section 5.5.

Remark. S is *not* usually left exact. This shows that the full exactness properties of a topos are required in this theorem. Various other combinations of exactness properties have been proposed as a non-additive analog of a category being Abelian (sets of properties which taken together with additivity would imply abelianness) but none of these proposals would appear to allow the proof of a theorem like Theorem 6 of Section 5.5.

Let $\mathcal{E}$ and $\mathcal{F}$ be toposes. Recall from Section 2.2 that a functor $u : \mathcal{E} \to \mathcal{F}$ is a **geometric morphism** if u has a left adjoint u^* and u^* is left exact. We will see in Section 6.4 that morphisms of sites induce geometric morphisms. At this point we wish merely to observe that the left adjoint F constructed above is left exact, so that

Corollary 7. *If* j *is a topology on the topos* E *and* E_{j} *the category of* j*-sheaves, then the inclusion* $\mathcal{E}_{\mathrm{j}} \to \mathcal{E}$ *is a geometric morphism.*

6.4. Left Exact Cotriples

A cotriple $\mathsf{G} = (G, \varepsilon, \delta)$ in which G is a left exact functor is called a **left exact cotriple.** In this section, we will prove

Theorem 1. *Let* $\mathcal{E}$ *be a topos and* G *a left exact cotriple in* $\mathcal{E}$. *Then the category* $\mathcal{E}_{\mathsf{G}}$ *of coalgebras of* G *is also a topos.*

The proof requires a sequence of propositions. In these propositions, $\mathcal{E}$ and G satisfy the requirements of the theorem. Note that G, being left exact, preserves pullbacks and subobjects.

It is not hard to show that when a cotriple is left exact we can speak of a subobject of a coalgebra being a subcoalgebra without ambiguity. See Exercise (Subco).

Proposition 2. *Let* (A, α) *be a coalgebra for* G *and* B *a subobject of* A. *Then* B *is a subcoalgebra if and only if the inverse image of* GB *along* α *is* B.

Proof. If the inverse image of GB along α is B, let the coalgebra structure $\beta : B \to GB$ be the restriction of α to B. This satisfies the required coalgebra identities because $G^2(B) \to G^2(A)$ is monic. To say that inclusion is a coalgebra map requires this diagram

(1)

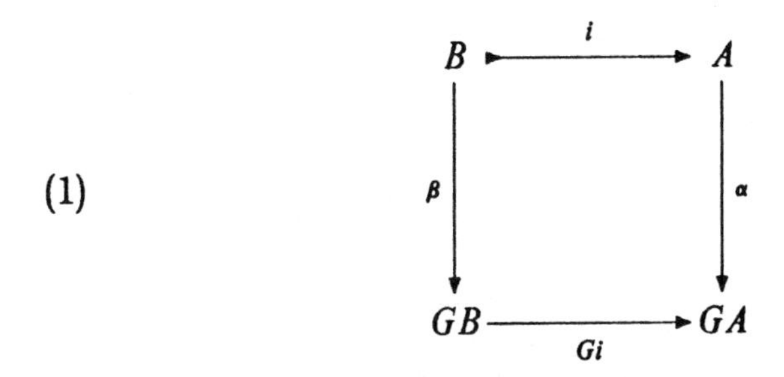

to commute, but in fact it is a pullback by assumption.

Conversely, suppose we are given β for which (1) commutes. We must show that any (variable) element $a : T \to A$ of A for which $\alpha(a) \in GB$ is actually in B. This follows from that fact that $a = \varepsilon A(\alpha(a))$ (where ε is the counit of the cotriple) which is an element of B because a natural transformation between left exact functors takes an element of a subobject to an element of the corresponding subobject.

Now let (A, α) and (B, β) be coalgebras. Let

$$\Phi(\alpha, \beta) : \mathrm{Sub}(A \times B) \to \mathrm{Sub}(A \times B)$$

be defined by requiring that for a subobject C of $A \times B$,

(2)

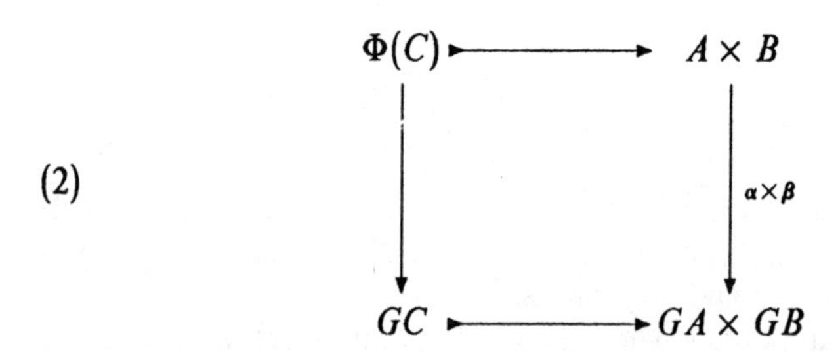

where we write $\Phi(C)$ for $\Phi(\alpha, \beta)(C)$, is to be a pullback.

Proposition 3. *Let (A, α) and (B, β) be coalgebras and C a subobject of $A \times B$. Then ΦC is a subcoalgebra of $A \times B$. It is the largest subcoalgebra contained in C; in particular, C is a subcoalgebra if and only if $C = \Phi C$.*

Proof. Since $\alpha \times \beta$ is a coalgebra structure on $A \times B$ (Exercise (PCA)), it suffices by Proposition 2 to show that we can fill in the upper left diagonal arrow in the diagram below so that the diagram commutes.

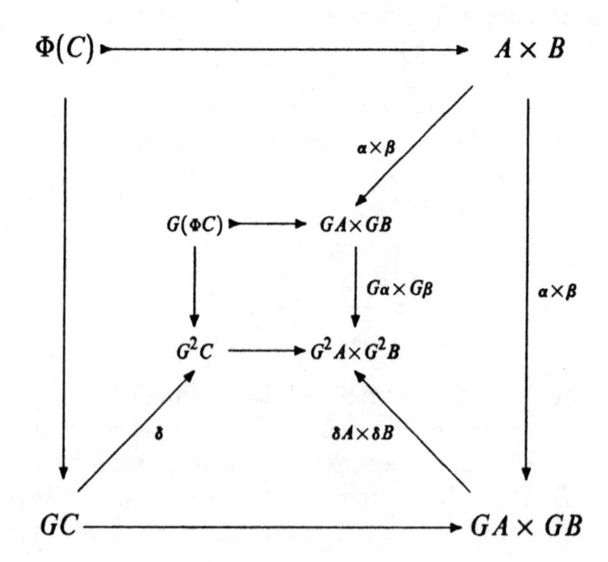

This follows immediately from the fact that all squares commute and the inner square is a pullback. The rest is left as Exercise (LSUB).

Proposition 4. *$\Phi(\alpha, \beta)$ is natural with respect to maps $f : (B', \beta') \to (B, \beta)$ in $\mathcal{E}_{\mathrm{G}}$.*

Proof. Naturality is equivalent to the requirement that the upper square in the diagram below must commute. Here, C is a subobject of $A \times B$ and C' is the inverse image of C along $A \times f$.

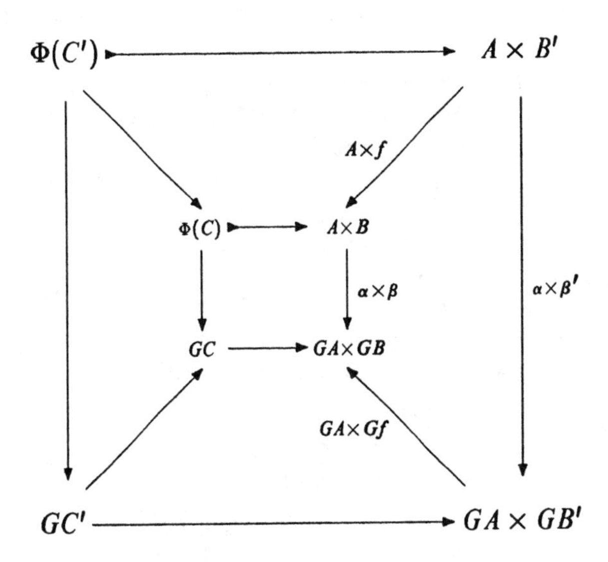

The inner and outer squares are pullbacks by definition. The bottom square is G applied to the pullback in the definition of the subobject functor and hence is a pullback because G is left exact. It follows that the upper square composed with the inner square is a pullback. Since the inner square is too, so is the top one, which therefore commutes.

Let $R : \mathcal{E} \to \mathcal{E}_{\mathrm{G}}$ be the right adjoint to $U : \mathcal{E}_{\mathrm{G}} \to \mathcal{E}$.

Corollary 5. *There is a map $\Phi(\alpha) : RPA \to RPA$ for which*

$$\mathrm{Hom}((B, \beta), \Phi(\alpha)) = \Phi(\alpha, \beta).$$

Proof. Yoneda.

We can now prove Theorem 1. Finite limits exist in $\mathcal{E}_{\mathrm{G}}$ because they are created by the underlying functor $U : \mathcal{E}_{\mathrm{G}} \to \mathcal{E}$. To prove this you use the same sort of easy argument as in proving Exercise (PCA).

The power object $\mathbf{P}(A, \alpha)$ for a coalgebra (A, α) is defined to be the equalizer of $\Phi(\alpha)$ and the identity map on $R\mathbf{P}A$. Since

$$\mathrm{Sub}(A \times B) \cong \mathrm{Hom}(B, \mathbf{P}A) \cong \mathrm{Hom}((B, \beta), R\mathbf{P}A),$$

the theorem follows from Proposition 3 and Corollary 5.

Observe that R is a geometric morphism. Its left adjoint is easily seen to be faithful, as well. We will see later that any geometric morphism with a faithful left adjoint arises from a cotriple in this way.

A nice application of the theorem is a new proof, much simpler than that in Section 2.1, that the functor category $\mathcal{S}et^{\mathcal{C}}$ is a topos for any small category $\mathcal{C}$. This follows from two observations: (i) If S is the set of objects of $\mathcal{C}$, then $\mathcal{S}et^{S} = \mathcal{S}et/S$, which is very easily seen to be a topos. (ii) The map $\mathcal{S}et^{\mathcal{C}} \to \mathcal{S}et^{S}$ induced by the forgetful functor is adjoint tripleable (Section 3.7), hence $\mathcal{S}et^{\mathcal{C}}$ is equivalent to the category of coalgebras of an (evidently) left exact cotriple on $\mathcal{S}et/S$.

Exercises 6.4

(SUBCO)°. Let $\mathbf{G} = (G, \varepsilon, \delta)$ be a left exact cotriple and (A, α) be a $\mathbf{G}$-coalgebra. Show that a subobject A_0 of A "is" a subcoalgebra if and only if there is a commutative square

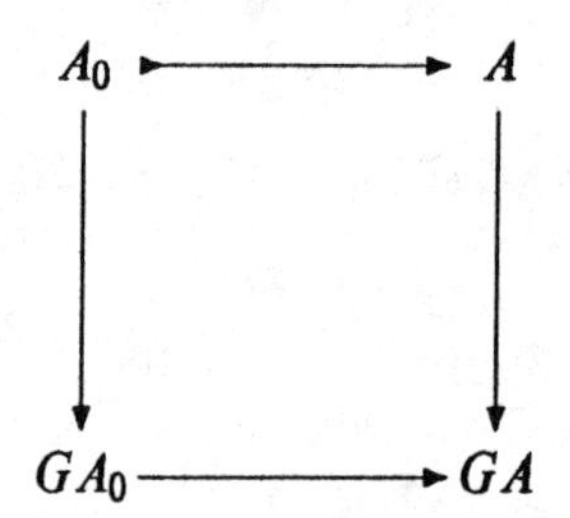

where the right arrow is the coalgebra structure map, in which case that square is a pullback. Conclude that there is at most one subcoalgebra structure on a subobject of an algebra, namely the left arrow in that square.

(PNT). Show that if F and G are product-preserving functors, $\lambda : F \to G$ a natural transformation, then for any objects A and B for which $A \times B$ exists, $\lambda(A \times B) = \lambda A \times \lambda B$.

(PCA)°. Show that if (G, ε, δ) is a left exact cotriple on a category with products and (A, α) and (B, β) are coalgebras, then so is $(A \times B, \alpha \times \beta)$.

(LSUB)°. Verify the third sentence of Proposition 3. (Hint: Follow the map $\Phi(C) \to GC$ by εC.)

6.5. Left Exact Triples

A **left exact triple** in a topos induces a topology on the topos for which the objects of the form TA are sheaves. We will use this construction and the topos of coalgebras of a cotriple discussed in the preceding section to obtain a factorization theorem for geometric morphisms.

Given a left exact triple $\mathsf{T} = (T, \eta, \mu)$ in a topos $\mathcal{E}$, define for each object A a function $\mathbf{j}A : \mathrm{Sub}\, A \to \mathrm{Sub}\, A$ in this way: for a subobject A_0 of A, $\mathbf{j}A(A_0)$ is the inverse image of TA_0 along ηA. In other words,

(1)

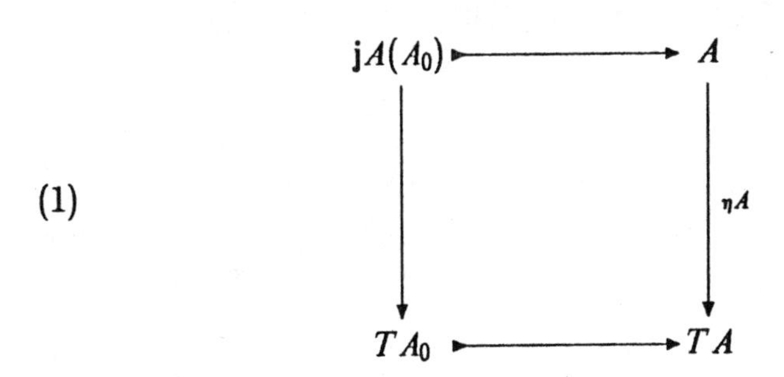

must be a pullback. (Note the lower arrow, hence the upper, must be monic because T is left exact.)

We will prove that these maps $\mathbf{j}A$ form a topology on $\mathcal{E}$. The following lemmas assume that T is a left exact triple, $\mathcal{E}$ is a category with finite limits, and $\mathbf{j}$ is defined as above.

Lemma 1. *Whenever*

(2)

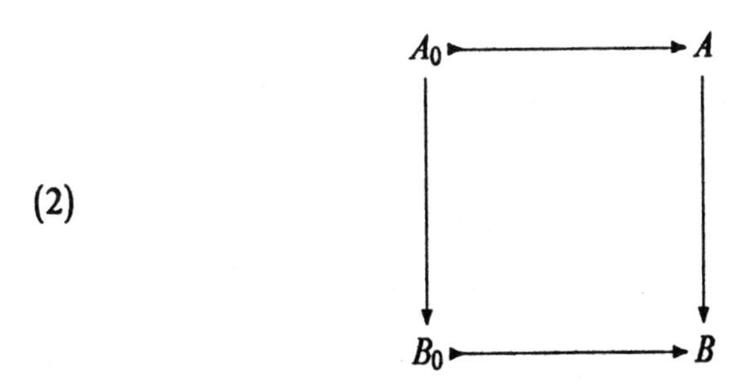

is a pullback, then so is

(3)

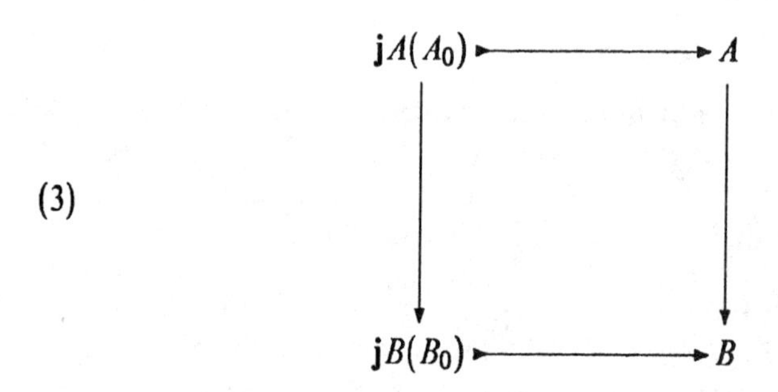

Proof. This can be read off the following square in much the same way that the naturality of Φ was deduced from (3) of Section 6.4.

(4)

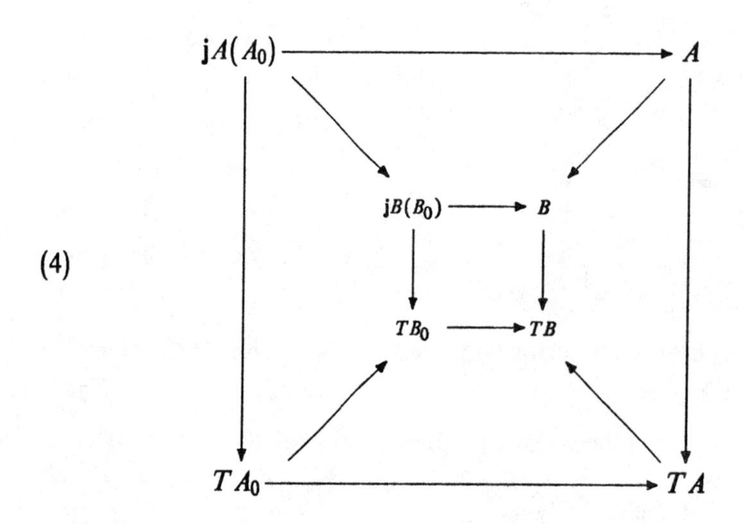

In this square, the inner and outer squares are pullbacks by definition and the bottom square because T is left exact.

Lemma 2. $A_0 \subseteq \mathbf{j}A(A_0)$ *for any subobject* A_0 *of an object* A.

Proof. Use the universal property of pullbacks on ηA_0 and the inclusion of A_0 in A.

Lemma 3. *For any subobject* A_0 *of an object* A, $T\mathbf{j}A(A_0) = \mathbf{j}TA(TA_0) = TA_0$.

Proof. To prove that $T\mathbf{j}A(A_0) = TA_0$, apply T to (1) and follow it by

μ, getting

(5)

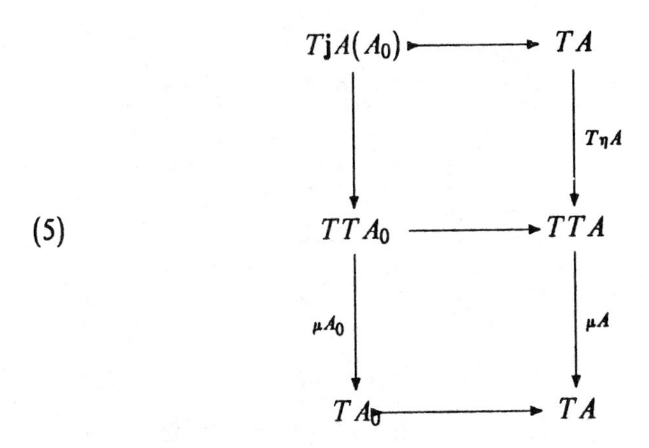

The right vertical map from TA to TA is the identity, so $T\mathbf{j}A(A_0) \to TTA_0 \to TA_0 \to TA$ is the inclusion. Cancelling the top and bottom arrows then shows that the left vertical arrow is an inclusion. This shows that $T\mathbf{j}A(A_0) \subseteq TA_0$, while the opposite inclusion is evident from Lemma 2.

To show that $\mathbf{j}TA(TA_0) = TA_0$, consider the following similar diagram.

(6)

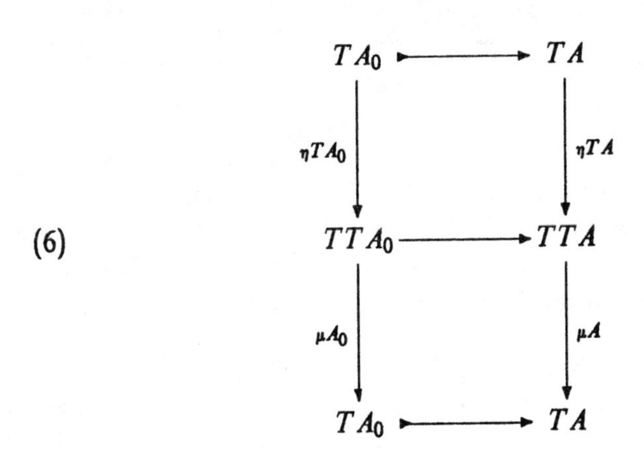

In the same way as for (5), the left and right vertical arrows from top to bottom are identities. This means the outer square is a pullback, so by Exercise (PBCC) of Section 2.2, the upper square is too, as required.

Lemma 4. *If* $A_0 \rightarrowtail A_1 \rightarrowtail A$ *then* $\mathbf{j}A(A_0) \subseteq \mathbf{j}A(A_1)$.

Proof. Easy consequence of the universal property of pullbacks.

Lemma 5. *$A_0 \rightarrowtail A$ is $\mathbf{j}$-dense if and only if there is an arrow from A to TA_0 making*

(7)

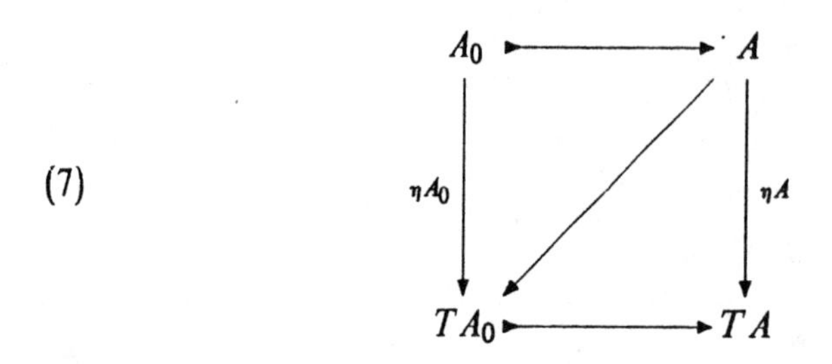

commute.

Proof. If $\mathbf{j}A(A_0) = A$, then there is a pullback diagram of the form

(8)

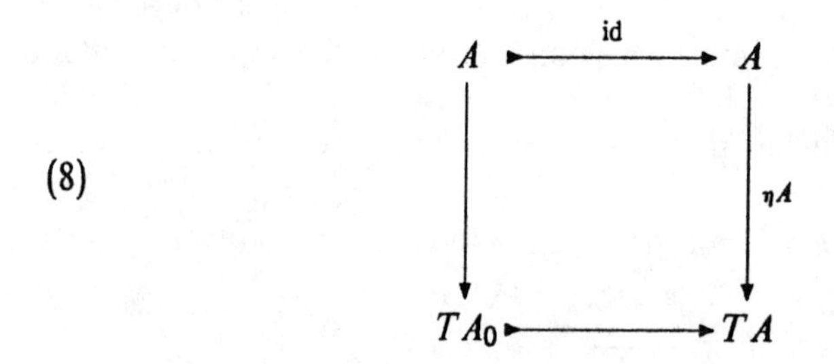

(Notice the subtle point here: $\mathbf{j}A(A_0)$ is defined *as a subobject of A*, which means that if it equals A the top arrow must be the identity). This gives the diagonal arrow in (7) and makes the lower triangle commute; but then the upper one does too since the bottom arrow is monic.

Conversely, if there is such a diagonal arrow, taking it as the left arrow in (8) is easily seen to make (8) a pullback, as required.

Theorem 6. *Given a left exact triple T in a topos $\mathcal{E}$, the maps $\mathbf{j}A$ defined above form a topology on $\mathcal{E}$ for which each object of the form TA is a $\mathbf{j}$-sheaf.*

Proof. That $\mathbf{j}$ is a natural transformation follows from Lemma 1 and Exercise (TPPB) of Section 2.3. Lemma 2 shows that $\mathbf{j}$ is inflationary, Lemma 3 that it is idempotent (because the diagrams corresponding to (1) for $\mathbf{j}A$ and $\mathbf{jj}A$ become the same) and Lemma 4 that it is order-preserving.

We use Proposition 8 of Section 6.2 to show that TA is a sheaf. Assume that B_0 is a dense subobject of an object B and $f : B_0 \to TA$ is given. We must find a unique extension $B \to TA$. This follows from the following diagram, in which g is the arrow given by Lemma 5.

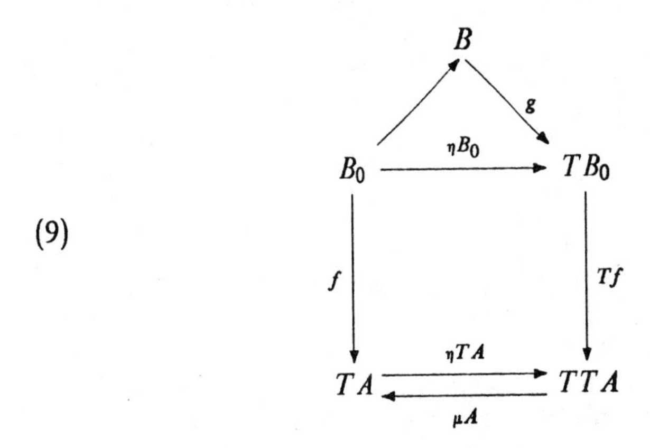

The required map is $\mu A \circ Tf \circ g$. It is straightforward to show that it is the unique map which gives f when preceded by the inclusion of B_0 in B.

Factorization of geometric morphisms

Now suppose that $U = U_* : \mathcal{E}' \to \mathcal{E}$ is a geometric morphism with inverse image map U^*. Like any adjoint pair, U determines a triple $\mathsf{T} = U_* \circ U^*, \eta, \mu)$ on $\mathcal{E}$. Let $\mathbf{j}$ be its topology induced as in Theorem 6, and $\mathcal{E}_\mathbf{j}$ the category of $\mathbf{j}$-sheaves.

Proposition 7. *For any object A of $\mathcal{E}'$, UA is a $\mathbf{j}$-sheaf.*

Proof. We again use Proposition 8 of Section 6.2. Suppose that $B_0 \rightarrowtail B$ is dense and $f : B_0 \to UA$. We get a diagram very much like (9):

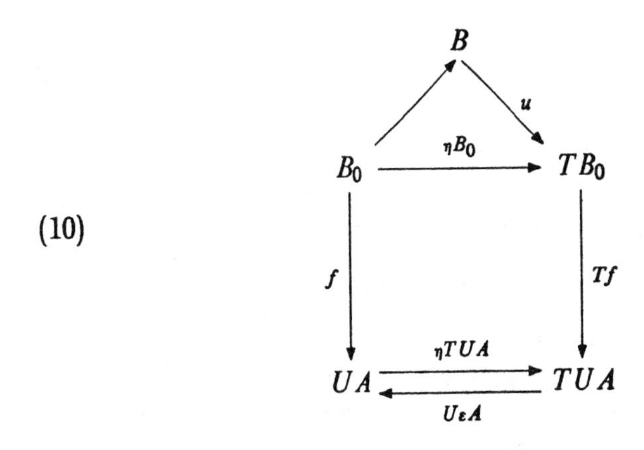

Here, u is given by Lemma 5 and $U\varepsilon A \circ \eta UA = \mathrm{id}$ by Exercise (Uco) of Section 1.9. It follows that $U\varepsilon A \circ Tf \circ u$ is the required arrow.

We now have a factorization of the given geometric morphism through $\mathcal{E}_j$ into two geometric morphisms.

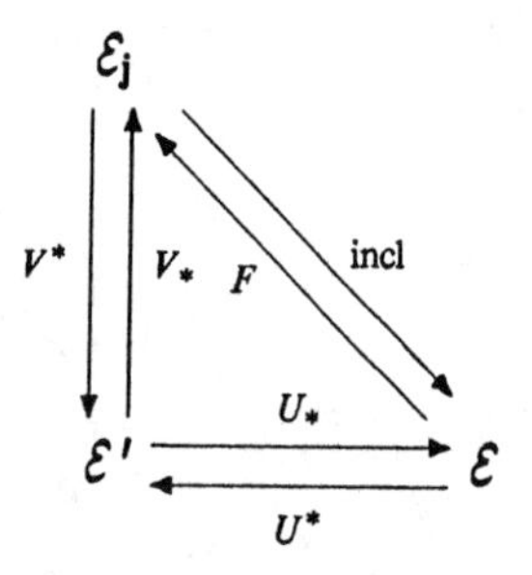

Here, V_* is U_* regarded as going into $\mathcal{E}_j$ and V^* is U^* composed with inclusion.

Theorem 8. *V^* is cotripleable.*

Proof. V^* is left exact by assumption, and so preserves all equalizers (and they exist because $\mathcal{E}_j$ is a topos). So all we need to show is that it reflects isomorphisms.

Suppose $f : A \to B$ in $\mathcal{E}_j$ is such that $V^*(f)$ is an isomorphism. In the following diagram, Δ is the diagonal map, $T' = V_* \circ V^*$, and d^0 and d^1 are the projections from the fiber product. All the vertical maps are components of the unit η corresponding to the adjunction of V^* and V_*.

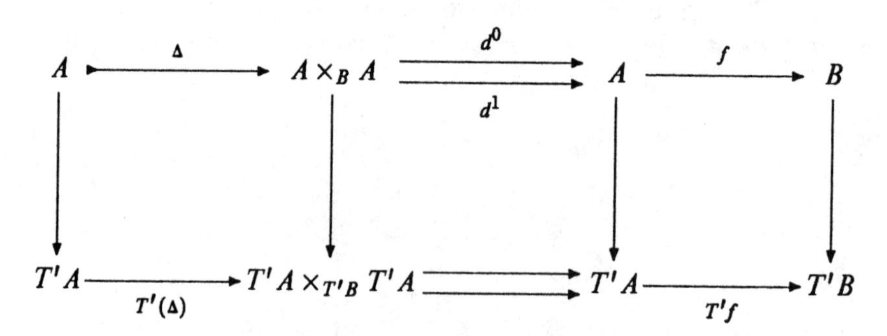

The composite across the top is f and $T'f$ is an isomorphism by assumption, so $T'd^0 = T'd^1$. This means $T'(\Delta)$ is an isomorphism. But A is a $\mathbf{j}$-sheaf, so it is $\mathbf{j}$-separated, meaning the left square is a pullback. (Note that as far as objects of $\mathcal{E}_j$ are concerned, T' is the triple determined by U and its left adjoint). That means that Δ is an isomorphism, so f is monic. *That* means that by the same argument the *right* square is a pullback, so f is an isomorphism.

Thus every sheaf category in a topos is the category of coalgebras for a left exact cotriple, and every geometric morphism is the composite of the cofree map for a left exact cotriple followed by the inclusion of its category of coalgebras as a sheaf category in the codomain topos.

Exercises 6.5

(SEP). Use Lemma 3 to given a direct proof that any object of the form TA is separated.

(ILET). If $\mathbf{j}$ is a topology on a topos $\mathcal{E}$, the inclusion of the category of sheaves in $\mathcal{E}$ and its left adjoint the sheafification functor F given by Theorem 9 of Section 6.2 produce a triple $\mathsf{T} = (T, \eta, \mu)$ in $\mathcal{E}$.

(a) Show that μ is an isomorphism. (A triple for which μ is an isomorphism is said to be **idempotent**).

(b) Show that the topology induced by that triple is $\mathbf{j}$.

(KEM). Show that the Kleisli and Eilenberg-Moore categories of an idempotent triple are equivalent.

6.6. Categories in a Topos

We will define category objects in a category $\mathcal{E}$ with finite limits by commutative diagrams as in Section 1.1. Functors between such category objects have a straightforward definition. What is more interesting is that functor from a category object C in $\mathcal{E}$ to $\mathcal{E}$ itself may be defined even though $\mathcal{E}$ is not itself a category object in $\mathcal{E}$. It turns out (as in $\mathcal{S}et$) that the category of such $\mathcal{E}$-valued functors is a topos when $\mathcal{E}$ is a topos.

Category objects

A **category object** in $\mathcal{E}$ is $C = (C, C_1, d^0, d^1, u, c)$, where C is the object of objects, C_1 is the object of morphisms, $d^0 : C_1 \to C$ is the domain map, $d^1 : C_1 \to C$ is the codomain map, $u : C \to C_1$ the unit map, and $c : C_2 \to C_1$ the composition. Here, C_2 is the fiber product $[(f, g) \mid d^0(f) = d^1(g)]$. In general, $C_n = [(f_1, \ldots, f_n) \mid d^1(f_i) = d^0(f_{i+1}), i = 1, \ldots, n-1]$, the object of composable n-tuples of maps of C. These objects and maps must satisfy the following laws:

(i) $d^0 \circ u = d^1 \circ u = \mathrm{id}_C$.

(ii) The following diagrams commute:

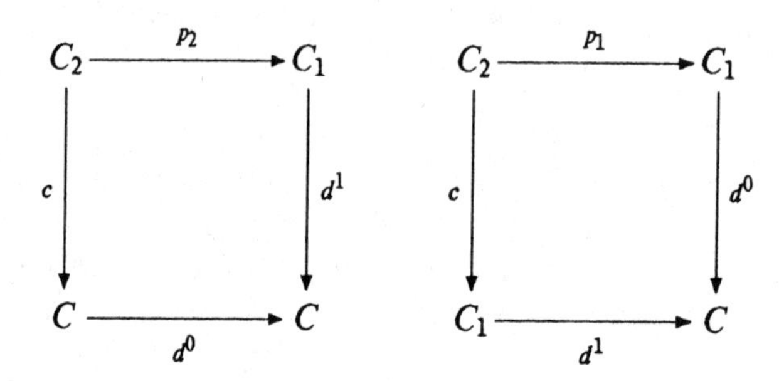

(in other words, $d^0(c(f,g)) = d^0(g)$ and $d^1(c(f,g)) = d^1(f)$ for all elements (f,g) of C_1),

(iii) $c(c \times \mathrm{id}) = c(\mathrm{id} \times c) : C_3 \to C_1$, and

(iv) $c(ud^1, \mathrm{id}) = c(\mathrm{id}, ud^0) = \mathrm{id} : C_1 \to C_1$.

In general, we will use the notation that when a letter denotes a category object, that letter with subscript 1 denotes its arrows. The maps d^0, d^1 and u will always be called by the same name.

An **internal functor** $F : \mathcal{C} \to \mathcal{D}$ between category objects of $\mathcal{E}$ is a pair of maps $F : C \to D$ and $F_1 : C_1 \to D_1$ which commutes with all the structure maps: $F \circ d^0 = d^0 \circ F_1$, $F \circ d^1 = d^1 \circ F_1$, $F_1 \circ u = u \circ F$, and $F_1 \circ c = c \circ (F_1 \times F_1)$. It is straightforward to show that the category of category objects and functors between them form a category $\mathcal{C}at\,(\mathcal{E})$.

$\mathcal{E}$-valued functors

There are two approaches to the problem of defining the notion of an $\mathcal{E}$-valued functor from a category object $\mathcal{C}$ in a topos $\mathcal{E}$ to $\mathcal{E}$. They turn out to be equivalent. One is a generalization of the algebraic notion of monoid action and the other is analogous to the topological concept of fibration. We describe each construction in $\mathcal{S}et$ first and then give the general definition.

The algebraic approach is to regard $\mathcal{C}$ as a generalized monoid. Then a $\mathcal{S}et$-valued functor is a generalization of the notion of monoid action. Thus if $F : \mathcal{C} \to \mathcal{S}et$ and $f : A \to B$ in $\mathcal{C}$, one writes fx for $Ff(x)$ when $x \in FA$ and shows that the map $(f,x) \to fx$ satisfies laws generalizing those of a monoid action: $(1_A)x = x$ and $(fg)x = f(gx)$ whenever fg and gx are defined. This map $(f,x) \to fx$ has a fiber product as domain: x must be an element of $d^0(f)$. Moreover, it is a map *over* C.

Guided by this, we say a **left** $\mathcal{C}$**-object** is a structure (A, φ, ψ) where $\varphi : A \to C$, and $\psi : C_1 \times_C A \to A$, where $C_1 \times_C A = [(g,a) \mid a \in A, g \in C_1$ and $\varphi(a) = d^0(g)]$, for which

(i) $\psi(u(\varphi(a)), a) = a$ for all elements a of A,
(ii) $\varphi\psi(g, a) = d^1(g)$ whenever $\varphi(a) = d^0(g)$, and
(iii) $\psi(c(f, g), a) = \psi(f, \psi(g, a))$ for all (f, g, a) for which $a \in A$, f, $g \in C_1$ and $\varphi(a) = d^0(g)$, $d^1(g) = d^0(f)$.

A morphism of left $\mathcal{C}$-objects is a map over C which commutes with φ and ψ in the obvious way. In $\mathcal{S}et$, given a functor $G : \mathcal{C} \to \mathcal{S}et$, A would be the disjoint union of all the values of G for all objects of $\mathcal{C}$, $\varphi(x)$ would denote the object C for which $x \in GC$, and $\psi(x, g)$ would denote $G(g)(x)$.

Contravariant functors can be handled by considering right $\mathcal{C}$-objects.

The other approach, via fibrations, takes the values of a functor $F : \mathcal{C} \to \mathcal{E}$ and joins them together in a category *over* $\mathcal{C}$. The result has a property analogous to the homotopy lifting property (Exercise (HLP)) and is a particular type of "opfibration". The general notion of opfibration (for $\mathcal{S}et$) is given in Exercise (OPF) and will not be used in this book (see Gray [1974]).

(The corresponding object for contravariant functors to $\mathcal{E}$—i.e. presheaves—is a "fibration." These ideas were discovered by Grothendieck, who was primarily interested in presheaves. What we call opfibrations he called cofibrations.)

The way this construction works in $\mathcal{S}et$ is this: Given $F : \mathcal{C} \to \mathcal{S}et$, construct the category $\mathcal{D}$ whose objects are the elements of the (disjoint) sets FC for objects C of $\mathcal{C}$. If you were explaining to someone the way F works, you might draw, for each element x of FC and $f : C \to C'$, an arrow from x to $Ff(x)$. These arrows are the arrows of $\mathcal{D}$. They compose in the obvious way, and there is an obvious map from $\mathcal{D}$ to $\mathcal{C}$. Then for $C \in \mathcal{O}b(\mathcal{C})$, FC is the inverse image of C under that map.

This has to be approached more indirectly in a topos. Given a topos $\mathcal{E}$ and a category object $\mathcal{C}$ of $\mathcal{E}$, a morphism $\varphi : \mathcal{D} \to \mathcal{C}$ of category objects is a **split discrete opfibration if**

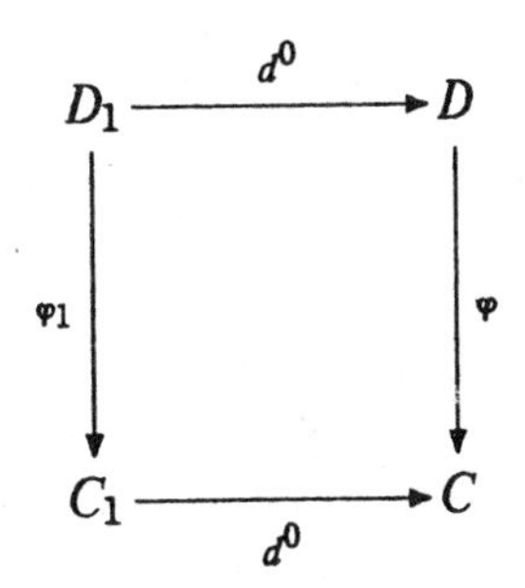

is a pullback. This says that the set of arrows of $\mathcal{D}$ is exactly the set

$$[(g, d) \mid g \in C_1, d \in D, \text{ and } d^0(g) = \varphi(d)].$$

Furthermore, the identification as a pullback square means that $d^0(g,d) = d$ (because the top arrow must be the second projection) and similarly $\varphi_1(g,d) = g$ (hence $\varphi(d^1(g,d)) = d^1(g)$).

It follows that for each object d of $\mathcal{D}$ and each arrow g out of $\varphi(d)$ in $\mathcal{C}$, there is exactly one arrow of $\mathcal{D}$ over g with domain d; we denote this arrow (g,d). This property will be referred to as the **unique lifting property** of opfibrations.

A split discrete opfibration over $\mathcal{C}$ is thus an object in $\mathcal{E}/\mathcal{C}$; we define a morphism of split discrete opfibrations over $\mathcal{C}$ to be just a morphism in $\mathcal{E}/\mathcal{C}$.

Proposition 1. *Let $\mathcal{E}$ be a left exact category and $\mathcal{C}$ a category object of $\mathcal{E}$. Then the category of split discrete opfibrations over $\mathcal{C}$ is equivalent to the category of left $\mathcal{C}$ objects. When $\mathcal{E} = \mathbf{Set}$, they are equivalent to the functor category $\mathbf{Set}^{\mathcal{C}}$.*

Proof. Suppose $\varphi : \mathcal{D} \to \mathcal{C}$ is a split discrete opfibration. Then $D_1 = C_1 \times_C D$ and $d^1 : D_1 \to D$. We claim that (D, φ, d^1) is a left $\mathcal{C}$ action. All the verifications, including that morphisms of split discrete opfibrations are taken to morphisms of left $\mathcal{C}$ actions, make use of the unique lifting property. We show two of the required properties and leave the others to you.

We show first that $d^1(u(\varphi d), d) = d$. Observe that $\varphi_1(u(\varphi d), d) = u(\varphi d) = \varphi_1(ud)$ and $d^0(u(\varphi d), d) = d = d^0(ud)$. Therefore by the unique lifting property, $ud = (u(\varphi d), d)$. The result follows from the fact that $d^1(ud) = d$.

We also need

$$d^1(c(f,g),d) = d^1(f, d^1(g,d)),$$

where $d^0(f) = d^1(g)$ and $d^0(g) = \varphi d$. Now the arrows $(c(f,g),d)$ and $c((f, d^1(g,d)), (g,d))$ (composition in $\mathcal{D}$) both have domain d and lie over $c(f,g)$. Therefore they are the same arrow, so d^1 of them is the same.

Going the other way, suppose that (A, φ, ψ) is a left $\mathcal{C}$-object. Let $D = C_1 \times_C A = [(g,a) \mid d^0(g) = \varphi(a)]$. Then the top part of the

following serially commutative diagram

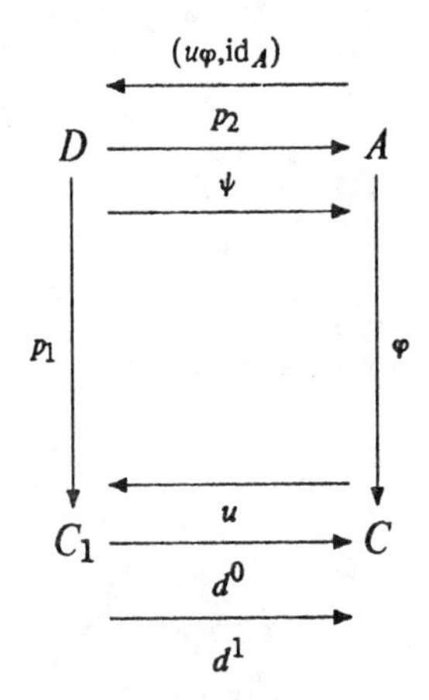

is a category object with composition taking $((g, a), (g', a'))$ to $(c(g, g'), a)$, and φ is a morphism of category objects. The verification of all the laws is tedious but straightforward. It is then immediate from the definition of D that (φ, p_1) is an split discrete opfibration.

Suppose (A, φ, ψ) is a left C-object in $\mathcal{S}et$. Then define a functor $F : C \to \mathcal{S}et$ by requiring that for an object C of C, $FC = \varphi^{-1}(C)$. If $x \in FC$ and $g : C \to D$, set $Fg(x) = \psi(g, x)$. If (A', φ', ψ') is another left C-object and $\lambda : A \to A'$ is a morphism, then λ corresponds to a natural transformation between the corresponding functors whose component at C is the restriction of λ to FC (which is a subset of A). That this construction gives an equivalence between the category of left C-objects and $\mathcal{S}et^{C}$ follows directly from (i)-(iii) in the definition of left C-object and the definition of morphism.

Left C-objects form a topos

We will now bring in heavy artillery from several preceding sections to show that the category of left C-objects and their morphisms form a topos.

Let C be a category object in the topos $\mathcal{E}$ and let $T : \mathcal{E}/C \to \mathcal{E}/C$ be the functor which takes $A \to C$ to $C_1 \times_C A \to C$ and $f : A \to B$ over C to $C_1 \times_C f$. Let η be the natural transformation from $\mathrm{id}_{\mathcal{E}}$ to T whose component at an object $s : A \to C$ is $\eta(s : A \to C) = (us, \mathrm{id}_A)$. Let $\mu : T^2 \to T$ be the natural transformation whose component at $A \to C$ is (c, id_A). It then follows directly from the definition of left C-objects and morphisms thereof that (T, η, μ) is a triple and that the category of Eilenberg-Moore algebras $(\varphi : A \to C, \psi)$ is the category of left C-objects.

Now come the heavy cannon. By the constructions of Theorem 6 and Corollary 7 of Section 5.3, T factors as the top row of

$$\mathcal{E}/C \underset{\prod d^0}{\overset{(d^0)^*}{\rightleftarrows}} \mathcal{E}/C_1 \underset{(d^1)^*}{\overset{\sum d^1}{\rightleftarrows}} \mathcal{E}/C$$

where the bottom row forms a right adjoint to T. Thus by Theorem 5 of Section 3.7, T is adjoint tripleable and the category of left C-objects is the category of coalgebras for a cotriple whose functor is $G = \prod d^0 \circ (d^1)^*$. G has a left adjoint, so gives a left exact cotriple. Hence by Theorem 1 of Section 6.4, we have:

Theorem 2. *For any category object C of a topos $\mathcal{E}$, the category of left C-objects (equivalently the category of split discrete opfibrations of C) is a topos.*

It is natural to denote the category of left C-objects by $\mathcal{E}^C$.

Theorem 3. *Any category object functor $f : C \to C'$ induces a functor $f^\# : \mathcal{E}^{C'} \to \mathcal{E}^C$ which is the restriction of the pullback functor f^*. Moreover, $f^\#$ has left and right adjoints.*

Proof. That pulling back a split discrete opfibration produces a split discrete opfibration is merely the statement that pulling back a pullback gives a pullback. The reader may alternatively define $f^\#$ on left C'-objects by stipulating that $f^\#(A', \varphi', \psi') = (A, \varphi, \psi)$, where $A = [(c, a') \mid fc = \varphi' a']$, φ is the first projection, and $\psi(g, (c, a'))$ (where necessarily $d^0(g) = c$ and $\varphi a' = fc$) must be $d^1(g), \psi'(f(g), a'))$.

The square

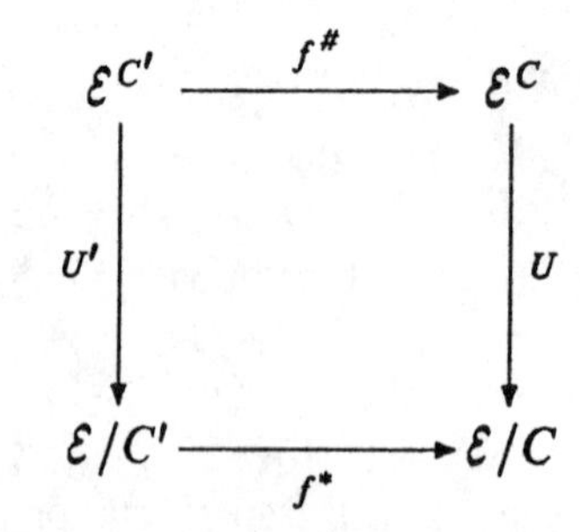

commutes, where the vertical arrows are the underlying functors from triple algebras. (They are in fact inclusions). Since the bottom arrow has both a left and a right adjoint, so does the top one by Butler's Theorems (Section 3.7, Theorem 3).

When C' is the trivial category object with $C' = C'_1 = 1$, then $\mathcal{E}^{C'}$ is $\mathcal{E}$ and the induced map $\mathcal{E} \to \mathcal{E}^{C}$ (where C is some category object) has left and right adjoints denoted $\lim_{\to} C$ and $\lim_{\leftarrow} C$ respectively. When $\mathcal{E}$ is $\mathcal{S}et$ they are in fact the left and right Kan extensions. Notice that this says that, given a set function $f : Y \to I$, then the existence of $\prod f^{-1}(i)$ (which is a completeness property—Y and I can both be infinite) depends only on the fact that in $\mathcal{S}et$ we have *finite* limits and a power object. Thus the existence of $\mathbf{P}$ is a powerful hypothesis.

Exercises 6.6

(Cso). A **simplicial object** in a category $\mathcal{E}$ is a functor $S : \Delta^{\text{op}} \to \mathcal{E}$ where Δ is the category whose objects are the finite sets $\{1, 2, \ldots, n\}$ for $n = 0, 1, 2, \ldots$ and whose arrows are the order-preserving maps between these sets. Prove that the category of category objects and functors between them in a left exact category $\mathcal{E}$ is equivalent to a full subcategory of the category of simplicial objects in $\mathcal{E}$.

(Opf). Let $F : \mathcal{D} \to \mathcal{C}$ be a functor between categories (in $\mathcal{S}et$). If A is an object of $\mathcal{C}$, the **fiber** $\mathcal{D}_A$ over $\mathcal{C}$ is the subcategory of $\mathcal{D}$ consisting of all objects mapping to A and all arrows mapping to 1_A. Its inclusion into $\mathcal{D}$ is denoted J_A. F is an **opfibration** if for each $f : A \to \mathcal{B}$ in $\mathcal{C}$ there is a functor $f^* : \mathcal{D}_A \to \mathcal{D}_B$ and a natural transformation $\theta_f : J_A \to J_B \circ f^*$ with every component lying over f, for which for any arrow $m : D \to E$ in $\mathcal{D}$ lying over f there is a unique arrow $f^*(D) \to E$ making

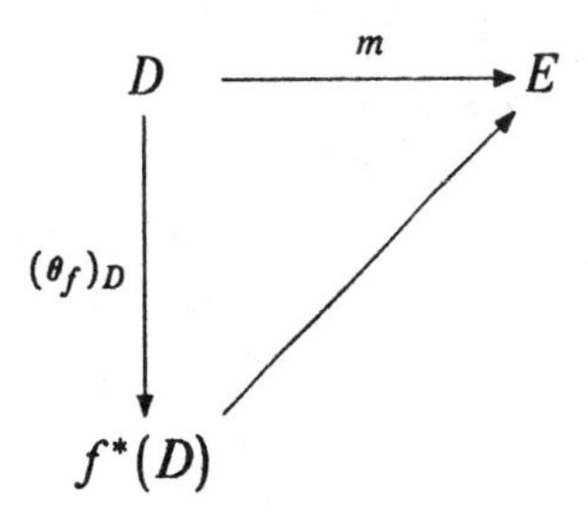

commute. The opfibration is **split** if $(1_A)^* = 1_{(\mathcal{D}_A)}$ and $g^* \circ f^* = (g \circ f)^*$ whenever $g \circ f$ is defined. It is **discrete** if the fibers $\mathcal{D}_A$ are all sets—i.e., their only arrows are identity arrows. ("Split" is also called "split normal," the normal referring to preservation of identities). A functor between opfibrations over $\mathcal{C}$ is a functor in $\mathcal{C}at/\mathcal{C}$ (it does *not* have to preserve $*$).

(a) Show that, for $\mathcal{S}et$, a split discrete opfibration as defined in the text is the same as that defined here.

(b) Show that the subcategory of split opfibrations over $\mathcal{C}$ and functors which commute with * is equivalent to the category of functors from $\mathcal{C}$ to $\mathcal{C}at$ and natural transformations.

(c) Show that opfibrations have the "homotopy lifting property": If G and H are functors from $\mathcal{A}$ to $\mathcal{C}$, $F : \mathcal{D} \to \mathcal{C}$ is an opfibration, $\lambda : G \to H$ is a natural transformation, and $G = F \circ G'$ for some functor $G' : \mathcal{A} \to \mathcal{D}$, then there is a functor $H' : \mathcal{A} \to \mathcal{D}$ such that $H = F \circ H'$ and a natural transformation $\lambda' : G' \to H'$.

(OPFC). (Categorical definition of opfibration). Let $F : \mathcal{D} \to \mathcal{C}$ be a functor. Let $\mathbf{2}$ denote the category with two objects 0 and 1 whose only nonidentity arrow u goes from 0 to 1. Let S be the functor from $\mathrm{Hom}(\mathbf{2}, \mathcal{D})$ to the comma category $(F, 1_{\mathcal{C}})$ which takes $M : \mathbf{2} \to \mathcal{D}$ to $(M(0), F(M(u)), F(M(1)))$ (you can figure out what it does to arrows of $\mathrm{Hom}(\mathbf{2}, \mathcal{D})$, which are natural transformations). Show that F is an opfibration if and only if S has a left adjoint which is also a right inverse of S.

6.7. Grothendieck Topologies

A Grothendieck topology on a category is a generalization of the concept of all open covers of all open sets in a topological space.

A **sieve** (called "crible" by some authors—"crible" is the French word for "sieve") on an object A is a family of arrows with codomain A. We will use the notation $\{A_i \to A\}$ for a sieve, the i varying over an unspecified index set. We will follow the convention that different sieves have possibly different index sets even if the same letter i is used, unless specifically stated otherwise.

The set of fiber products $\{A_i \times_A A_j\}$, where i and j both run over the index set of the sieve, will be used repeatedly in the sequel. If $f : A \to B$ and $\{A_i \to A\}$ is a sieve on A, we write $f|A_i$ for the composite of the projection $A_i \to A$ followed by f, and $f|A_i \times_A A_j$ for the composite of $A_i \times_A A_j \to A$ followed by f.

A sieve $\{A_i \to A\}$ **refines** a sieve $\{B_j \to A\}$ if every arrow in the first factors through at least one arrow in the second.

A **Grothendieck topology** in a left exact category $\mathcal{A}$ is a family of sieves, called **covers**, with the following properties.

(i) For each object A of $\mathcal{A}$, id_A is a sieve.

(ii) (Stability) If $\{A_i \to A\}$ is a cover and $B \to A$ is an arrow, then $\{B \times_A A_i \to B\}$ is a cover.

(iii) (Composability) If $\{A_i \to A\}$ is a cover, and for each i, $\{A_{ij} \to A_i\}$ is a cover, then $\{A_{ij} \to A\}$ is a cover.

A Grothendieck topology is **saturated** if whenever $\{A_i \to A\}$ is a sieve and for each i, $\{A_{ij} \to A_i\}$ is a sieve for which $\{A_{ij} \to A\}$ (doubly indexed!) is a cover, then $\{A_i \to A\}$ is a cover. It is clear that each Grothendieck topology is contained in a unique saturated Grothendieck topology. It follows from (ii) (stability) that the saturation contains all the sieves which have a refinement in the original topology.

Refinement is also defined for topologies: one topology refines another if every cover in the saturation of the second is in the saturation of the first.

A **site** is a left exact category together with a specific saturated Grothendieck topology. A **morphism of sites** is a left exact functor between sites which takes covers to covers.

Some examples of sites:

(a) The category of open sets of a fixed topological space together with all the open covers of all open sets.

(b) Any left exact category together with all the universal regular epimorphisms, each regarded as a sieve containing a single arrow. ("Universal" or "stable" means preserved under pullbacks.)

(c) A sieve $\{f_i : A_i \to A\}$ is an epimorphic family if and only if whenever $g \neq h$ are two maps from A to B, there is at least one index i for which $g \circ f_i \neq h \circ f_i$. If epimorphic families are stable under pullbacks, they form a site.

If $\mathcal{C}$ is any category, an object of $\mathcal{S}et^{\mathcal{C}^{\mathrm{op}}}$ is called a **presheaf**, a terminology used particularly when $\mathcal{C}$ is a site. If $f : A_1 \to A_2$ in $\mathcal{A}$ and $a \in A_2$, then, motivated by the discussion of sheaves on a topological space in Section 2.2, we write $a|A_1$ for $Ff(a)$. (We recommend this notation—it makes the theory much more manageable.)

Let $S = \{A_i \to A\}$ be a sieve in a category $\mathcal{C}$. If $F : \mathcal{C}^{\mathrm{op}} \to \mathcal{S}et$ is a presheaf, we say that F is S**-separated** if $FA \to \prod FA_i$ is injective and that F is an S**-sheaf** if

$$FA \longrightarrow \prod FA_i \rightrightarrows \prod F(A_i \times_A A_j)$$

is an equalizer. This generalizes the sheaf condition for topological spaces given by Proposition 1, Section 2.2. Stated in terms of restrictions, F is S-separated if whenever $a, a' \in FA$ with the property that for every i, $a|A_i = a'|A_i$, then $a = a'$. F is an S-sheaf if in addition for every tuple of elements $a_i \in A_i$ with the property that $a_i|A_i \times_A A_j = a_j|A_i \times_A A_j$, there is a (unique) element $a \in FA$ such that $a|A_i = a_i$.

It is straightforward to see that a sieve S is an epimorphic family if every representable functor $\mathrm{Hom}(-, B)$ is S-separated, and we say that S is a

regular or **effective epimorphic family** if every representable functor is an S-sheaf. Both epimorphic and regular epimorphic families are called **stable** or **universal** if they remain epimorphic (respectively regular epimorphic) when pulled back.

In any category, stable epimorphic families, and also stable regular epimorphic families, form a Grothendieck topology.

Note that since every epimorphism in a topos is universal and regular by Corollary 5 of Section 5.4 and Proposition 3 of Section 5.5, the class of all epimorphisms in a topos forms a Grothendieck topology on the topos. (See Exercise (CANON)). However, epimorphic families in a topos need not be regular unless the topos has arbitrary sums.

A site $\mathcal{A}$ determines a topology $\mathbf{j}$ on $\mathcal{Set}^{\mathcal{A}^{\text{op}}}$ as follows. If F is a presheaf in $\mathcal{Set}^{\mathcal{A}^{\text{op}}}$ and F_0 is a subpresheaf, then $\mathbf{j}F_0 = \mathbf{j}F(F_0)$ is the functor whose value at A consists of all $a \in FA$ for which there is a cover $\{A_i \to A\}$ such that $a|A_i \in F_0A_i$ for all i.

Proposition 1. $\mathbf{j}$ *as constructed above is a topology on* $\mathcal{Set}^{\mathcal{A}^{\text{op}}}$.

Proof. (i) Naturality translates into showing that if the left square below is a pullback then so is the right one.

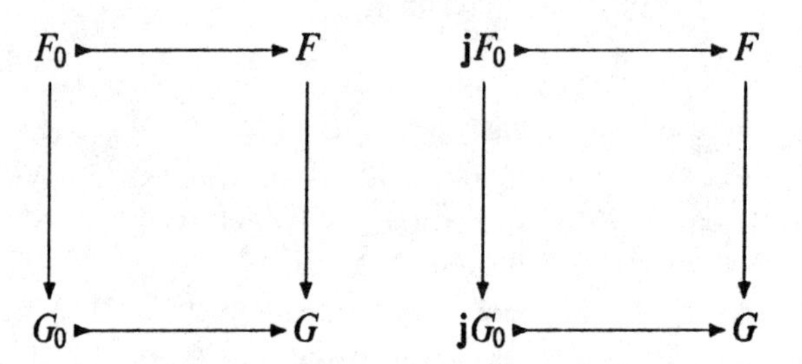

Suppose $a \in \mathbf{j}G_0A$, $b \in FA$ have the same image in GA. We must show $b \in \mathbf{j}F_0A$. Let $\{A_i \to A\}$ witness that $a \in \mathbf{j}G_0(A)$. In other words, $a|A_i \in G_0A_i$ for all i. Now consider this pullback diagram:

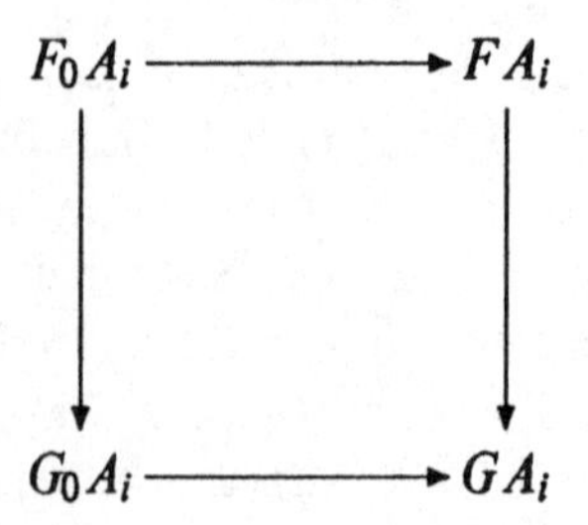

We know $a|A_i \in G_0A_i$ and that $b|A_i \in FA_i$ for all i. They must have the same image in GA_i, so $b|A_i \in F_0A_i$. Therefore $b \in \mathbf{j}F_0(A)$.

(ii) The inflationary property follows from the fact that id_A is a cover.

(iii) The monotone property is a trivial consequence of the definition.

(iv) Idempotence follows from the composition property for covers.

Conversely, given a topology $\mathbf{j}$ on $\mathcal{S}et^{\mathcal{A}^{\mathrm{op}}}$, we can construct a Grothendieck topology on $\mathcal{A}$ as follows. Any sieve $\{A_i \to A\}$ determines a subfunctor R of $\mathrm{Hom}(-, A)$ defined for an object B by letting RB be the set of $f : B \to A$ for which for some i, there is a factorization $B \to A_i \to A$ of f. This is the same as saying R is the union of the images of $\mathrm{Hom}(-, A_i)$ in $\mathrm{Hom}(-, A)$. Then we say $\{A_i \to A\}$ is a covering sieve for the Grothendieck topology if R is $\mathbf{j}$-dense.

Proposition 2. *For any topology* $\mathbf{j}$, *the definition just given produces a Grothendieck topology on* $\mathcal{A}$.

The proof is straightforward and will be omitted.

The two constructions given above produce a one to one correspondence between saturated Grothendieck topologies on $\mathcal{A}$ and topologies on $\mathcal{S}et^{\mathcal{A}^{\mathrm{op}}}$. We make no use of this fact and the proof is uninteresting, so we omit it.

Proposition 3. *Let* $\mathcal{A}$ *be a site and* $\mathbf{j}$ *the corresponding topology on* $\mathcal{S}et^{\mathcal{A}^{\mathrm{op}}}$. *Then a presheaf* F *is a* $\mathbf{j}$*-sheaf if and only if it is an* S*-sheaf for every sieve* S *of the topology.*

Proof. Let F be an S-sheaf for every cover S of the site. Let $G_0 \to G$ be a dense inclusion of presheaves. We must by Theorem 8 of Section 6.2 construct for any $\alpha_0 : G_0 \to F$ a unique extension $\alpha : G \to F$.

Let A be an object of $\mathcal{A}$. Since G_0 is dense in G, for all $a \in GA$ there is a cover $\{A_i \to A\}$ with the property that $a|A_i \in G_0A_i$ for each i. Thus $\alpha_0A(a|A_i) \in FA_i$ for each i. Moreover,

$$(a|A_i)|A_i \times_A A_j = a|A_i \times_A A_j = (a|A_j)|A_i \times_A A_j,$$

so applying α_0A, it follows that $\alpha_0A(a|A_i) \in FA$ and thus determines the required arrow α.

To prove the converse, first note that the class of covers in a Grothendieck topology is filtered with respect to refinement: if $\{A_j \to A\}$ and $\{A_k \to A\}$ are covers then $\{A_j \times_A A_k \to A\}$ is a cover refining them.

Now suppose F is a sheaf. For each object A, define F^+A to be the colimit over all covers of A of the equalizers of

$$\prod FA_i \rightrightarrows \prod F(A_i \times_A A_j)$$

It is easy to see that F^+ is a presheaf. We need only show that $F = F^+$. To do that it is sufficient to show that F^+ is separated and $F \to F^+$ is dense. The latter is obvious. As for the former, it follows from the definition of $\mathbf{j}$ that F^+ is separated if and only if whenever $\{A_i \to A\}$ is a cover and $a_1, a_2 \in F^+A$ with $a_1|A_i = a_2|A_i$ for all i, then $a_1 = a_2$. Since F is a sheaf, F^+ satisfies this condition.

Now suppose $a_1, a_2 \in F^+A$ and there is a cover $\{A_i \to A\}$ for which $a_1|A_i = a_2|A_i$ for all i. (In this and the next paragraph, restriction always refers to F^+). By definition of F^+, there is a cover $\{A_j \to A\}$ for which $a_1|A_j \in FA_j$ and a cover $\{A_k \to A\}$ for which $a_2|A_k \in FA_k$ for all k.

Next let $\{A_l \to A\}$ be a cover simultaneously refining $\{A_i \to A\}$, $\{A_j \to A\}$ and $\{A_k \to A\}$. Then $a_1|A_l = a_2|A_l$ for all l because $\{A_l \to A\}$ refines $\{A_i \to A\}$. $a_1|A_l \in FA_l$ and $a_2|A_l \in FA_l$ because $\{A_l \to A\}$ refines both $\{A_j \to A\}$ and $\{A_k \to A\}$. Thus the two elements are equal in F^+A because they are equal at a node in the diagram defining F^+A. Hence F^+ is separated, as required.

Special types of Grothendieck topologies

A topology for which every representable functor is a sheaf is called **standard**, or **subcanonical**. Note that for a standard topology, if L is sheafification and Y is the Yoneda embedding, then $LY = Y$. Thus when a site is standard, it is fully embedded by y into its own category of sheaves. This embedding plays a central role in the proof of Giraud's Theorem in Section 6.8 and in the construction of cocone theories in Chapter 8.

The **canonical** Grothendieck topology on a left exact category is the finest (most covers) subcanonical topology.

Proposition 4. *The following are equivalent for a Grothendieck topology.*

(a) *The covers are effective epimorphic families.*
(b) *The representable functors are sheaves for the topology.*
(c) *The functor $y = LY$, where L is sheafification, is full and faithful.*

Proof. (a) implies (b) is clear from the definitions. If (b) is true, $LY = Y$ and Y is full and faithful, so (c) is true.

If (c) is true, for a given cover let R be the subfunctor of YA constructed above. Then R is j-dense in YA. Apply L to the diagram

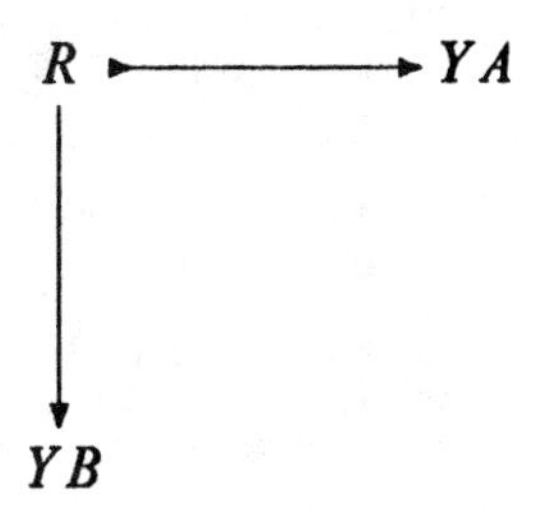

Then the top line becomes equality so we have, from (c),

$$\begin{aligned}\mathrm{Hom}(LYA, LYB) &\cong \mathrm{Hom}(A, B)\\ &\cong \mathrm{Hom}(YA, YB) \to \mathrm{Hom}(R, YB) \to \mathrm{Hom}(LR, LYB)\\ &\cong \mathrm{Hom}(LYA, LYB)\end{aligned}$$

with composites all around the identity. Hence (a) is true.

Exercises 6.7

(Mty)°. Prove that the empty sieve is an effective epimorphic family over the initial object, and that it is universally so if and only if the initial object is strict.

(Epi)°. Show that the following are equivalent for a Grothendieck topology.

(a) The covers are epimorphic families.
(b) The representable functors are separated presheaves for the topology.
(c) The functor $y = LY$, where L is sheafification, is faithful.

6.8. Giraud's Theorem

Any topos $\mathcal{E}$ is a site with respect to the canonical topology. Giraud's Theorem as originally stated says that certain exactness conditions on a category plus the requirement that it have a set of generators (defined below) are equivalent to its being the category of sheaves over a *small* site. It is stated in this way, for example, in Makkai and Reyes [1977], p. 53 or in Johnstone [1977], p. 17.

Our Theorem 1 below is stated differently, but it is in essence a strengthening of Giraud's Theorem.

If $\mathcal{C}$ is a category, a subset U of the objects of $\mathcal{C}$ is a set of **(regular) generators** for $\mathcal{C}$ if for each object C of $\mathcal{C}$ the set of all morphisms from objects of U to C forms a (regular) epimorphic family. We say that U is a **conservative** generating family if for any object C and proper subobject $C_0 \subsetneqq C$, there is an object $G \in U$ and an element $c \in^G C$ for which $c \notin^G C_0$. See Exercise (GEN) for the relationships among these conditions.

Recall from Exercise (RGFAC) of 5.5 that a regular category is a category with finite limits in which regular epis are stable under pullback.

A **Grothendieck topos** is a category which

(i) is complete;
(ii) has all sums and they are disjoint and universal;
(iii) is regular with effective equivalence relations; and
(iv) has a small set of regular generators.

We do not yet know that a Grothendieck topos is a topos, but that fact will emerge.

Theorem 1 (Giraud). *The following are equivalent:*

(a) *$\mathcal{E}$ is a Grothendieck topos;*
(b) *$\mathcal{E}$ is the category of sheaves for a small site;*
(c) *$\mathcal{E}$ is a topos with arbitrary sums and a small set of generators.*

Proof. That (b) implies (c) is clear because a sheaf category is a reflective subcategory of a functor category. Earlier results, taken together with Exercise (GEN), show that (c) implies (a). For example, we have shown in Theorem 7 of 2.3 that equivalence relations are effective. In Corollary 8 of 5.3 we showed that pullbacks have adjoints from which the disjoint stable sums and epis follow.

The proof that (a) implies (b) is immediate from the following proposition.

Proposition 2. *Let $\mathcal{E}$ be a Grothendieck topos, and $\mathcal{C}$ a left exact subcategory containing a (regular) generating family of $\mathcal{E}$, regarded as a site on the topology of all sieves which are regular epimorphic families in $\mathcal{E}$. Then $\mathcal{E}$ is equivalent to $\mathsf{Sh}(\mathcal{C})$. Moreover, if $\mathcal{C}$ is closed under subobjects, then the topology on $\mathcal{C}$ is canonical.*

Proof. The proof consists of a succession of lemmas, in which $\mathcal{E}$ and $\mathcal{C}$ satisfy the hypotheses of the Proposition. We begin by showing that the regular epimorphic families form a topology.

Lemma 3. *Every regular epimorphic family in a Grothendieck topos is stable.*

Proof. A regular epimorphic family $\{E_i \to E\}$ in a category with all sums is characterized by the fact that

$$\sum(E_i \times_E E_j) \rightrightarrows \sum E_i \longrightarrow E$$

is a coequalizer. Given the stable sums, this is equivalent to the assertion that

$$(\sum E_i) \times_E (\sum E_i) \rightrightarrows \sum E_i \longrightarrow E$$

is a coequalizer, which implies that $\sum E_i \to E$ is a regular epi. Since both sums and regular epis are stable, so is this condition.

In the rest of this section, we use Y, L, and $y = LY$ as defined in 6.7. We show eventually that y is left exact and cocontinuous and deduce that it is an equivalence.

Corollary 4. *For every object E of $\mathcal{E}$, $Y(E)$ is a sheaf.*

Proof. Since every cover is a regular epimorphic family in $\mathcal{E}$, it follows from the definition of covers that representable functors are sheaves.

Lemma 5. $y(0) = 0$.

Proof. The stable sums in $\mathcal{E}$ imply that 0 is a strict initial object. For any object C of $\mathcal{C}$, $Y(0)(C) = \mathrm{Hom}(C, 0) = \emptyset$ unless $C = 0$ in which case it is a singleton. If 0 is not an object of $\mathcal{C}$, this is the constantly null functor which is initial in the functor category, so that its associated sheaf $y(0)$ is initial in the sheaf category. If 0 is an object of $\mathcal{C}$, then $y(0)$ can readily be seen to be initial, from the definition of initial, as soon as we observe that $F(0) = 0$ for every sheaf F. Since 0 is covered by the empty sieve, we have, for any sheaf F, an equalizer

$$F(0) \longrightarrow \prod(\,) \rightrightarrows \prod(\,)$$

in which the empty products have a unique element so that $F(0)$ does too.

Lemma 6. *y preserves sums.*

Proof. We must show that if $E = \sum E_i$, then $\sum YE_i \to YE$ is a dense inclusion (in the presheaf category) so that the sheaf associated to the first is the second. It is an inclusion because $E_i \times_E E_j = 0$, the injections into the sum are mono, and Y preserves limits. Evaluating this at C in $\mathcal{C}$ (remember that Y is Hom-functor-valued), we must show that for any $f : C \to E$, there is a cover $\{C_k \to C\}$ such that $f|C_k \in \sum \mathrm{Hom}(C_k, E_i)$. This means that for each k there is an i with $f|C_k \in \mathrm{Hom}(C_k, E_i)$.

Given f, let

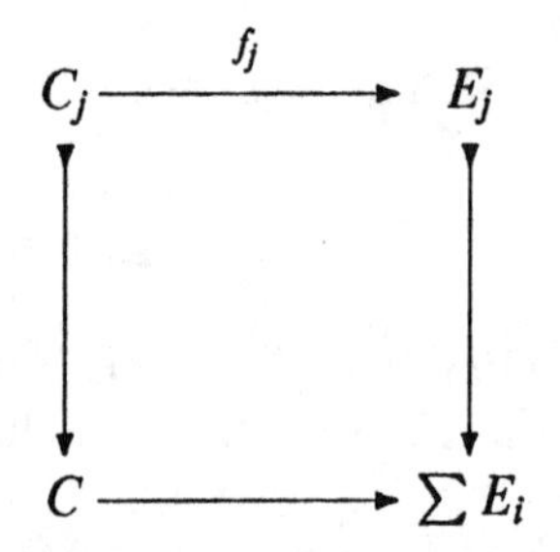

be a pullback. The objects C_j may not be in $\mathcal{C}$, but for each one there is a regular epimorphic family (hence cover) $\{C_{jk} \to C_j\}$ with all C_{jk} in $\mathcal{C}$ because $\mathcal{C}$ contains a generating family. (Observe that the second subscript k varies over an index set which depends on the first subscript j). Since $C = \sum C_j$, $\{C_j \to C\}$ is a cover of C. Hence the doubly indexed sieve $\{C_{jk} \to C\}$ is a cover of C in $\mathcal{C}$. Furthermore, $f|C_{jk} = f_j|C_{jk}$, which is in $\mathrm{Hom}(C_{jk}, E_j)$.

Lemma 7. *y preserves regular epis.*

Proof. The argument is similar to the above. Given $E' \twoheadrightarrow E$ an element $f \in \mathrm{Hom}(C, E)$ will not necessarily lift to an element of $\mathrm{Hom}(C, E')$, but will do so on a cover of C, namely a cover $\{C_i \to C'\}$ by objects in $\mathcal{C}$ of the object C' gotten by pulling f back

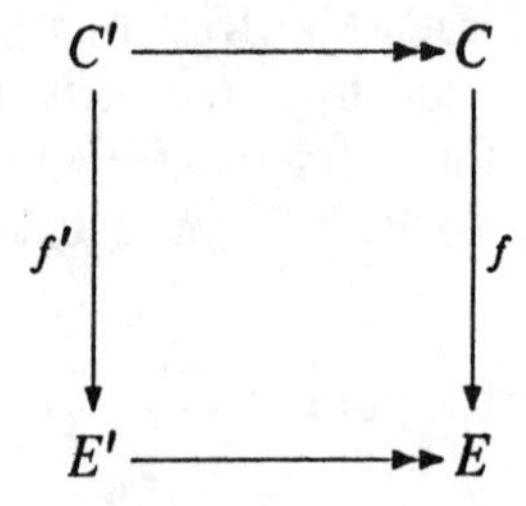

Lemma 8. *y preserves coequalizers.*

Proof. It preserves limits because L and Y do. We also know that it preserves sums and images and hence unions (even infinite unions). The constructions used in Exercise (GEQ) are therefore all preserved by y, so y preserves the construction of equivalence relations. Thus if

$$A \underset{k}{\overset{h}{\rightrightarrows}} B \xrightarrow{c} C$$

is a coequalizer, y takes the kernel pair of c (which is the equivalence relation generated by h and k) to the kernel pair of yc. Since yc is regular epi, it is the coequalizer of yh and yk.

Lemma 9. *y is full and faithful.*

Proof. The faithfulness follows immediately from the fact that $\mathcal{C}$ contains a set of regular generators for $\mathcal{E}$ and that the covers are regular epimorphic families. As for the fullness, the definition of y implies that when C is an object of $\mathcal{C}$, $\mathrm{Hom}(yC, yE) \cong \mathrm{Hom}(C, E)$. The universal property of colimits, together with the fact that these are preserved by y allows one to extend this conclusion easily to the case that C belongs to the colimit closure of $\mathcal{C}$ which is $\mathcal{E}$.

Lemma 10. *Every sheaf F is a coequalizer of a diagram of the kind*

$$\Sigma(yC'_k) \rightrightarrows \Sigma(yC_i) \longrightarrow F$$

Proof. This is a trivial consequence of the fact that every functor, hence every sheaf is a colimit of a diagram of representables.

Lemma 11. *If $\mathcal{C}$ is closed in $\mathcal{E}$ under subobjects, then every cover in $\mathcal{C}$ is also a cover in $\mathcal{E}$.*

Proof. Let $\{C_i \to C\}$ be a cover in $\mathcal{C}$. As seen in Exercise (RGFAC) of 5.5, a regular category has a factorization system using regular epis and monos. The regular image in $\mathcal{E}$ of $\sum C_i \to C$ is a subobject $C_0 \subseteq C$ which, by hypothesis, belongs to $\mathcal{C}$. If $C_0 \neq C$, the family is not regular epimorphic.

Now we can finish the proof of Proposition 2. We already know that y is full and faithful. Let F be a sheaf and represent it by a diagram as in Lemma 10 above. The fact that y preserves sums implies that there is a coequalizer

$$yE_1 \rightrightarrows yE_0 \longrightarrow F$$

and the fact that y is full means the two arrows from E_1 to E_0 come from maps in $\mathcal{E}$. Letting E be the coequalizer of those maps in $\mathcal{E}$, it is evident that $yE \cong F$. Hence y is an equivalence.

The last sentence of the proposition follows from Lemma 11.

It should remarked in connection with this theorem that if $\mathcal{E}$ has a small regular generating set, $\mathcal{C}$ can be taken to be small by beginning with the generating set and closing it under finite products and subobjects. However, nothing in the proof requires $\mathcal{C}$ to be small and it could even be taken to be $\mathcal{E}$ itself. Of course, in that case, it may be thought that the functor category and the reflector L may not necessarily exist, but this is a philosophical, not a mathematical objection. For those who care, we remark that it is a topos in a larger universe. Moreover, the "reflection principle" guarantees that the theorem is valid even in the category of *all* sets.

Theorem 12. *Any left exact cocontinous functor between Grothendieck toposes is the left adjoint part of a geometric morphism.*

Proof. We need only show that the right adjoint exists. But Grothendieck toposes are complete with generators so the Special Adjoint Functor Theorem guarantees the existence.

A category whose Yoneda embedding has a left adjoint is called **total.** If the left adjoint is also left exact, the category is called **lex total.** Freyd and Street have characterized Grothendieck toposes as lex total categories satisfying a mild size restriction [Street, 1981]. Street [1983] characterizes lex total categories in terms of conditions on epimorphic families generalizing the requirements in Giraud's theorem of having universal effective epimorphisms. Total categories are cocomplete in a strong sense. Street [1983] says of them that they are "...precisely the the algebraic and topological categories at which traditional category theory was aimed."

Some of the more obvious relations among these various notions appear in the theorem below.

Theorem 13. *Let $\mathcal{E}$ be a category. Then each of the following properties implies the next. Moreover, (iv) plus the existence of a generating set implies (i).*

(i) *$\mathcal{E}$ is a Grothendieck topos,*
(ii) *$\mathcal{E}$ is lex total,*
(iii) *$y : \mathcal{E} \to \mathrm{Sh}(\mathcal{E})$ is an equivalence,*
(iv) *$\mathcal{E}$ is a complete topos.*

Proof. We prove that (ii) $\Rightarrow$ (iii), which requires the most argument, and leave the others as exercises.

Suppose $\mathcal{E}$ is lex total. Then $Y : \mathcal{E} \to \mathcal{P}sh(\mathcal{E})$ has a left exact left adjoint $Y^{\#}$. It follows from results of Section 6.5 that $\mathcal{E}$ is a cocomplete topos. Let $I : \mathcal{S}h(\mathcal{E}) \to \mathcal{P}sh(\mathcal{E})$ be the inclusion with left adjoint L, and as before, let $y = LY$. We will use the fact that y preserves colimits; this follows from the fact that it preserves sums and (regular because $\mathcal{E}$ is a topos) epis, and, as we will see in Section 7.6, Propositions 1 and 2, that is enough in a countably complete topos. To show that y is an equivalence, it is enough to show that every sheaf F is isomorphic to yE for some object E of $\mathcal{E}$. Since every presheaf is a colimit of representables $IF = \operatorname{colim} YE_i$, so

$$F = LIF = \operatorname{colim} LYE_i = \operatorname{colim} yE_i = y \operatorname{colim} E_i,$$

as required.

We make no attempt here to investigate the reverse implications. Exercises (GRTOP) and (BIGACT) give an example of a complete topos that lacks a small generating set.

Exercises 6.8

(GSV). We have defined a sieve and thus a cover in terms of collections $\{A_i \to A\}$. There is another way. Let us say that a **Giraud sieve** on A is a subfunctor of the representable functor $\operatorname{Hom}(-, A)$. Say that a collection of Giraud sieves forms a **Giraud topology** if it includes the identity sieve on every object and is invariant under pullback and composition. Further say that a sieve (in the sense used previously in this book) $\{A_i \to A\}$ is **saturated** if whenever an arrow $B \to A$ factors through one of the arrows in the sieve, it already is one of the arrows in the sieve. Show that there is a one-to-one correspondence between saturated sieves and Giraud sieves on an object A. Conclude that topologies and Giraud topologies are equivalent.

(OMT)°. Let $\mathcal{C}$ be a site and $\mathcal{E} = \mathcal{S}h(\mathcal{C})$.

(a) Show that there is a presheaf which assigns to each object the set of Giraud sieves on that object and that that is the subobject classifier Ω in the presheaf category.
(b) Show that the presheaf $\Omega_{\mathbf{j}}$ that assigns to each object the set of Giraud covers is a sheaf and in fact the subobject classifier in the sheaf category.
(c) Show that the topology $\mathbf{j}$ (which, recall, can be viewed as an endomorphism of Ω) is the classifying map of $\Omega_{\mathbf{j}}$.

(DNC)°. Let $\mathcal{C}$ be a site and $\mathcal{E} = \mathcal{Shv}(\mathcal{C})$. Prove that a Grothendieck topology on $\mathcal{E}$ is contained in the $\neg\neg$ topology if and only if no cover in the Grothendieck topology is empty.

(GEN)°. (a) Show that in a category $\mathcal{C}$ a set U of objects is a set of generators if and only if when $f, g : C \to B$ are distinct then there is an element $c \in^G C$ for some $G \in U$ for which $f(c) \neq g(c)$.

(b) Show that any regular generating family is conservative.

(c) Show that in a category with equalizers, any conservative generating family is a generating family.

(d) Show that in a topos any generating family is conservative. (Hint: every mono is regular.)

(e) Show that in a complete topos, every epimorphic family is regular; hence the converse of (b) is true and every generating family is regular.

(GEQ)°. Let $\mathcal{E}$ be a countably complete (hence countably cocomplete) topos and $R \subseteq A \times A$ a relation on an object A. If S_1 and S_2 are two relations on A, then the **composite** of S_1 and S_2 is, as usual, is the image of the pullback $S_1 \times_A S_2$ where S_1 is mapped to A by the second projection and S_2 by the first.

(a) Let $S = R \cup \Delta \cup R^{\mathrm{op}}$. Show that S is the reflexive, symmetric closure of R.

(b) Let E be the union of the composition powers of S. Show that E is the equivalence relation generated by R.

(Hint: Pullbacks, unions and countable sums are all preserved by pulling back.)

(GRTOP). By a measurable cardinal α is meant a cardinal number for which there is an ultrafilter $\mathfrak{f}$ with the property that if any collection $\{U_i\}, i \in I$ is given for which $\#(I) < \alpha$, and each $U_i \in \mathfrak{f}$, then $\cap U_i \in \mathfrak{f}$. Call such an $\mathfrak{f}$ an α-measure. It is not known that measurable cardinals exist, but for this exercise, we will assume not only that they exist, but that there is a proper class of them. So let $\alpha_1, \alpha_2, \ldots, \alpha_\omega, \ldots$ (indexed by all ordinals) be an increasing sequence of measurable cardinals and $\mathfrak{f}_1, \mathfrak{f}_2, \ldots, \mathfrak{f}_\omega, \ldots$ a corresponding sequence of ultrafilters. Assume the following (known) property of measurable cardinals: they are strongly inaccessible, meaning they cannot be reached by operations of product, sum or exponentiation involving fewer, smaller cardinals.

Define, for any set S and ultrafilter $\mathfrak{f}$ a functor $FS = \operatorname{colim} S^U, U \in \mathfrak{f}$. Note that the diagonal map $S \to S^U$ induces a function $S \to FS$, which is, in fact, the component at S of a natural transformation.

(a) Show that the functor F is left exact.

(b) Show that if α is measurable, $\mathfrak{f}$ an α-measure and $\#(S) < \alpha$, then $S \to FS$ is an isomorphism.

(c) Show that if F_i is defined as above, using the ultrafilter $\mathfrak{f}_i$ on α_i, then there is a sequence H_i of left exact endofunctors on $\mathcal{Set}$ defined by $H_{j+1} = F_j \circ H_j$ and $H_j = \operatorname{colim} H_k$, $k < j$ when j is a limit ordinal.

(d) Show that for any set S there is an i dependent on S such that for $j > i$, $G_j(S) \to G_i(S)$ is an isomorphism.

(e) Conclude that there is a left exact functor $H : \mathcal{Set} \to \mathcal{Set}$ whose values are not determined by the values on any small subcategory.

(f) Show that there is a left exact cotriple on $\mathcal{Set} \times \mathcal{Set}$ whose functor part is given by $G(S,T) = (S \times HT, T)$ for which the category of algebras does not have a set of generators and hence is not a Grothendieck topos.

(BIGACT). Let C be a proper class and G be the free group generated by C. (If you don't like this, we will describe an alternate approach later. Meantime continue.) Let $\mathcal{E}$ be the category of those G-sets which have the property that all but a small subset of the elements of C act as the identity automorphism. Show that $\mathcal{E}$ is a complete topos which does not have a small generating set.

An alternate approach to the same category is to take as an object a 3-tuple (S, C_0, f) in which S is a set, C_0 a sub-*set* of C and $f : C_0 \to \operatorname{Aut}(S)$ a function. Morphisms are defined so as to make this the category of G-sets as defined above. Although set-theoretically unassailable, this approach seems conceptually much less clear.

As a matter of historical interest, this category was one of the earliest known examples to show the necessity of the solution set condition in the GAFT. The evident underlying set functor is easily seen to lack an adjoint while satisfying all the other conditions. We believe it is due to Freyd.

(RGCO)°. Prove that if U is a set of regular generators for a category $\mathcal{C}$, then every object of $\mathcal{C}$ is a colimit of an indexed family of objects of U.

Chapter 7.

Representation Theorems

7.1. Freyd's Representation Theorems

In this section we prove a number of theorems due to Freyd representing toposes into various special classes of toposes. The development follows Freyd [1972] very closely.

Terminology

We define several related concepts which will be used in Chapters 7 and 8. We have put all the definitions here, although they are not all used in this section, because some of the terminology varies in the literature.

A category is **regular** if it is left exact and has coequalizers, and every regular epimorphism is stable (preserved under pullbacks). Weaker definitions are sometimes used in the literature. A functor is **regular** if it is left exact and preserves regular epis (but it need not preserve all coequalizers.) An **exact category** is a category which has finite limits and finite colimits. An **exact functor** is one which preserves finite limits and colimits, i.e., it is left and right exact. In the past the term "exact category" has been used to denote a regular category which has effective equivalence relations (every equivalence relation is the kernel pair of some arrow), and regular functors have also been called exact functors, but we will not use that terminology.

A **pretopos** is a left exact category with effective equivalence relations which has finite sums which are disjoint and stable and in which every morphism factors as a composite of a stable regular epimorphism and a monomorphism. The corresponding type of morphism is a **near exact** functor which is left exact and preserves regular epimorphisms and finite sums. The main import of the work of Makkai and Reyes [1977] is that pretoposes correspond to a broad class of theories in the sense of model theory in mathematical logic.

Booleanness

A topos $\mathcal{E}$ is **Boolean** if for every subobject A of an object B, $A \vee \neg A \cong B$. In the following proposition, 2 denotes $1 + 1$.

Proposition 1. *The following are equivalent for a topos* $\mathcal{E}$ *:*

(a) $\mathcal{E}$ *is Boolean.*
(b) *Every subobject of an object in* $\mathcal{E}$ *has a complement.*
(c) *The Heyting algebra structure on* Ω *as defined in Section 5.6 is a Boolean algebra.*
(d) *If* false : $1 \to \Omega$ *is the classifying map of the zero subobject, then* $\langle\text{true}, \text{false}\rangle : 2 \to \Omega$ *is an isomorphism.*

Proof. (a) implies (b) because for any subobject A, $A \wedge \neg A = 0$. (b) and (c) are clearly equivalent by Theorem 4 of Section 5.6.

To see that (b) implies (d), observe that true: $1 \to \Omega$ has a complement $A \to \Omega$. But $A \to \Omega$ classifies subobjects just as well as true because there is a one to one correspondence between subobjects and their complements so that Proposition 4 of Section 2.3 shows that $A = 1$.

Finally, if (d) holds, any map $f : E \to 2$ defines a complemented subobject since E is the sum of the inverse images of the two copies of 1 (sums are stable under pullback).

Warning: Even when 2 is the subobject classifier there may be other global sections $1 \to 2$. In fact, the global sections of 2 can form any Boolean algebra whatever. However, see Proposition 2 and Theorem 4 below. A topos is said to be **2-valued** if the only subobjects of 1 are 0 and 1.

A topos is **well-pointed** if it is nondegenerate (that is $0 \neq 1$) and 1 is a generator. By Exercise (GEN) of Section 6.8, this is equivalent to saying that for each object A and proper subobject $A_0 \subsetneq A$, there is global element of A that does not factor through A_0. In particular, every non-zero object has a global element.

Proposition 2. *Let* $\mathcal{E}$ *be a well-pointed topos. Then*

(a) $\mathcal{E}$ *is* 2*-valued.*
(b) $\mathcal{E}$ *is Boolean.*
(c) $\text{Hom}(1, -)$ *preserves sums, epimorphisms, epimorphic families and pushouts of monomorphisms.*
(d) *Every nonzero object is injective.*
(e) *Every object not isomorphic to* 0 *or* 1 *is a cogenerator.*

Proof. (a) If U were a proper nonzero subobject of 1, the hypothesis would force the existence of a map $1 \to U$, making $U = 1$.

(b) There is a map $(\text{true}, \text{false}) : 1 + 1 \to \Omega$. This induces a map from $\text{Hom}(1, 1 + 1)$ to $\text{Hom}(1, \Omega)$ which is an isomorphism by (a). Since 1 is a generator, it follows that $(\text{true}, \text{false})$ is an isomorphism. Booleanness now follows from Proposition 1.

(c) A map from 1 to ΣA_i induces by pulling back on the inclusions of A_i into the sum a decomposition $1 = \Sigma U_i$. By (a), all but one of the U_i must be zero and the remaining one be 1, which means that the original map factors through exactly one A_i.

Given an epi $A \to B$ and a global element of B, the pullback

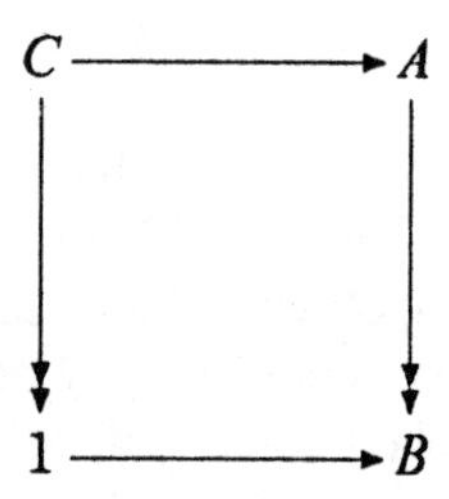

and the map $1 \to C$ provides a map $1 \to A$ which shows that $\text{Hom}(1, A) \to \text{Hom}(1, B)$ is surjective, as required.

The preservation of epimorphic families follows from the preservation of sums and epimorphisms.

Finally, given a pushout

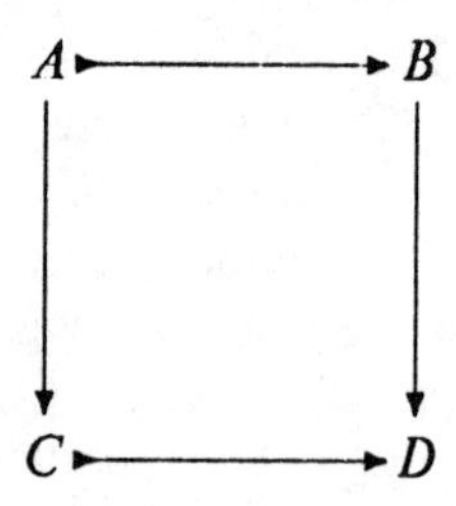

the Booleanness implies that $B = A + A'$ for some subobject A' of B, whence $D = C + A'$, which we know is preserved by $\text{Hom}(1, -)$.

(d) If $A \subseteq B = A + A'$ and $A \neq 0$ then the existence of $A' \to 1 \to A$ provides a splitting for the inclusion $A \to B$. (See Exercise (INJ) of Section 2.1.)

As for (e), let A be an object different from 0 and 1. Since $A \neq 0$, there is a map $1 \to A$ and since $A \neq 1$, that map is not an isomorphism so that there is a second map $1 \to A$ that does not factor through the first. Any map out of 1 is a monomorphism. Since 1 has no non-zero subobjects, these determine disjoint subobjects of A each of which is isomorphic to 1. Since they are disjoint, their sum gives a mono of 2 into A. Hence it is sufficient to show that 2 is a cogenerator. Given a parallel pair

$$B \overset{f}{\underset{g}{\rightrightarrows}} C$$

with $f \neq g$, let $v : 1 \to B$ be a map with $f \circ v \neq g \circ v$. Then we have the map $[f \circ v, g \circ v] : 1 + 1 \to C$ which is a mono by exactly the same reasoning and, since 2 is injective, splits. But this provides a map $h : C \to 2$ for which $h \circ f \circ v$ is the one injection of 1 into 2 and $h \circ g \circ v$ is the other, so that they are different. Hence $h \circ f \neq h \circ g$ which shows that 2 is a cogenerator.

Embedding theorems

Theorem 3. *Every small topos has an exact embedding into a Boolean topos. This embedding preserves epimorphic families and all colimits.*

Proof. Let $\mathcal{E}$ be the topos. For each pair $f, g : A \to B$ of distinct arrows we will construct a left exact, colimit preserving embedding of $\mathcal{E}$ into a Boolean topos which keeps f and g distinct, and then take the product of all the Boolean toposes so obtained. It is an easy exercise (Exercise (PBOOL)) that the product (as categories) of Boolean toposes is a Boolean topos.

The map from $\mathcal{E}$ to $\mathcal{E}/A$ which takes an object C to $C \times A \to A$ certainly distinguishes f and g In the category $\mathcal{E}/A$, the diagonal arrow $A \to A \times A$ followed by $A \times f$ (respectively $A \times g$) gives a pair of distinct global elements of B whose equalizer is a proper subobject U of 1. The topology j induced by U as in Example (c) of Section 6.1 makes the equalizer 0 in $Sh_{\mathrm{j}}(\mathcal{E}/A)$, which is clearly a nondegenerate topos. The double negation sheaves in that category is a Boolean topos with the required property (Exercise (DN2)). The arrows that f and g go to are still distinct because $\neg\neg 0 = 0$.

The limit preservation properties follow from the fact that the map $\mathcal{E} \to \mathcal{E}/A$ has both adjoints and the associated sheaf functor is left exact and has a right adjoint.

Theorem 4. *Every small Boolean topos $\mathcal{B}$ has a logical embedding to a product of small well-pointed toposes.*

Proof. The argument goes by constructing, for each nonzero object A of $\mathcal{B}$, a logical morphism $T : \mathcal{B} \to \mathcal{C}$ (where $\mathcal{C}$ depends on A) with $\mathcal{C}$ well-pointed and $TA \neq 0$. This will show that the the mapping of $\mathcal{B}$ into the

product of all the categories $\mathcal{C}$ for all objects A is an embedding (Exercise (FAITH)).

The proof requires the following lemma.

Lemma 5. *For every small Boolean topos $\mathcal{B}$ and nonzero object A of $\mathcal{B}$ there is a small topos $\mathcal{B}'$ and a logical morphism $T : \mathcal{B} \to \mathcal{B}'$ with $TA \neq 0$ and such that for all objects B of $\mathcal{B}$ either $TB = 0$ or TB has a global element.*

Proof. Well order the objects of $\mathcal{B}$ taking A as the first element. Let $\mathcal{B}_0 = \mathcal{B}$ and suppose that for all ordinal numbers $\beta < \alpha$, $\mathcal{B}_\beta$ has been constructed, and whenever $\gamma < \beta$, a family of logical morphisms $u_{\beta\gamma} : \mathcal{B}_\gamma \to \mathcal{B}_\beta$ is given such that

(i) $u_{\beta\beta} = 1$ and
(ii) for $\delta \leq \gamma \leq \beta, u_{\beta\gamma} \circ u_{\gamma\delta} = u_{\beta\delta}$.

(Such a family is nothing but a functor on an initial segment of ordinals regarded as an ordered category and is often referred to as a **coherent family**).

If α is a limit ordinal, let $\mathcal{B}_\alpha$ be the direct limit of the $\mathcal{B}_\beta$ for $\beta < \alpha$. If α is the successor of β then let B be the least object of $\mathcal{B}$ which has not become 0 nor acquired a global section in $\mathcal{B}_\beta$. Let $\hat{B}$ be its image in $\mathcal{B}_\beta$ and let $\mathcal{B}_\alpha$ be $\mathcal{B}_\beta/\hat{B}$. Stop when you run out of objects. Since toposes are defined as models of a left exact theory and logical functors are morphisms of that theory, it follows from Theorem 4 of Section 4.4 that the direct limit is a topos. By Exercise (BWP), the last topos constructed by this process is the required topos. It is easy to see that the functors in the cone are logical.

To prove Theorem 4, form the direct limit $\mathcal{C}$ of $\mathcal{B}$, $\mathcal{B}'$, $\mathcal{B}''$ (forming $\mathcal{B}''$ using the image of A in $\mathcal{B}'$) and so on. The image of A in $\mathcal{C}$ will be nonzero, and every nonzero object of $\mathcal{C}$ has a global element.

The product of all these categories $\mathcal{C}$ for all objects A of $\mathcal{B}$ is the required topos.

Theorem 6. *Every small topos has an exact embedding into a product of well-pointed toposes.*

Theorem 7. *Every small topos has an embedding into a power of* $\mathbf{Set}$

that preserves finite limits, finite sums, epimorphisms, and the pushout of a monomorphism.

Proof. Every well-pointed topos has a functor to $\mathcal{S}et$, namely $\mathrm{Hom}(1, -)$, with those properties.

Exercises 7.1

(BWP)°. Prove that a Boolean topos is well-pointed if and only if every nonzero object has a global section.

(PWP). Prove that well-pointed toposes are not the models of an LE theory.

(DN2)°. Show that the category of sheaves for the topology of double negation (Exercise (DN) of Section 6.1) is a Boolean topos. (Hint: In any Heyting algebra $\neg\neg a = \neg\neg\neg\neg a$.)

(FAITH)°. (a) Show that an exact functor from a Boolean topos is faithful if and only if it takes no non-zero object to zero.
(b) Show that an exact functor from a 2-valued topos to any non-degenerate topos is faithful.

(RFI)°. Show that the embeddings of Theorems 6 and 7 reflect all limits and colimits which they preserve. (Hint: First show that they reflect isomorphisms by considering the image and the kernel pair of any arrow which is not an isomorphism.)

(PBOOL)°. (a) Show that the product as categories of toposes is a topos.
(b) Show that the product of Boolean toposes is Boolean.

(LOG). A category has **stable sups** if the supremum of any two subobjects exists and is preserved by pullbacks. It has **stable images** if for any arrow $f : A \to B$, $\mathrm{Sub} f : \mathrm{Sub}\, B \to \mathrm{Sub}\, A$ has left adjoint which is preserved by pullbacks. A left exact category with stable sups and stable images is called a **logical category**. Show that a category is a pretopos if and only if it is logical, has finite disjoint sums and effective equivalence relations. (This comes from [Makkai and Reyes, 1977, pp.121–122].)

7.2. The Axiom of Choice

The Axiom of Choice

If $f : A \to B$ is an arrow in a category, we say that a map $g : B \to A$ is a **section of** f if it is a right inverse of f, i.e. $f \circ g = 1$. If f has a section, we say that it is a **split epi** (it is necessarily epi), although the second word is often omitted when the meaning is clear. It is easy to see that the Axiom of Choice in ordinary set theory is equivalent to the statement that in the category of sets, all epis are split. A section of the map $() : A \to 1$ is a global element. A global element of A is thus often called a **global section** of A.

We say a topos satisfies the Axiom of Choice (AC) if every epi splits. It is often convenient to break this up into two axioms:

SS (Supports Split): Every epimorphism whose codomain is a subobject of 1 splits.

IAC (Internal Axiom of Choice): If $f : A \to B$ is an epi, then for every object C, $f^C : A^C \to B^C$ is an epi.

The name "Supports Split" comes from the concept of the **support** of an object X, namely the image of the map $X \to 1$ regarded as a subobject of 1. An object has **global support** if its support is 1.

It is an easy exercise that AC implies SS and IAC. It will emerge from our discussion that SS and IAC together imply AC.

We say $f : A \to B$ is a **powerful epi** if it satisfies the conclusion of IAC. We define $\S f$ by the pullback

(1)

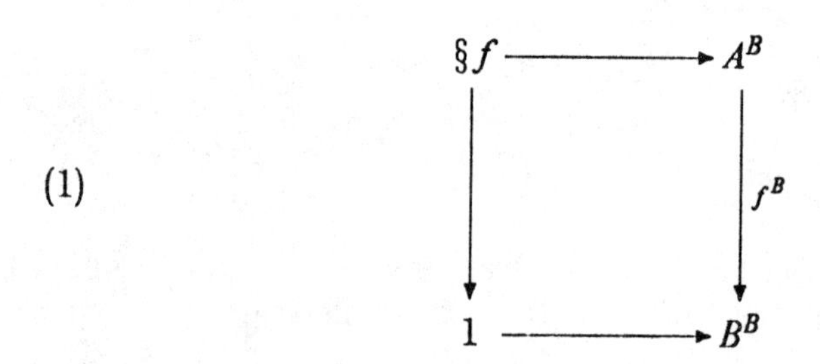

where the lower map is the transpose of the identity. Intuitively, $\S f$ is the set of sections of f.

It is clear that if f^B is epi then $\S f$ has global support (the converse is also true, see Exercise (SEC)) and that $\S f$ has a global section if and only if f has a section. In fact, global sections of $\S f$ are in one to one correspondence with sections of f.

Proposition 1. *For a morphism $f : A \to B$ in a topos $\mathcal{E}$, the following are equivalent:*

(a) *f is a powerful epi;*

(b) *f^B is epi;*

(c) *$\S f$ has global support.*

(d) *There is a faithful logical embedding $L : \mathcal{E} \to \mathcal{F}$ into some topos $\mathcal{F}$ such that Lf is split epi.*

Proof. (a) implies (b) by definition. (b) implies (c) because a pullback of an epi is epi. To see that (c) implies (d) it is sufficient to let $\mathcal{F}$ be $\mathcal{E}/\S f$ and L be $\S f \times -$. By Theorem 6 of Section 5.3, L is faithful and logical. L therefore preserves the constructions of diagram (1). In the corresponding diagram in $\mathcal{F}$, $\S Lf$ has a global section (the diagonal) which corresponds to a right inverse for Lf. For (d) implies (a), let C be an object and g a right inverse for Lf. Then g^{LC} is a right inverse for $L(f^C)$, which is isomorphic to $(Lf)^C$. Thus $L(f^C)$ is epi, which, because L is faithful, implies that f^C is epi.

Proposition 2. *Given a topos $\mathcal{E}$ there is a topos $\mathcal{F}$ and a logical, faithful functor $L : \mathcal{E} \to \mathcal{F}$ for which, if f is a powerful epi in $\mathcal{E}$, then Lf is a split epi in $\mathcal{F}$.*

Proof. Well order the set of powerful epis. We construct a transfinite sequence of toposes and logical morphisms as follows: If $\mathcal{E}_\alpha$ is constructed, let $\mathcal{E}_{\alpha+1}$ be $\mathcal{E}_\alpha/\S f_\alpha$ where f_α is the powerful epi indexed by α and the logical functor that constructed by Proposition 1. At a limit ordinal α, let $\mathcal{E}_\alpha$ be the direct limit of all the preceding logical functors. The required topos $\mathcal{F}$ is the direct limit of this family. As observed in the proof of Lemma 4 of Section 7.1, the direct limit of toposes and logical morphisms is a topos and the cone functors are logical.

Corollary 3. *Given a topos $\mathcal{E}$, there is a topos $\mathcal{F}$ in which every powerful epi splits and a faithful, logical functor L from $\mathcal{E}$ to $\mathcal{F}$.*

Proof. Repeat the above process countably often.

Corollary 4. *A topos satisfies the IAC if and only if it has a faithful, logical embedding into a topos that satisfies the Axiom of Choice.*

Proof. The "if" part is very easy. For if a topos has such an embedding, it is immediate that every epi is powerful. So suppose $\mathcal{E}$ is a topos that satisfies the IAC. If we show that every slice and any colimit of such slices (called a **limit slice** because it is a limit in the category of geometric morphisms) satisfies the IAC, then the topos constructed above will have every epi powerful and every powerful epi split. The limit part is trivial since every epi in the colimit is an epi before the colimit is reached (since the functors are all faithful). Thus it is sufficient to show that slicing

preserves the IAC. If

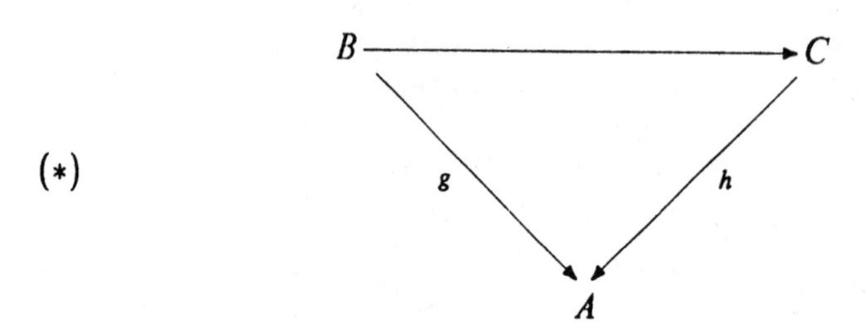

is an epimorphism in $\mathcal{E}/A$, then $f : B \to C$ is epi in $\mathcal{E}$. Since $\mathcal{E}$ satisfies the IAC, the object $\S f$ has global support. Now consider the diagram of toposes and logical functors:

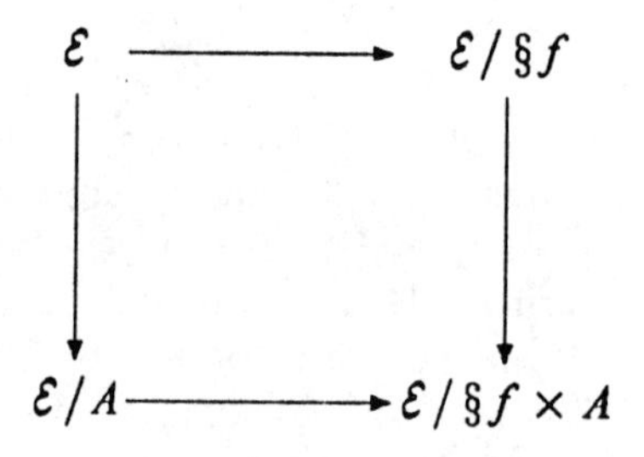

If we apply $\S f \times -$ to (*), we get

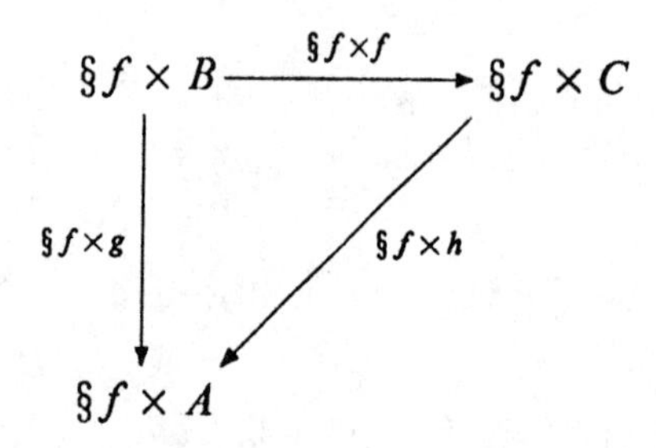

The map $\S f \times f$ has a splitting in $\mathcal{E}/\S f$. It is immediate, using the fact that it *is* a section, to see that it makes the triangle commute and is hence a section in $\mathcal{E}/\S f \times A$. But then $\mathcal{E}/A$ has a faithful (because $\S f \times A \to A$ has global support), logical functor into a topos in which f has a section and hence by Proposition 1, f is a powerful epi.

AC and Booleanness

Our goal is to show that a topos satisfying AC, and ultimately a topos satisfying IAC, is Boolean. We require a lemma:

Lemma 5. *If*

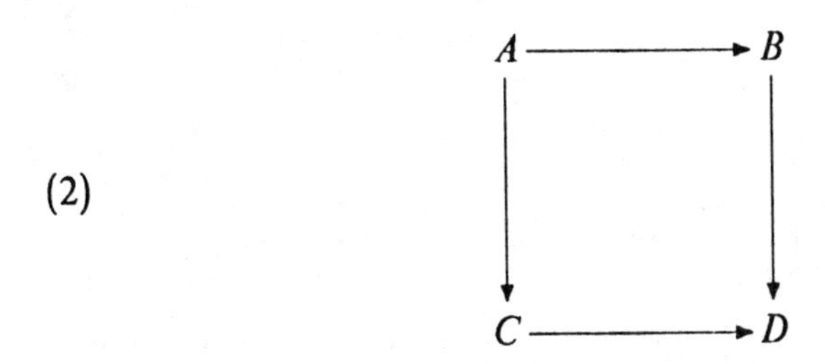

is a pullback and a pushout with all arrows monic, then its image under any near exact functor Φ *from* $\mathcal{E}$ *to a topos* $\mathcal{F}$ *is a pushout as well as a pullback.*

Proof. Since (2) is a pushout,

$$A \rightrightarrows B + C \longrightarrow D$$

is a coequalizer. The kernel pair of the arrow from $B + C$ to D is $(B + C) \times_D (B + C)$, which is isomorphic to

$$B \times_D B + B \times_D C + C \times_D B + C \times_D C.$$

$B \to D$ monic implies that $B \times_D B = B$ and similarly $C \times_D C = C$. The fact that (2) is a pullback implies that the other two summands are each A. The result is that the kernel pair is the reflexivized, symmetrized image of A in $B + C$. The same considerations will apply to Φ applied to (2). Thus

$$\Phi B + \Phi A + \Phi A + \Phi C \rightrightarrows \Phi B + \Phi C \longrightarrow \Phi D$$

is a kernel pair and since the right arrow is epi, is also a coequalizer. It follows that

$$\Phi A \rightrightarrows \Phi B + \Phi C \longrightarrow \Phi D$$

is a coequalizer and hence that Φ of (2) is a pushout.

Theorem 6. *A topos* $\mathcal{E}$ *which satisfies the Axiom of Choice is Boolean.*

Proof. We must show that every subobject A of an object B has a complement. Since a slice of a topos satisfying AC is a topos satisfying AC and subobjects of B in $\mathcal{E}$ are the same as subobjects of 1 in $\mathcal{E}/B$, it is sufficient to consider the case $B = 1$. So let A be a subobject of 1. Form the coequalizer

$$A \rightrightarrows 1 + 1 \longrightarrow 1 +_A 1$$

and find a right inverse f for the right arrow. Take the pullbacks C_1 and C_2 of f along the two inclusions of 1 in $1 + 1$; then we claim that the complement of A is $C = C_1 \cap C_2$ as subobjects of 1. We know that A and C are subobjects of 1, so we have a map $A + C \to 1$, which we want to prove is an isomorphism. By Freyd's Theorem 7 and Exercise (RFI) of Section 7.1, there is a family of near-exact functors $\Phi : \mathcal{E} \to \mathbf{Set}$ which collectively reflect isomorphisms. By Lemma 5, everything in the construction of $A + C$ is preserved by near-exact functors, so it is sufficient to prove that $A + C \to 1$ is an isomorphism in $\mathbf{Set}$. In $\mathbf{Set}$, either $A = \emptyset$ in which case $C_1 = C_2 = C = 1$ which is the complement of A, or $A = 1$ in which case one of C_1 or C_2 is empty, so the intersection is empty and is therefore the complement of A. In each case the map $A + C \to 1$ is an isomorphism.

Corollary 7. *A topos satisfying IAC is Boolean.*

Proof. Suppose A is a subobject of B in a topos satisfying IAC. According to Proposition 4, the topos has a logical, faithful embedding into a topos satisfying the Axiom of Choice. Since it is logical, it preserves the construction of $\neg A$. Thus the equation $A + \neg A = B$ is true in the original topos if and only if it is true after the embedding. This follows from Theorem 6.

Exercises 7.2

(SEC)°. Show that if $f : A \to B$ is a map for which $\S f$ has global support, then f^B is epi. (Hint: Slice by $\S f$.)

(GAC). Show that if G is a nontrivial group, $\mathbf{Set}^G$ satisfies IAC but not AC.

(COMP). If A is a subobject of B, then A has a complement in B if and only if the epimorphism $B + B \to B +_A B$ is split.

(COMP2). If A is a subobject of B and C is any object whose support includes the support of B, then A is complemented in B if and only if $A \times C$ is complemented in $B \times C$. (Hint: Adapt the argument used in Corollary 7.)

7.3. Morphisms of Sites

In this section, we state a theorem about extensions of left exact cover-preserving functors to the sheaf category which will play the same role for theories with cocones (treated in Chapter 8) that Theorem 4 of Section 4.3 plays for theories with only cones. This theorem is also used in the proof of Deligne's Theorem in the next section.

We need some preliminary results.

Proposition 1. *Let $\mathcal{B}$ and $\mathcal{D}$ be complete toposes, $\mathcal{A}$ a left exact generating subcategory of $\mathcal{B}$, and $f : \mathcal{B} \to \mathcal{D}$ a colimit-preserving functor whose restriction to $\mathcal{A}$ is left exact. Then f is left exact.*

Proof. Since $\mathcal{A}$ is left exact, it contains the terminal object 1, so $f(1) = 1$. Thus we need "only" show that f preserves pullbacks.

In the proof below we systematically use A and B with or without subscripts to refer to objects in $\mathcal{A}$ or $\mathcal{B}$ respectively. The proof requires several steps.

(i). Let $B = \sum A_i$. *Then for any A, f preserves the pullback*

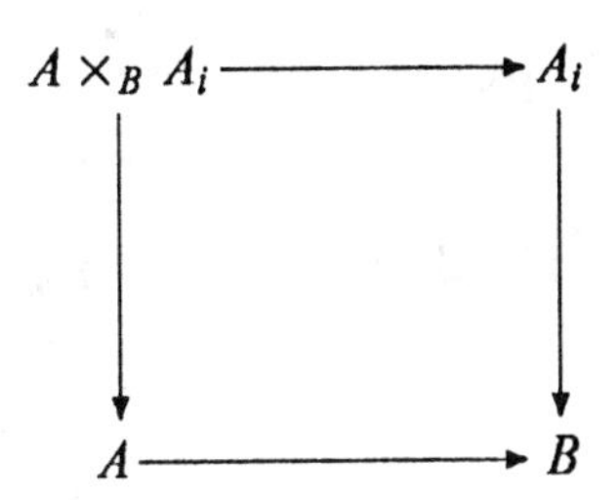

where the right arrow is coordinate inclusion.

Proof. We have

$$\sum A \times_B A_i \cong A$$

and

$$\sum f(A \times_B A_i) \cong fA \cong \sum f(A) \times_{fB} f(A_i),$$

because f commutes with sums and so do pullbacks in a topos. Both these isomorphisms are induced by the coordinate inclusions $A \times_B A_i \to A$. Thus the sum of all the maps $f(A \times_B A_i) \to fA \times_{fB} fA_i$ is an isomorphism, so all the individual ones are.

(ii). *If $\sum \mathcal{A}$ is the full subcategory of $\mathcal{B}$ consisting of sums of objects of $\mathcal{A}$, then $\sum \mathcal{A}$ has finite limits and the restriction of f to $\sum \mathcal{A}$ is left exact.*

Proof. Suppose we are given

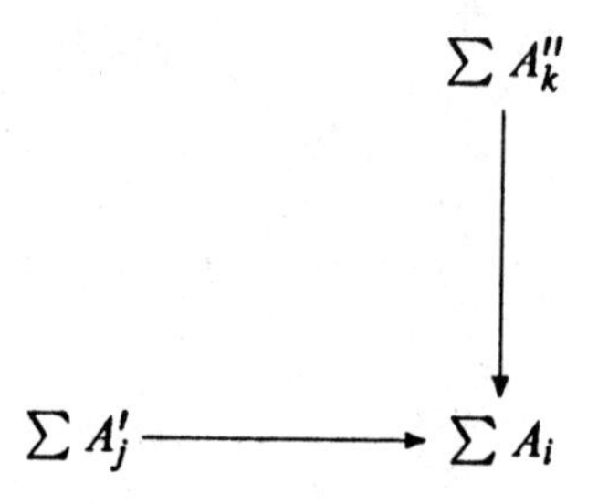

For each j, define A'_{ji} so that

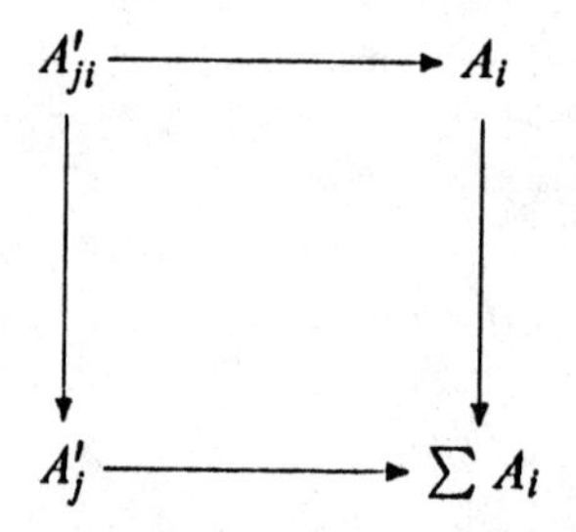

is a pullback, and similarly define A''_{ki} for each k. Because pullbacks distribute over sums,

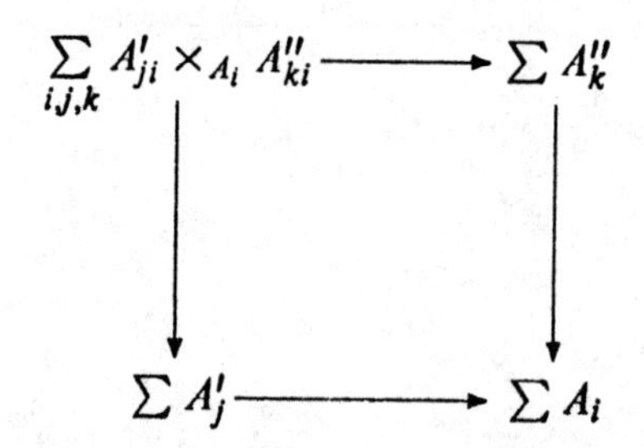

is a pullback. This proves that pullbacks and hence all limits exist in $\sum \mathcal{A}$. Since sums and all the pullbacks used in the preceding construction are preserved by f, f preserves these pullbacks, and therefore all finite limits in $\sum \mathcal{A}$.

We will henceforth assume that $\mathcal{A}$ is closed under sums.

(iii). *f preserves pullbacks of diagrams of this form:*

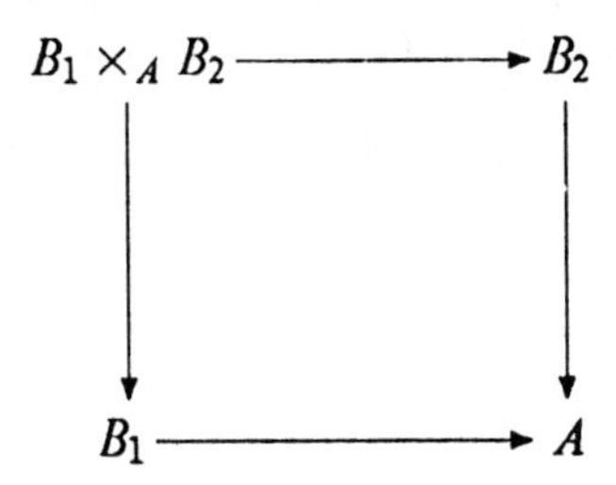

In particular, f preserves monos whose target is in $\mathcal{A}$.

Proof. By Exercise (RGCO), Section 6.8, B_1 is the colimit of objects A_i, i running over some index set, and B_2 is the colimit of objects A'_j, j running over a different index set. For each i and j, form the pullback $A_i \times_A A'_j$. Then $B_1 \times_A B_2$ is the colimit of these pullbacks for i and j ranging over their respective index sets. Then we calculate

$$\begin{aligned} f(B_1 \times_A B_2) &\cong f(\operatorname{colim}(A_i \times_A A'_j) \cong \operatorname{colim} f(A_i \times_A A'_j) \\ &\cong \operatorname{colim} f A_i \times_{fA} f(A'_j) \cong f(B_1) \times_{fA} f(B_2). \end{aligned}$$

(iv). *f preserves pullbacks*

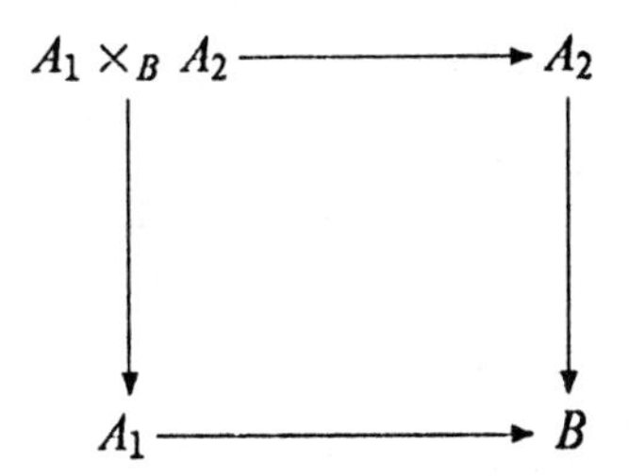

in which the two maps to B are joint epi.

Proof. Let $A = A_1 + A_2$. Under the hypothesis, $A \twoheadrightarrow B$ is an epi whose kernel pair is the disjoint sum

$$E = A_1 \times_B A_1 + A_1 \times_B A_2 + A_2 \times_B A_1 + A_2 \times_B A_2.$$

E is an equivalence relation whose transitivity is equivalent to the existence of an arrow $E \times_A E \to E$ (the fiber product being the pullback of one projection against the other) satisfying certain equations. This pullback is of the type shown in (iii) to be preserved by f. It follows that fE is transitive on fA. The fact that E is a subobject of $A \times A$ implies in a similar way that fE is a subobject of $f(A \times A) \cong f(A) \times f(A)$. Symmetry and reflexivity are preserved by any functor, so fE is an equivalence relation on fA. Hence it is the kernel pair of its coequalizer. Since f preserves colimits, that coequalizer is $fA \to fB$. So letting $A_{ij} = A_i \times_B A_j$ and $C_{ij} = fA_i \times_{fB} fA_j$ for $i,j = 1,2$, both rows in

$$\begin{array}{ccccc}
f(A_{1,1}) + f(A_{1,2}) + f(A_{2,1}) + f(A_{2,2}) & \rightrightarrows & fA & \longrightarrow & fB \\
\downarrow & & \downarrow = & & \downarrow = \\
C_{1,1} + C_{1,2} + C_{2,1} + C_{2,2} & \rightrightarrows & fA & \longrightarrow & fB
\end{array}$$

are kernel pairs, so the left vertical arrow is an isomorphism, from which it follows that each component is an isomorphism. The second component is the one we were interested in.

(v). *Given the pullback*

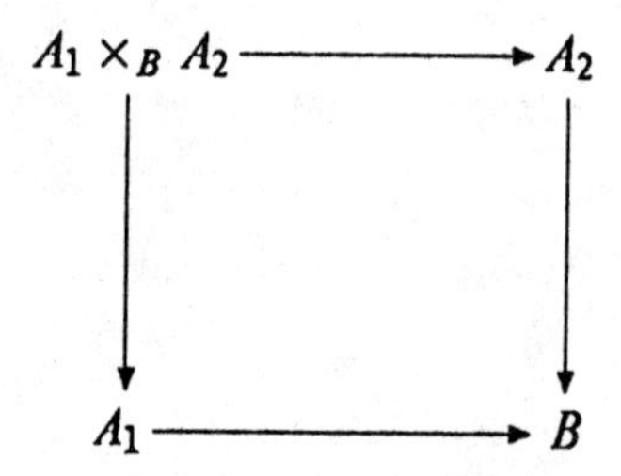

the induced map $f(A_1 \times_B A_2) \to f(A_1) \times_{fB} f(A_2)$ *is monic.*

Proof. This follows from the diagram

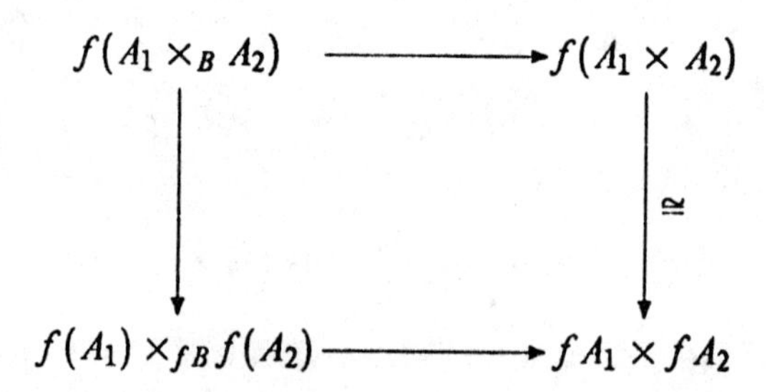

where the horizontal arrows are induced by the mono $A_1 \times_B A_2 \to A_1 \times A_2$ (and so are mono by (iii)) and the fact that if the composite of two arrows is monic then so is the first one.

(vi). *Given the pullback*

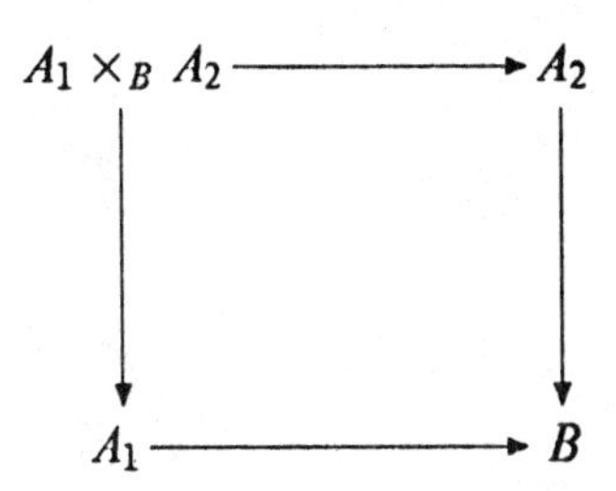

the induced map $f(A_1 \times_B A_2) \to f(A_1) \times_{fB} f(A_2)$ *is epic.*

Proof. Since $\mathcal{A}$ is closed under sums, every object B has a presentation $A \to B$ (epi). Form the following diagram

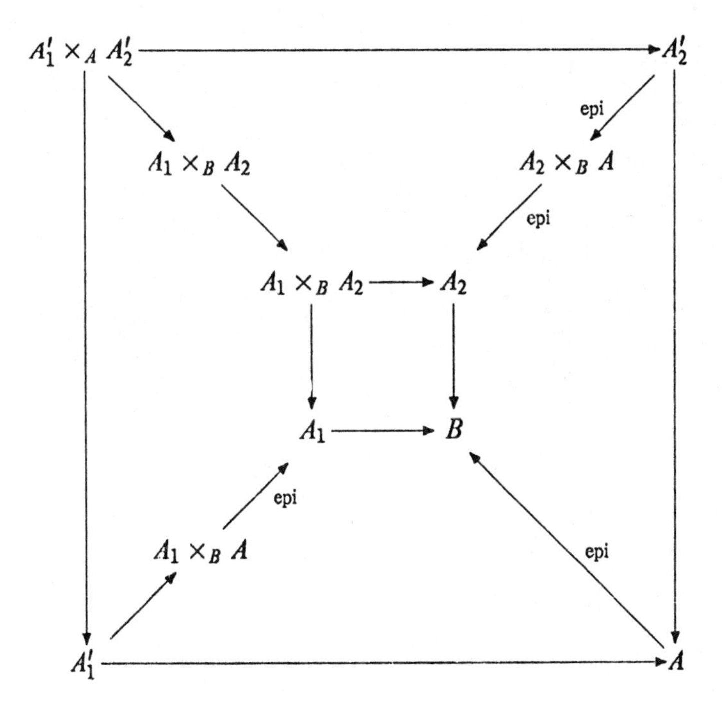

in which $A \to B$ is a presentation, and then $A_i' \to A_i \times_B A$ are presentations for $i = 1, 2$. Furthermore, $A_i' \times_B A$ is a pullback which by (iv) is preserved by f. The composite arrow in the upper left corner can be seen to be epi using Freyd's Theorem 5 of 7.2 (the near-exact embedding into a power of $\mathcal{S}et$) and a diagram chase in $\mathcal{S}et$.

In the target category $\mathcal{D}$ we have a similar diagram:

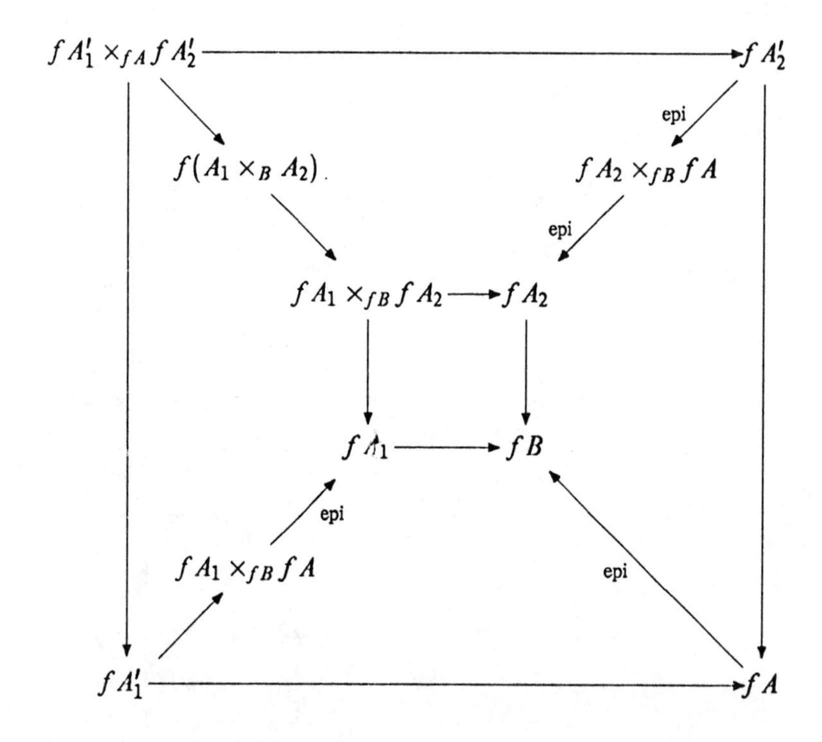

The maps $fA'_i \to fA_i \times_{fB} fA$ $(i = 1, 2)$ are epi because they are f of the corresponding arrows in the diagram preceding this one. By the same argument using Freyd's representation theorem as was used in the earlier diagram,

$$fA'_1 \times_{fA} fA'_2 \to fA_1 \times_{fA} fA_2$$

is epi, so the second factor

$$f(A_1 \times_B fA_2) \to fA_1 \times_{fB} fA_2$$

is also epi.

(vii). *f preserves arbitrary pullbacks.*

Proof. It follows from (v) and (vi) that $f(A_1 \times_B A_2) \to fA_1 \times_{fB} fA_2$ is an isomorphism. Thus we can apply the argument used in (iii) above to

the diagram

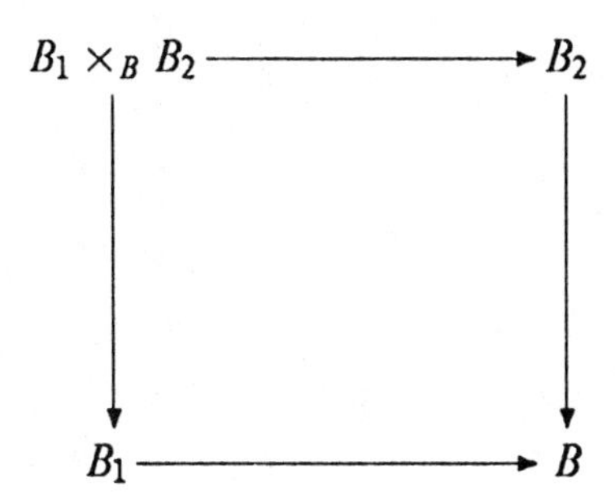

by replacing A by B throughout.

There are other proofs of Proposition 1 known, not quite as long, which depend on analyzing the form of the Kan extension. See for example Makkai and Reyes [1977, Theorem 1.3.10].

In the application below, the categories $\mathcal{B}$ and $\mathcal{D}$ are both functor categories, and the use of Freyd's theorem can be avoided in that case by a direct argument.

Theorem 2. *Suppose $\mathcal{A}$ and $\mathcal{C}$ are sites, and $f : \mathcal{A} \to \mathcal{C}$ is a morphism of sites. Then there is a functor $f_\# : \mathrm{Sh}(\mathcal{A}) \to \mathrm{Sh}(\mathcal{C})$ which is left exact and has a right adjoint for which*

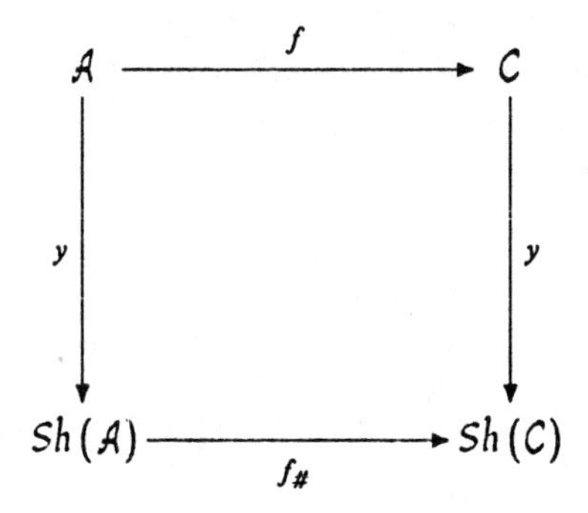

commutes.

Proof. Form the diagram

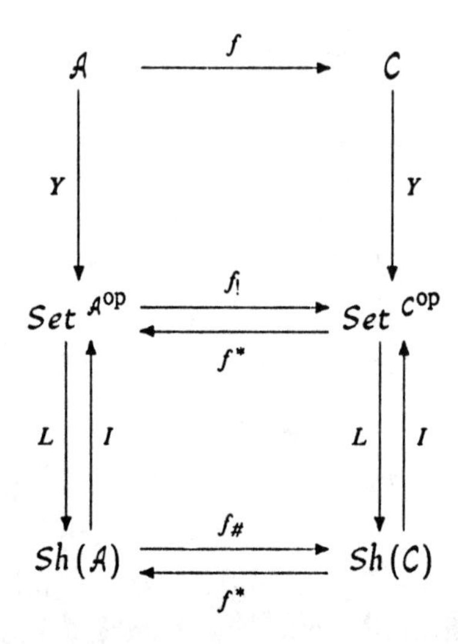

in which f^* is the functor composing with f and $f_!$ is the left Kan extension. Y is Yoneda, I is inclusion, and L is sheafification. $y = L \circ Y$. The fact that f is cover-preserving easily implies that f^* takes sheaves to sheaves and so induces a functor which we also call f^* on the sheaf categories. Then by Theorem 2 of 4.3, $f_\# = L \circ f_! \circ I$ is left adjoint to f^*.

The commutativity follows from the following calculation:

$$\begin{aligned}\mathrm{Hom}(f_\# yA, F) &\cong \mathrm{Hom}(yA, f^*F) = \mathrm{Hom}(LYA, f^*F)\\ &\cong \mathrm{Hom}(YA, If^*F) \cong If^*F(A) = IF(fA)\\ &\cong \mathrm{Hom}(YfA, IF) \cong \mathrm{Hom}(LYfA, F) = \mathrm{Hom}(yfA, F).\end{aligned}$$

Since $f_!$ is an instance of $f_\#$, it commutes with Y. Since Y is left exact, $f_!Y = Yf$ is left exact and Proposition 1 forces $f_!$ to be left exact. Thus $f_\# = Lf_!I$ is the composite of three left exact functors.

Exercise 7.3

(Univ). Let $\mathcal{A}$ denote the category whose objects are Grothendieck toposes and whose morphisms are left exact functors with a right adjoint (that is the adjoints to geometric morphisms). Let $\mathcal{B}$ denote the category whose objects are essentially small sites, meaning those sites which possess a small subcategory with the property that every object of the site can be covered by covering sieves with domains in that subcategory. There is an underlying functor $U : \mathcal{A} \to \mathcal{B}$ which associates to each Grothendieck topos the site which is the same category equipped with the category of epimorphic families (which, in a Grothendieck topos, is the same as the topology of regular epimorphic families). Show that the category of sheaves functor is left adjoint to U.

7.4. Deligne's Theorem

A topos $\mathcal{E}$ is **coherent** if it has a small full left exact generating subcategory $\mathcal{C}$ such that every epimorphic family $E_i \to C$ (for any object C) contains a finite epimorphic subfamily. Johnstone [1977] gives a proof of a theorem due to Grothendieck that characterizes coherent Grothendieck toposes as those which are categories of sheaves on a site which is a left exact category with a topology in which all the covers are finite. In Chapter 8, we will see that coherent toposes classify theories constructed from left exact theories by adding some finite cocones. (In general, geometric theories allow cocones of arbitrary size).

Theorem 1 (Deligne). *Let $\mathcal{E}$ be a coherent Grothendieck topos. Then $\mathcal{E}$ has a left exact embedding into a product of copies of the category of sets which is the left adjoint of a geometric morphism.*

Proof. Let $\mathcal{E}_0$ be the smallest subtopos of $\mathcal{E}$ which contains $\mathcal{C}$ as well as every $\mathcal{E}$-subobject of every object of $\mathcal{C}$. $\mathcal{E}_0$ is small because each object has only a set of subobjects (they are classified by maps to Ω), and we need only repeatedly close under $\mathbf{P}$, products and equalizers at most a countable number of times, and take the union.

According to Theorem 7 of Section 7.1, there is a faithful near-exact embedding T from $\mathcal{E}_0$ to a power of $\mathcal{S}et$. $f = T|\mathcal{C}$ is left exact and preserves finite epimorphic families which, given the nature of $\mathcal{C}$, implies that it preserves covers (by cover we mean in the topology of epimorphic families). We interrupt the proof for

Lemma 2. *f preserves noncovers; that is, if a sieve $\{C_i \to C\}$ is not a cover, then $\{fC_i \to fC\}$ is not a cover.*

Proof. $E = \bigcup \operatorname{Im}(C_i \to C)$ exists in $\mathcal{E}$ because $\mathcal{E}$ is complete. Since $\mathcal{E}_0$ is closed under subobjects, that union belongs to $\mathcal{E}_0$; moreover, by hypothesis, it cannot be all of C. Therefore since T is faithful, it follows that TE is a proper subobject of TC, through which all the TC_i factor. Consequently all the fC_i factor through this same proper subobject of fC.

Let $\mathcal{S}$ be the codomain of f. By theorem 2 of Section 7.3, $f : \mathcal{C} \to \mathcal{S}$

extends to a left exact functor $f_\# : \mathcal{E} = \mathcal{S}h(\mathcal{C}) \to \mathcal{S}$ which has a right adjoint f^*. We claim that $f_\#$ is faithful. It is enough to show that given a proper subobject E_0 of an object E, then $f_\#(E_0)$ is a proper subobject of $f_\#(E)$.

Since $\mathcal{C}$ generates, there is an object C of $\mathcal{C}$ and an arrow $C \to E$ which does not factor through E_0. Form the diagram

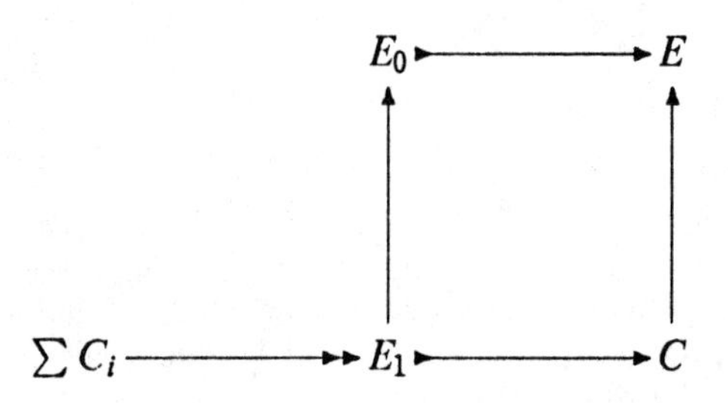

in which the square is a pullback and the C_i are a cover of E_1 and belong to $\mathcal{C}$. Since $C \to E$ does not factor throught E_0, E_1 is a proper subobject and so the sieve $\{C_i \to C\}$ is a noncover. By Lemma 2 and the fact that $f_\#$ is a left adjoint, we have

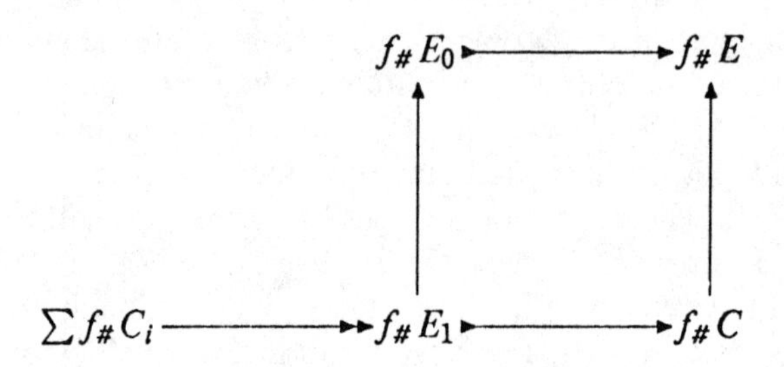

where the square is still a pullback and the sieve is a noncover. It follows that the top map cannot be an isomorphism.

7.5. Natural Number Objects

In a topos, (A, a, t) is a **pointed endomorphism structure**, or **PE-structure**, if $a : 1 \to A$ is a global element of A and $t : A \to A$ is an endomorphism. PE-structures are clearly models of an FP-theory, and $f : (A, a, t) \to$

(A', a', t') is a morphism of PE-structures if

(1)

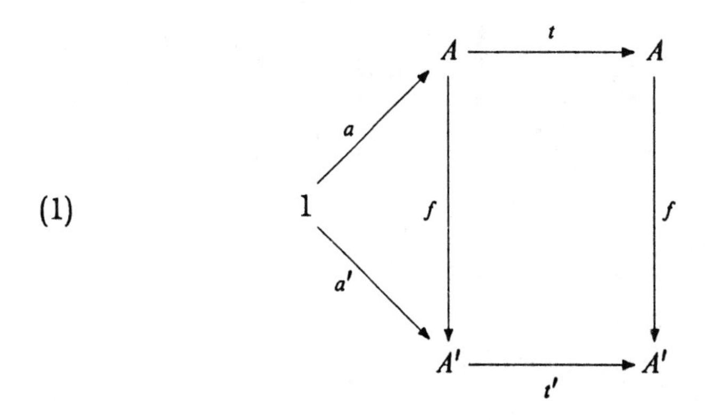

commutes.

A PE-structure $\mathsf{N} = (N, 0, s)$ is a **natural number object** or **NNO** (or object of natural numbers or natural numbers object) if for any PE-structure (A, a, t) there is a unique morphism $t^{(-)}a : (N, 0, s) \to (A, a, t)$. If we write (suggestively) $t^n(a)$ for $t^{(-)}a(n)$ when $n \in N$, then the defining properties of a morphism of PE-structure means that

(i) $t^0(a) = a$, and
(ii) $t^{sn}(a) = t \circ t^n(a)$.

It follows immediately that if we identify the natural number n with the global element $s \circ s \circ s \cdots s \circ 0$ (s occurring in the expression n times) of N, then expressions like $t^n(a)$ are not ambiguous. However, now $t^n(a)$ is defined for *all* elements n of N, not merely those global elements obtained by applying s one or more times to 0.

In this section, we will derive some basic properties of natural number objects and prove a theorem (Theorem 6 below) due to Freyd that characterizes them by exactness properties. The proof is essentially the one Freyd gave; it makes extensive use of his embedding theorems (Section 7.1).

We begin with Proposition 1 below, which says in effect that any PE-structure contains a substructure consisting of all elements $t^n(a)$. Note that this is a statement about closure under a "countable" union in *any* topos, so it will not be a surprise that the proof is a bit involved. Mikkelson [1976] has shown that internal unions in $\mathbf{P}A$, suitably defined, always exist. (Of course, *finite* unions always exist). That result and a proof of Proposition 1 based on it may be found in Johnstone [1977].

Proposition 1. *If (A, a, t) is a PE-structure, then there is a substructure A' of A for which the restricted map $\langle a, t\rangle : 1 + A' \to A'$ is epi.*

The notation $\langle a, t\rangle$ is defined in Section 1.8. A PE-structure (A', a', t') for which $\langle a', t'\rangle$ is epi will be called a **Peano system.**

Proof. We begin by defining a natural transformation $r : \text{Sub}(- \times A) \to \text{Sub}(- \times A)$ which takes $U \subseteq A \times B$ to

$$U \cap (\text{Im}(\text{id}_B \times a) \cup (\text{id}_B \times t)(U)),$$

where $(\text{id}_B \times t)(U)$ means the image of

$$U \rightarrowtail B \times A \xrightarrow{\text{id}_B \times t} B \times A$$

That r is natural in B follows easily from the fact that pullbacks commute with coequalizers, hence with images. Note that if $A' \subseteq A$ then $rA' = A'$ if and only if $A' \subseteq \text{Im}\, a \cup tA'$.

The function r induces an arrow also called $r : \text{P}A \to \text{P}A$. Let E be the equalizer of r and $\text{id}_{\text{P}A}$. Define C by the pullback

(2)

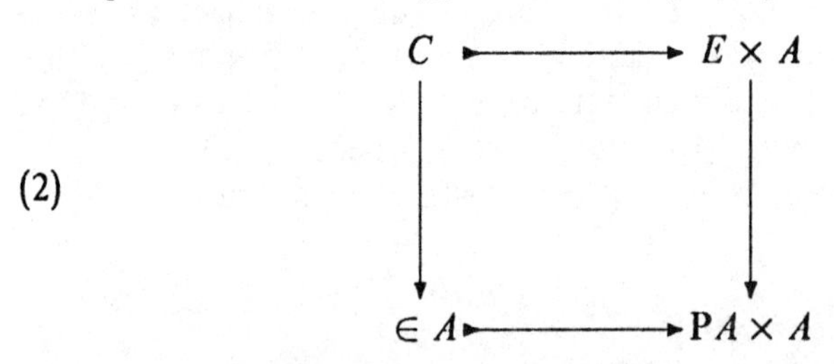

Here $\in A$ is defined in Exercise (Eps) of Section 2.1. In rest of the proof of Proposition 1, we will repeatedly refer to the composites $C \to E$ and $C \to A$ of the inclusion with the projections.

In the following lemma, $A' \subseteq A$ corresponds to $cA' : 1 \to \text{P}A$.

Lemma 2. *The following are equivalent for a subobject* $A' \rightarrowtail A$.

(i) $A' \subseteq \text{Im}\, a \cup tA'$.
(ii) $rA' = A'$
(iii) $cA' : 1 \to \text{P}A$ *factors through* E *by a map* $u : 1 \to E$.
(iv) A' *can be defined as a pullback*

(3)

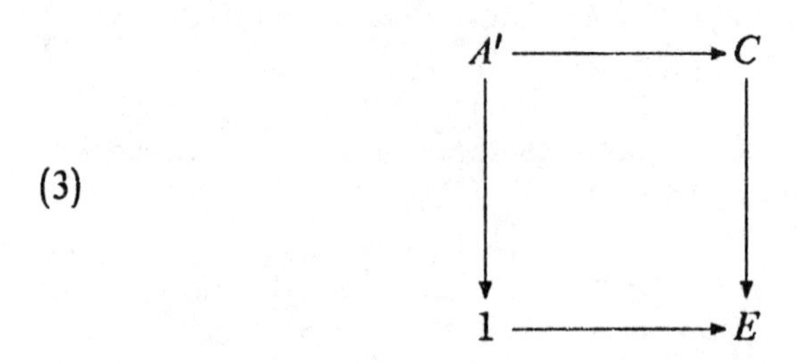

for which the inclusion $A' \rightarrowtail A = A' \to C \to E \times A \to A$, *the last map being projection.*

Proof. That (i) is equivalent to (ii) follows from the fact that in a lattice, $A \cap B = B$ if and only if $B \subseteq A$. That (ii) is equivalent to (iii) follows from the definition of E: take $B = 1$ in the definition of r.

Assuming (iii), construct $v : A' \to C$ by the following diagram, in which the outer rectangle is a pullback, hence commutative, and the bottom trapezoid is the pullback (2).

(4)

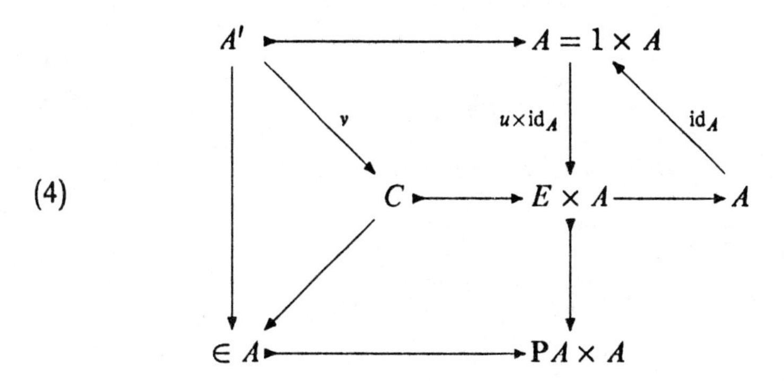

The last part of (iv) follows immediately from the preceding diagram.

Now in the following diagram, II, III, IV and the left rectangle are pullbacks and III is a mono square. Hence I and therefore the top rectangle are pullbacks by Exercise (PBCC), Section 2.2.

(5)

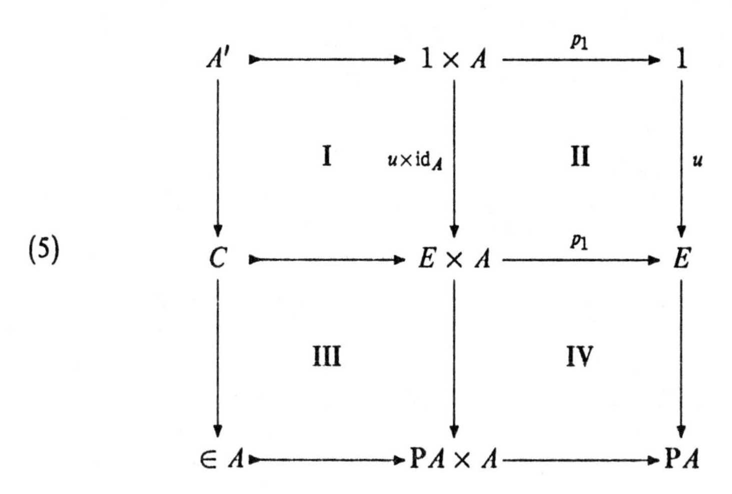

Thus (iii) implies (iv).

Given (iv), let $u : 1 \to E$ be the bottom arrow in (3). Then in (5), III is a pullback and so is the rectangle I+II, so I is a pullback because II is a mono square. Hence the rectangle I+III is a pullback. Clearly rectangle II+IV is a pullback, so the outer square is a pullback as required.

Let D be the image of $C \to E \times A \to A$. We will show that D is the subobject required by Proposition 1.

Lemma 3. *The statement*

(A) $$D \subseteq \operatorname{Im} a \cup \iota D$$

is true if and only if

(B) *the map $e : P_1 \to C$ in the following pullback is epi.*

(6)

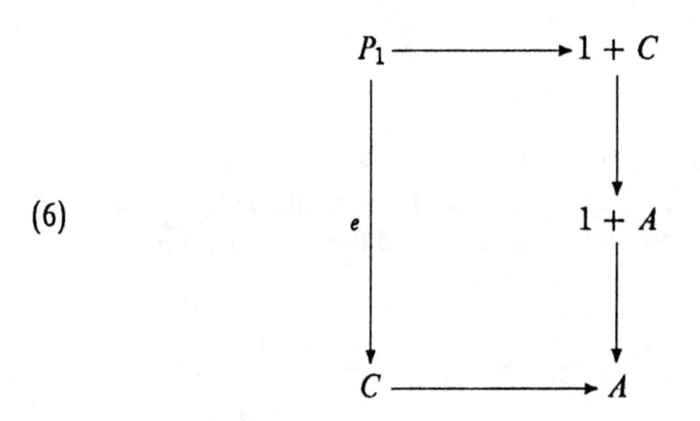

Proof. We use Freyd's near-exact embedding Theorem 7 of Section 7.1. By Lemma 2, the statement (A) is the same as saying that two subobjects of A are equal, a statement both preserved and reflected by a near-exact faithful functor. (Note that the definition of r involves almost all the constructions preserved by a near-exact functor). Since statements (A) and (B) are equivalent in the category of sets (easy), they are equivalent in a power of the category of sets since all limits and colimits are constructed pointwise there. It thus follows from the near-exact embedding theorem that (A) is equivalent to (B).

Now we will do another transference.

Lemma 4. *Suppose that for every global element $c : 1 \to C$ the map*

$f : P_2 \to 1$ in the pullback

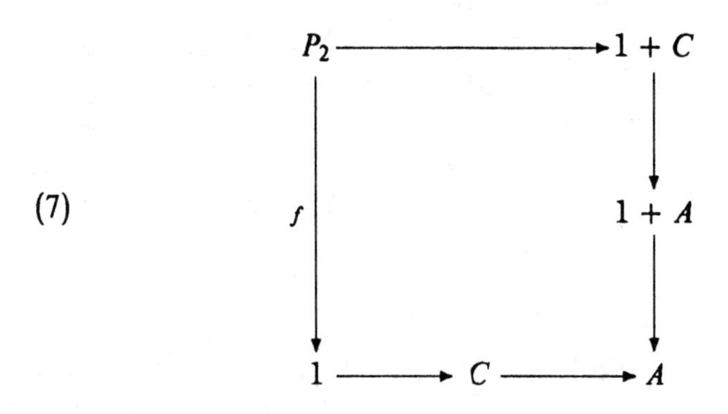

is epic. Then (B) *is true.*

Proof. We first observe that by a simple diagram chase, if $e : P \to C$ is not epi, then neither is the left vertical arrow in the following pullback diagram.

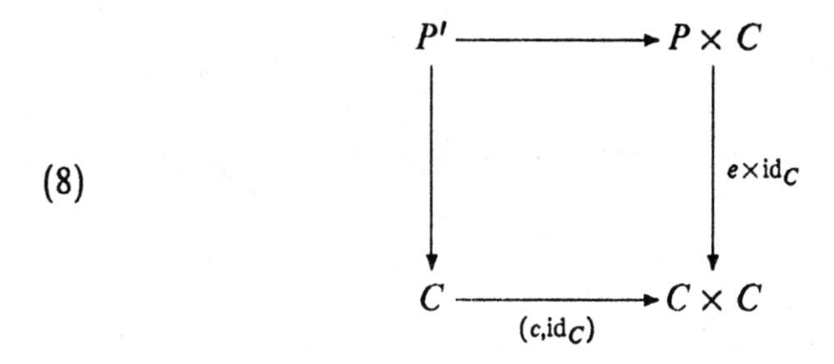

Now if e in (6) is not epi, its image in the slice $\mathcal{E}/C$ is not epi either. The observation just made would then provide a global element of C in a diagram of the form (7) in which the map $P \to 1$ is not epic. The lemma then follows from the fact that all constructions we have made are preserved by the logical functor $\mathcal{E} \to \mathcal{E}/C$.

Lemma 5. (B) *is true.*

Proof. We prove this by verifying the hypothesis of Lemma 4. Let c be a global element of C. Define C' by requiring that

(9)

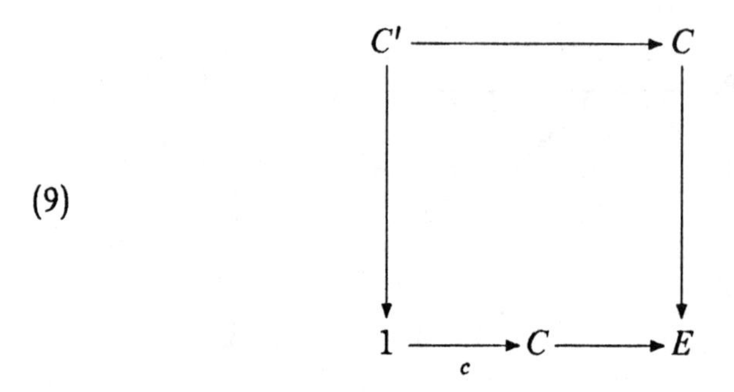

be a pullback, and P_3 by requiring that the top square in

(10)

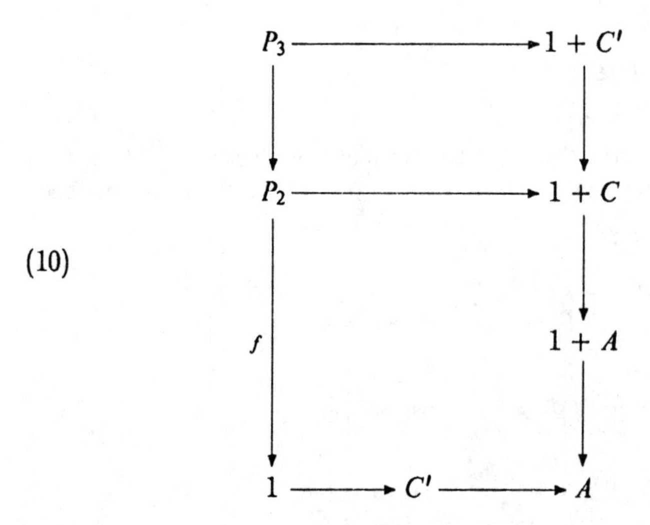

be a pullback. The bottom square is (7) with c replaced by the global element of C' induced by c and the definition of C'. This square is easily seen to be a pullback, so the big rectangle is a pullback.

Because $g \circ h$ epi implies g epi, it suffices to show that $P_3 \to 1$ is epic. We can see that by factoring the big rectangle in (10) vertically:

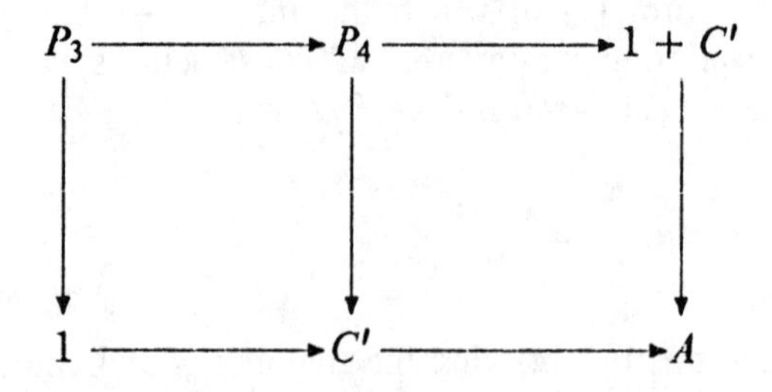

Here P_4 is defined so that the right square is a pullback. The middle arrow is epi by Lemma 2 and Lemma 3, so the left arrow is epi as required.

By Lemma 5, D satisfies property (i) of Lemma 2. By Lemma 2, any subobject A' which has that property factors through C, hence through its image D. Since $\operatorname{Im} a \cup tD$ also has property (i) of Lemma 2 (easy), it must be that $D = \operatorname{Im} a \cup tD$. This proves Proposition 1.

Theorem 6 (Freyd). *A PE-structure (A, a, t) for which*

(i) *$\langle a, t\rangle$ is an isomorphism and*
(ii) *the coequalizer of id_A and t is 1*

is a natural number object, and conversely.

The proof will make use of

Proposition 7 (The Peano Property). *A PE structure (A, a, t) which satisfies requirements (i) and (ii) of Theorem 6 has no proper PE-substructures.*

Proof. Let (A', a', t') be a substructure. By going to a subobject if necessary we may assume by Proposition 1 that the restricted $\langle a', t'\rangle : 1 + A' \to A'$ is epi. Since this proposition concerns only constructions preserved by exact functors, we may assume by Corollary 6 of Section 7.1 that the topos is well-pointed, hence by Proposition 2 of the same section that it is Boolean.

Let A'' be the complement of A' in A. If the topos were ***Set***, it would follow from the fact that $\langle a, t\rangle$ is an isomorphism on $1 + A$ and an epimorphism to A' on $1 + A'$ that

$$(*) \qquad tA'' \subseteq A''.$$

Since sums, isomorphisms, epimorphisms and subobjects are preserved by near-exact functors, using the near-exact embedding of Theorem 7 of Section 7.1, (*) must be true here. Thus $t = t' + t''$ where t' and t'' are the restrictions of t to A' and A'', respectively. Since colimits commute with colimits, the coequalizer of id_A and t is the sum of the coequalizers of $\mathrm{id}_{A'}$ and t' and of $\mathrm{id}_{A''}$ and t''. Such a sum cannot be 1 unless one of the terms is 0 (Proposition 2(a) of Section 7.1). Since A' is a substructure it contains the global element a and so cannot be 0; hence it must be A, as required.

We now have all the ingredients to prove Theorem 6. Suppose we are given a natural number object $\mathsf{N} = (N, 0, s)$. We first show (i) of the theorem. A straightforward diagram chase shows that if $i_1 : 1 \to 1 + N$ is the inclusion and $t = i_2 \circ \langle 0, s\rangle : 1 + N \to 1 + N$, then $(1 + N, i_1, t)$ is a PE-structure and $\langle 0, s\rangle$ is a morphism from this structure to $\mathbf{N}$. Thus the composite $\langle 0, s\rangle \circ t^{(-)}(i_1)$ (we remind you just this once that the last morphism, by definition, is the unique morphism from $\mathbf{N}$ to $(1 + N, i_1, t)$) is an endomorphism of $\mathbf{N}$ as a PE-structure, so by definition of NNO must be the identity.

If we can show that the opposite composite $t^{(-)}(i_1) \circ \langle 0, s\rangle$ is the identity on $1 + N$ we will have shown that $\langle 0, s\rangle$ is an isomorphism. That follows from this calculation, in which we use the notation of (ii) at the beginning of this section:

$$\begin{aligned} t^{(-)}(i_1) \circ \langle 0, s\rangle &= \langle t^0(i_1), t^s(i_1)\rangle \\ &= \langle i_1, t^{s \circ \mathrm{id}_N}(i_1)\rangle = \langle i_1, t \circ t^{(-)}(i_1)\rangle. \end{aligned}$$

But this last term is $\langle i_1, i_2\rangle = \mathrm{id}_{1+N}$, where i_2 is the inclusion $N \to 1 + N$, since $t = i_2 \circ \langle 0, s\rangle$ and we have already shown that $\langle 0, s\rangle \circ t^{(-)}(i_1)$ is the identity.

To show that 1 is the coequalizer of s and id_N, we need to show that given any $f : N \to X$ with $f \circ s = f$, there is an arrow $x : 1 \to X$ with $f = x()$. (As before, $()$ denotes the unique map from something to 1—here the something must be N). We define $x = f(0)$. It is easy to see that both f and $x()$ are PE-morphisms from $\mathbf{N}$ to the PE-structure (X, x, id_X), hence must be the same. Uniqueness follows because N has a global element 0, so the map $N \to 1$ is epic.

For the converse, let $(N, 0, s)$ satisfy (i) and (ii), so $\langle 0, s\rangle$ is an isomorphism and the coequalizer of id_N and s is 1. If (A, a, t) is any PE structure and $f, g : N \to A$ two PE-morphisms, then the equalizer of f and g would be a PE-substructure of $(N, 0, s)$, so must be all of N by Proposition 7. Hence $(N, 0, s)$ satisfies the uniqueness part of the definition of NNO.

Now suppose (B, b, u) is a PE-structure. Then so is $(N \times B, 0 \times b, s \times u)$. By Proposition 1, this structure must contain a substructure (A, a, t) with $\langle a, t\rangle$ epic. Moreover, the projection maps composed with inclusion give PE-morphisms $A \to N = A \rightarrowtail N \times B \to N$ and $A \to B = A \rightarrowtail N \times B \to B$. Thus if we can show that the map $A \to N$ is an isomorphism, we will be done.

This map $A \to N$ is epic because its image must be a PE-substructure of $(N, 0, s)$, so must be all of it by the Peano property. To see that it is monic (in a topos, monic + epic = iso), it is sufficient to prove:

(**) If $\langle a, t\rangle$ is epi, $\langle 0, s\rangle$ is an isomorphism, the coequalizer of id_N and s is 1, and $f : A \to N$ is a PE-morphism, then f is monic.

The constructions involved in this statement are all preserved by exact functors, so it is enough to prove it in a well-pointed topos. Let $K \rightarrowtail A \times A$ be the kernel pair of f, K' the complement of the diagonal Δ in K, and N' the image of the map $K' \to A \to N$ (take either projection for $K' \to A$—the kernel pair is symmetric). Let M be the complement of N'. We must show $M = N$, for then $N' = 0$, so $K' = 0$ (the only maps to an initial object in a topos are isomorphisms), so $K = \Delta$ and f is then mono.

We show that $M = N$ by using the Peano property.

If a global element n of N is in N' it must lift to at least two distinct global elements a_1, a_2 of A for which $fa_1 = fa_2$. (This is because $\mathrm{Hom}(1, -)$ preserves epis by Proposition 2 of Section 7.1, so an element of N' must lift to an element of K'.) If it is in M it must lift to a unique element of A which f takes to n. Since any element of N must be in exactly one of N' or M, the converse is true too: An element which lifts uniquely must be in M.

Suppose the global element $0 : 1 \to N$ lifted to some a_1 other than a. (Of course it lifts to a.) Since $\langle a, t \rangle$ is epi, $a_1 = t(a_2)$. Thus $0 = ft(a_2) = sf(a_2)$ which would contradict the fact that $\langle 0, s \rangle$ is an isomorphism. Thus by the argument in the preceding paragraph, $a \in M$. A very similar argument shows that if the global element n of N has a unique lifting then so does sn. Hence $sM \subseteq M$, so by the Peano property, $M = N$ as required.

Corollary 8. *An exact functor between toposes takes a NNO to a NNO.*

Exercise 7.5

(NNOP). (a) Show that $\mathcal{S}et \times \mathcal{S}et$ is a topos with natural number object $\mathsf{N} \times \mathsf{N}$.

(b) Show more generally that if $\mathcal{E}_1$ and $\mathcal{E}_2$ are toposes with natural number objects N_1 and N_2, respectively, then $\mathcal{E}_1 \times \mathcal{E}_2$ is a topos with natural number object $\mathsf{N}_1 \times \mathsf{N}_2$.

7.6. Countable Toposes and Separable Toposes

A Grothendieck topos is called **separable** if it is the category of sheaves on a site which is countable as a category and which, in addition, has the property that there is a countable base for the topology. By the latter condition is meant that there is a countable set of covering sieves such that a presheaf is a sheaf if and only if it satisfies the sheaf condition with respect to that set of sieves. Makkai and Reyes [1977] have generalized Deligne's theorem to the case of separable toposes. In the process of proving this, we also derive a theorem on embedding of countable toposes due to Freyd [1972].

Standard toposes

A topos $\mathcal{E}$ is called **standard** if for any object A of $\mathcal{E}$ and any reflexive, symmetric relation $R \subseteq A \times A$ the union of the composition powers of R exists. That is if $R^{(n)}$ is defined inductively by letting $R^{(1)} = R$ and $R^{(n+1)}$ be the image in $A \times A$ of the pullback $R^{(n)} \times_A R$, then we have

$$R \subseteq R^{(1)} \subseteq R^{(2)} \subseteq \cdots \subseteq R^{(n)} \subseteq \cdots$$

and we are asking that this chain have a union. By Exercise (GEQ) of Section 6.8, this union, when it exists, is the least equivalence relation containing by R. Such a least equivalence relation always exists, being the kernel pair of the coequalizer of the two projections of R, so this condition is equivalent to requiring that $\{R^{(n)}\}$ be an epimorphic family over that least equivalence relation. Of course if the topos has countable sums the union may be formed as the image in $A \times A$ of the sum of those composition powers. Hence we have,

Proposition 1. *A countably complete topos is standard.*

Proposition 2. *Let $\mathcal{E}$ be a standard topos and $u : \mathcal{E} \to \mathcal{F}$ be a near exact functor that preserves countable epimorphic families. Then u is exact.*

Proof. It is sufficient to show that u preserves coequalizers. In any regular category the coequalizer of a parallel pair of maps is the same as the coequalizer of the smallest equivalence relation it generates. The reflexive, symmetric closure of a relation $R \subseteq B \times B$ is $R \vee \Delta \vee R^{\text{op}}$, a construction which is preserved by near exact functors. Thus it is sufficient to show that such a functor u preserves the coequalizer of a reflexive, symmetric relation.

So let R be a reflexive, symmetric relation and E be its transitive closure. Then by hypothesis the composition powers, $R^{(n)}$ of R, are dense in E. Hence the images $uR^{(n)} \cong (uR)^{(n)}$ are dense in uE. The isomorphism comes from the fact that all constructions used in the building the composition powers are preserved by near exact functors. Moreover, uE is an equivalence relation for similar reasons: the transitivity comes from a map $E \times_B E \to E$ and this arrow is simply transported to $\mathcal{F}$. The least equivalence relation on uB generated by uR contains every composition power of uR, hence their union which is uE. Since uE is an equivalence relation, this least equivalence relation is exactly uE.

To finish the argument, we observe that near exact functors preserve coequalizers of effective equivalence relations. This is a consequence of the facts that they preserve regular epis, that they preserve kernel pairs and that every regular epi is the coequalizer of its kernel pair.

Proposition 3. *If $u : \mathcal{E} \to \mathcal{F}$ is an exact embedding (not necessarily full) between toposes and $\mathcal{F}$ is standard, so is $\mathcal{E}$.*

Proof. Let R be a reflexive, symmetric relation and E be the least equivalence relation containing it. If the composition powers of R do not form an epimorphic family over E, there is a proper subobject $D \subseteq E$ which contains every $R^{(n)}$. Applying u and using the fact that it is faithful, we find that uD is a proper subobject of uE that contains every $uR^{(n)}$. But the construction of E as the kernel pair of a coequalizer is preserved by exact functors, so the $uR^{(n)}$ must cover uE in the standard topos $\mathcal{F}$. Thus $\mathcal{E}$ must be standard as well.

Proposition 4. *A small topos is standard if and only if it has an exact embedding into a complete topos that preserves epimorphic families. If the domain is Boolean, resp. 2-valued, the codomain may be taken to be Boolean, resp. 2-valued.*

Proof. The "only if" part is a consequence of Propositions 1 and 3. As for the converse, it suffices to take the category of sheaves for the topology of epimorphic families. The functor y of Section 6.8 is an embedding that preserves epimorphic families, and Proposition 2 gives the conclusion.

If $\mathcal{B}$ is a Boolean topos, this embedding may be followed by the associated sheaf functor into the category of double negation sheaves. An exact functor on a Boolean topos is faithful if and only if it identifies no non-zero object to zero (Exercise (Faith) of Section 7.1). But an object B is identified to 0 if and only if $0 \to B$ is dense, which is impossible for $B \neq 0$ as 0 certainly has a complement. If $\mathcal{B}$ is 2-valued, every map to 1 with a non-0 domain is a cover. If $F \neq 0$ is a presheaf and $B \to F$ is an arbitrary element of F, then $B \to F \to 1$ is epi, whence so is $F \to 1$. Thus 1 has no proper subobjects except 0 and so the topos is 2-valued.

A topos with an NNO is called N-standard if the ordinary natural numbers (that is, $0, 1, 2, 3, \cdots$) form an epimorphic family over N. It follows from Theorem 6 of Section 7.5 that an exact functor preserves the NNO, if any. It is clear that any countably complete topos is N-standard since N is then the sum of the ordinary natural numbers. If $\mathcal{E}$ is a standard topos with an NNO, then from Proposition 4 it follows that $\mathcal{E}$ has an exact embedding into a complete topos that preserves epimorphic families. As in the proof of Proposition 3, a faithful exact functor reflects epimorphic families. Thus we have,

Proposition 5. *A standard topos has an exact embedding into an* N-*standard topos that preserves epimorphic families; moreover a standard topos with an NNO is* N-*standard.*

Proposition 6. *An* N-*standard topos has an exact embedding into an* N-*standard Boolean topos that preserves epimorphic families.*

Proof. Apply Theorem 3 of Section 7.1. The embedding preserves N and preserves epimorphic families.

Proposition 7. *A countable* N-*standard Boolean topos has a logical functor to a* 2-*valued* N-*standard Boolean topos.*

Proof. We begin by observing that if U is a subobject of 1 in a topos $\mathcal{E}$, then the induced maps from subobject lattices in $\mathcal{E}$ to those of $\mathcal{E}/U$ are surjective. Moreover a slice functor has a right adjoint and so preserves epimorphic families. Now let $\mathcal{B}$ be a Boolean topos with a standard NNO and let $U_1, U_2, \cdots$ enumerate the subobjects of N. Having constructed the sequence

$$\mathcal{B} \to \mathcal{B}_1 \to \mathcal{B}_2 \to \cdots \to \mathcal{B}_n$$

of toposes and logical functors gotten by slicing by subobjects of 1, we continue as follows. Let m be the least integer for which the image of U_m in $\mathcal{B}_n$ is non-0. Since the natural numbers are an epimorphic family over N, there is at least one natural number, say $p : 1 \to \mathsf{N}$ for which the pullback

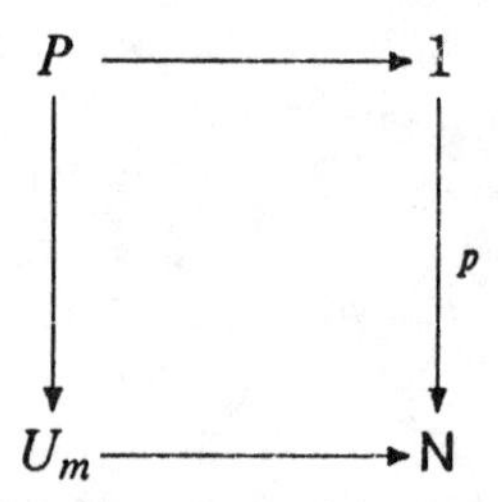

is non-0. The process can even be made constructive by choosing the least such. Then let $\mathcal{B}_{n+1} = \mathcal{B}_n/P$. In the limit topos, every non-0 subobject of N has a global section which is a natural number. There are two conclusions from this. First, the natural numbers cover and second, since

1 is subobject of N, every non-0 subobject of 1 also has a global section which implies that the topos is 2-valued.

It is worth mentioning that the transfinite generalization of this argument breaks down because at the limit ordinals you will lose the property of being N-standard; the epimorphic families may cease being so at the limits (where the transition functors are not faithful; see Lemma 9 below). The proof above uses an explicit argument to get around that.

Corollary 8. *Let $\mathcal{B}$ be a countable N-standard Boolean topos. Then for every non-0 object A of $\mathcal{B}$, there is a logical functor from $\mathcal{B}$ into a countable 2-valued N-standard topos in which A has a global section.*

Proof. Just apply the above construction to $\mathcal{B}/A$ and replace, if necessary, the resultant topos by a countable subtopos that contains A, N and the requisite global sections.

Lemma 9. *Let $\mathcal{E}_\alpha$ be a directed diagram of toposes and faithful logical morphisms which preserve epimorphic families. Let $\mathcal{E} = \operatorname{colim} \mathcal{E}_\alpha$. Then for any α the canonical functor $\mathcal{E}_\alpha \to \mathcal{E}$ is faithful, logical and preserves epimorphic families.*

Proof. We use the notation $T_{\beta,\alpha} : \mathcal{E}_\alpha \to \mathcal{E}_\beta$ for $\alpha \leq \beta$ and $T_\alpha : \mathcal{E}_\alpha \to \mathcal{E}$ for the element of the cocone. Let $f_i : E_i \to E$ be an epimorphic family in $\mathcal{E}_\alpha$. Let $g, h : T_\alpha E \to E'$ be distinct maps in $\mathcal{E}$. Directedness implies the existence of $\beta \geq \alpha$, an object E'' of E_β and (necessarily distinct) maps

$$g', h' : T_{\beta,\alpha}E \to E''$$

such that $T_\beta(g') = g$ and $T_\beta(h') = h$. Since $\{T_{\beta,\alpha}f_i\}$ is also an epimorphic family, there is an index i for which

$$g' \circ T_{\beta,\alpha}f_i \neq h' \circ T_{\beta,\alpha}f_i$$

Since for all $\gamma \geq \beta$, $T_{\gamma,\beta}$ is faithful, it follows that

$$T_\beta g' \circ T_\alpha f_i \neq T_\beta h' \circ T_\alpha f_i,$$

which shows that the $\{T_\alpha f_i\}$ are an epimorphic family. The fact that the T_α are faithful and logical is implicit in what we have done.

Proposition 10. *Every countable N-standard Boolean topos has a logical embedding into a product of N-standard well-pointed toposes.*

Proof. For a non-zero object A of the topos there is, by Corollary 8, a logical functor into a countable 2-valued N-standard topos $\mathcal{B}$ in which A is not sent to zero (in fact has a global section). Enumerate the elements of $\mathcal{B}$ and define a sequence of toposes $\mathcal{B} = \mathcal{B}_0 \subseteq \mathcal{B}_1 \subseteq \mathcal{B}_2 \subseteq \cdots$ in which $\mathcal{B}_{i+1}$ is obtained from $\mathcal{B}_i$ by applying that Corollary to the ith object in the enumeration. If $\mathcal{B}_\omega$ is the direct limit of this process, then every object of $\mathcal{B}$ has a global section in $\mathcal{B}_\omega$. Moreover, it follows from Exercise (FAITH) of Section 7.1 and Lemma 9 that the resultant functor $\mathcal{B} \to \mathcal{B}_\omega$ is logical, faithful and preserves epimorphic families from which it follows that $\mathcal{B}_\omega$ is N-standard. Applying the same argument to the sequence $\mathcal{B} \subseteq \mathcal{B}_\omega \subseteq \mathcal{B}_{\omega\omega} \subseteq \mathcal{B}_{\omega\omega\omega} \subseteq \cdots$, and repeating the above argument gives the required result.

Proposition 11. *An* N*-standard topos is standard.*

Proof. The conclusion is valid as soon as it is valid in every countable subtopos. Propositions 6 and 7 reduce it to the case of a 2-valued N-standard Boolean topos. Accordingly, let $\mathcal{B}$ be such a topos, B be an object of $\mathcal{B}$ and $R \subseteq B \times B$ be a reflexive, symmetric relation on B. Using the mapping properties of N, let $f : \mathrm{N} \to 2^{B\times B}$ be the unique map such that the diagram

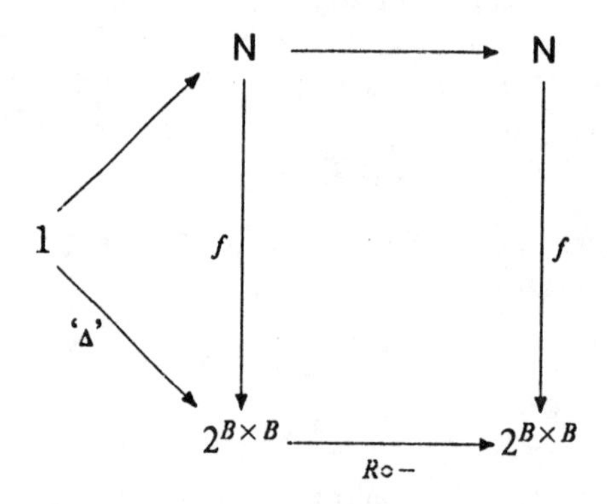

commutes. Here 'Δ' corresponds to the diagonal of $B\times B$ and $R\circ -$ is the internalization of the operation on subobjects of $B \times B$ which is forming the composition with R. It is constructed using the Yoneda lemma.

The transpose of f is a map $g : \mathrm{N} \times B \times B \to 2$ which classifies $Q = [(n, b_1, b_2) \mid (b_1, b_2) \in R^{(n)}]$. The image E of Q in $B \times B$ under the projection is intuitively the set of all $[(b_1, b_2)]$ which are in some $R^{(n)}$.

The diagram

(*)

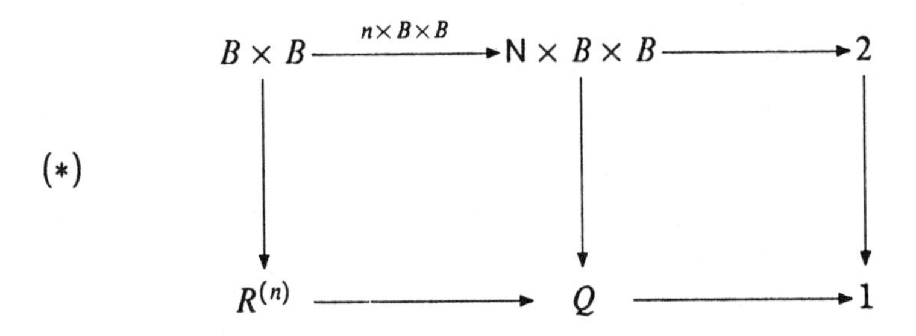

shows that every $R^{(n)}$ is contained in E. We claim that E is an equivalence relation. First observe that in a 2-valued N-standard Boolean topos the only global sections of N are the standard ones (that is are either 0 or one its successors; this is, of course, where the present use of "standard" comes from). For when 1 has no proper non-0 subobjects, any two global sections are equal or disjoint. Moreover, in an N-standard topos, no subobject of N, hence no global section can be disjoint from every standard global section. Hence, every global section is a standard one. Second observe that in a well-pointed topos, two subobjects of an object are equal if and only if they admit the same global sections. Thus to show that $E \circ E \subseteq E$, it is sufficient to show it on global sections. A global section of E lifts to one on Q and a global section on $E \circ E$ lifts to a pair of sections (n, b_1, b_2) and (m, b_2, b_3) such that $(b_1, b_2) \in R^{(n)}$ and $(b_2, b_3) \in R^{(m)}$. We can easily show by induction that

$$R^{(n)} \circ R^{(m)} \subseteq R^{(m+n)} \subseteq E$$

from which it follows that E is an equivalence relation.

To complete the argument, we observe that since the natural numbers cover N and products have adjoints, the maps $n \times B \times B$ cover $\mathsf{N} \times B \times B$ as n varies over the standard natural numbers. Since (*) is a pullback, it follows that the various $R^{(n)}$ cover Q and *a fortiori* cover E. Hence $\mathcal{B}$ is standard.

Theorem 12 (Freyd). *Every countable standard topos can be embedded exactly into a power of* **Set**.

Proof. By Proposition 5 a countable standard topos has an exact embedding into an N-standard topos which by Proposition 6 can be exactly embedded into an N-standard Boolean topos. By replacing the target by a countable subtopos, we can apply Proposition 10 to embed it logically—hence exactly—into a product of N-standard well-pointed toposes. In a well-pointed topos, the functor Hom$(1, -)$ is an exact $\mathcal{S}et$ -valued functor. Putting this all together, we have the desired conclusion.

Recall that a separable topos is the category of sheaves on a site which is both countable and for which there is a set of countable covers that generate the topology.

Theorem 13 (Makkai and Reyes). *If $\mathcal{E}$ is a separable Grothendieck topos, then there is a faithful family of $\mathcal{S}et$ -valued left exact functors on $\mathcal{E}$ which have right adjoints.*

$\mathcal{S}et$ -valued functors on a topos which are left exact and have a right adjoint are called **points.** This theorem says that a separable Grothendieck topos has enough points.

Proof. Let $\mathcal{E}$ be a separable topos. For each object C of $\mathcal{E}$ and proper subobject E of C, we construct inside $\mathcal{E}$ a countable subcategory $\mathcal{C}$ with certain properties. It should be a topos; it should contain the image of $E \rightarrowtail C$, the NNO of $\mathcal{E}$ (so $\mathcal{C}$ is standard), and a generating set for $\mathcal{E}$. It should contain all the objects used in the countably many sieves $\{A_i \to A\}$ that generate the topology. It should also include the sum S of the objects that appear in each such sieve, together with the coproduct injections, the induced map $S \to A$ and the unique arrow $S \to \mathsf{N}$ for which

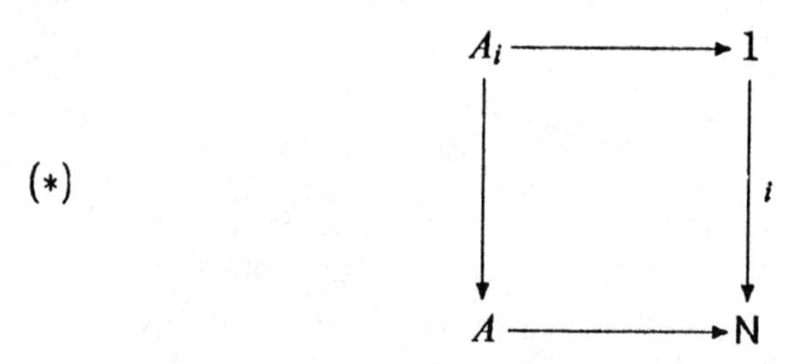

is a pullback. Since we are closing a countable category under finitary and countable operations, the resulting category is countable, by exactly the same kind of argument that you use to show that the free group generated by a countable set is countable. Note that S is not in general the categorical sum in $\mathcal{C}$.

Now apply Freyd's Theorem 12 to get a faithful family of exact functors $\mathcal{C} \to \mathcal{S}et$. Each such functor takes (*) to a similar diagram from which it is clear that each such functor preserves the sum $\sum A_n$. Since that sum maps epimorphically onto A, each of the functors preserves the covers which generate the topology and hence extends to a point of $\mathcal{E}$.

7.7. Barr's Theorem

By adapting the proof of Theorem 4 of Section 7.1, we can obtain the following proposition, from which we can deduce an embedding theorem for any Grothendieck topos.

Proposition 1. *For every small Boolean topos $\mathcal{B}$ there is a small Boolean topos $\overline{\mathcal{B}}$ in which subobjects of* 1 *generate, and a logical morphism $\mathcal{B} \to \overline{\mathcal{B}}$ which preserves epimorphic families.*

Well-order the set of objects of $\mathcal{B}$ which have global support. Given a Boolean topos $\mathcal{B}$ and an object A of global support, the map $\mathcal{B} \to \mathcal{B}/A$ is faithful by Exercise (SF) of Section 5.2, and preserves epimorphic families because it has a right adjoint. Moreover, the image of A has a global section. Thus we can construct a well-ordered sequence of Boolean toposes $\mathcal{B}_\alpha$ together with faithful logical morphisms $T_{\beta,\alpha} : \mathcal{B}_\alpha \to \mathcal{B}_\beta$ whenever $\alpha \leq \beta$ as follows. Begin by letting $\mathcal{B}_0 = \mathcal{B}$. Having constructed $\mathcal{B}_\alpha$ together with the appropriate transition functor, let A be the least object in the well-ordering which has global support and lacks a global section. Take $\mathcal{B}_{\alpha+1} = \mathcal{B}_\alpha/A$. At a limit ordinal take a direct limit. In the latter case preservation of epimorphic families follows from Lemma 9 of Section 7.6. Taking the direct limit of this sequence, we get a Boolean topos $\tilde{\mathcal{B}}$ and a faithful logical morphism $\mathcal{B} \to \tilde{\mathcal{B}}$ which preserves epimorphic families, again by Lemma 9. Moreover, every object of $\mathcal{B}$ with global support has a global section in $\tilde{\mathcal{B}}$.

By iterating this construction a countable number of times, and taking the directed limit of the resulting sequence, we get a topos $\overline{\mathcal{B}}$ and a logical morphism $\mathcal{B} \to \overline{\mathcal{B}}$ which is faithful and preserves epimorphic families. In $\overline{\mathcal{B}}$, every object with global support has a global section. If A is an arbitrary object of $\overline{\mathcal{B}}$ with support S, let T be the complement of S in the subobject lattice of 1. Then the object $A + T$ has a global section which restricts to a section of $A \to S$. Hence the subobjects of 1 generate.

Theorem 2 (Barr). *Every Grothendieck topos has a left exact cotripleable embedding into the topos of sheaves on a complete Boolean algebra.*

Proof. To show that an exact functor is cotripleable, it is sufficient to show that it is faithful and has a right adjoint. Let $\mathcal{E}$ be a Grothendieck topos. Let $\mathcal{C}$ be a small topos contained in $\mathcal{E}$ which contains a set of generators for $\mathcal{E}$ so that $\mathcal{E}$ is the category of canonical sheaves on $\mathcal{C}$ as in Theorem 1 of 6.8. Combining Theorem 3 of 7.1 with the above proposition we conclude that $\mathcal{C}$ can be embedded into a Boolean topos $\mathcal{B}^{\#}$ which is generated by its subobjects of 1 so that the embedding is left exact and preserves epimorphic families. If $\hat{B}$ is the completion of the Boolean algebra B of subobjects of 1 in $B^{\#}$, we claim that $\mathcal{B}^{\#}$ is embedded in the category $Sh(\hat{B})$ of sheaves on $\hat{B}$ and this embedding is exact and preserves epimorphic families. Assuming this true, the result follows from Theorem 2 of 7.3 plus Theorem 1 of 6.8.

To complete the argument, we must show that $\mathcal{B}^{\#}$ is embedded in $Sh(\hat{B})$. There is a functor $\mathcal{B}^{\#} \to Sh(\hat{B})$ given by representing an element $b \in \hat{B}$ as $\sup b_i$, with b_i in B and defining, for an object C of $\mathcal{B}^{\#}$ the presheaf on $\hat{B}$ given by $\operatorname{colim} \operatorname{Hom}(b_i, C)$. These are not necessarily sheaves, so we must follow this by the associated sheaf functor. To see that the resultant composite is faithful, let $C \rightarrowtail D$ be a non-isomorphic mono in $\mathcal{B}^{\#}$ and let E be a complement of C in D. Then E is not the initial object of $\mathcal{B}^{\#}$ so that it has a section over a non-zero object of B and hence is not the initial presheaf. The associated sheaf contains a quotient of E and cannot be the initial sheaf. But it remains a complement of the sheaf associated to C so that the inclusion of C into D does not induce an isomorphism and the functor is faithful.

Note that the existence of the right adjoint to this left exact embedding means that the embedding is the left adjoint part of a geometric morphism. Part of the significance of this result arises from the following.

Theorem 3. *Let $\mathcal{E}$ be the category of sheaves on a complete Boolean algebra. Then $\mathcal{E}$ satisfies the Axiom of Choice.*

Proof. We begin by observing that if B is a complete Boolean algebra, a functor $F : B^{\mathrm{op}} \to Set$ is a sheaf if and only if whenever $b = \bigcup b_i$ is a *disjoint* union, $Fb \cong \prod Fb_i$, the isomorphism being the canonical map. The condition is necessary for then the b_i cover b and the intersection of any two of them is empty. To see that the condition is sufficient it is enough to note that every cover has a refinement which is disjoint. For if $b = \bigcup b_i$ is not disjoint, choose a simple ordering of the index set and replace b_i by $b_i - \bigcup b_j$, the union taken over the $j < i$.

Next we note that an epimorphism is onto. Let $f : F \twoheadrightarrow G$ be an epimorphism of sheaves and $G_0 \rightarrowtail G$ be the presheaf image of the map. We claim that $G_0 = G$. For let $b = \bigcup b_i$ be a disjoint union. In the

diagram

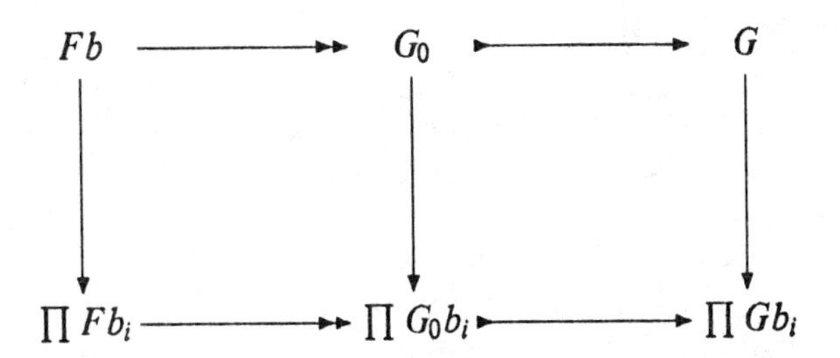

it is easy to see that when the outer vertical arrows are isomorphisms, so is the middle one. Thus G_0 is a sheaf, which shows that $G_0 = G$. Now to split the epi f, we choose a maximal element (J, h) among the partially ordered set of pairs (I, g) in which I is an ideal of B and g is a splitting of $f|I$. If J is not the whole of B, we consider separately the cases that J is or is not principal. In the former case, let $J = (b)$ and suppose $b' \notin J$. Then we may replace b' by $b' - b$ and suppose that b' is disjoint from b. Given an element of $x \in Gb'$, choose an arbitrary element of $y \in Fb'$ mapping to it and extend h by $h(b' \cup c) = y \cup hc$. It is easy to see that this gives a morphism of presheaves, hence of sheaves. If J is not principal, let $b = \bigcup J$ and extend h to b by $h(b) = \bigcup h(J)$. We see, using the fact that F and G are sheaves that this extends h. This completes the construction of a section of f.

7.8. Notes to Chapter 7

The results of Section 7.1, Section 7.5, and the parts of Section 7.6 pertaining to standard toposes appeared in a remarkable paper that Peter Freyd wrote during a visit to Australia in 1971 [Freyd, 1972]. These results form the basis for the modern representation theory of toposes. We have followed Freyd's exposition quite closely.

Although Barr's theorem [Barr, 1974] appeared two years later than Freyd's paper, the work was done in ignorance of Freyd's work with a proof quite different from that presented here. The latter is, of course, based on the ideas used by Freyd. There is an entirely different proof of this theorem due to Diaconescu which is given in Johnstone [1977]. The result was in response to a question of Lawvere's as to whether the example of a Boolean-valued model of set theory for which the Boolean algebra lacked points (complete 2-valued homomorphisms) was essentially the most general example of a topos without points. The result, that every topos has a faithful point in a suitable Boolean topos, showed that the Lawvere's surmise was correct.

Diaconescu [1975] was the first to show that AC implied Boolean.

Deligne's theorem [SGA 4, 1972] was proved for the purposes of algebraic geometry. The original proof was completely different. So far as we know, this is the first place in which it has been derived as a simple consequence of Freyd's theorem.

Makkai and Reyes [1977] proved the theorem credited to them without reference to Freyd's work. Again, this seems to be the first place in which the proof is done using Freyd's results.

Johnstone [1977] has put down Freyd's representation theorems as "... something of a blind alley". This chapter clearly demonstrates the utility of the theorems. It is possible, of course, to want to avoid the use of Freyd's theorems out of dislike of the use of representation theory for proving things, or from a more general preference for elementary or constructive methods. We do not share those attitudes. We feel it is a matter of taste whether, for example, the proof we have given of the fact that AC implies Boolean is better or worse than a harder, but elementary proof. We generally prefer a proof which is more readily understood. (That is not necessarily the same as easier—see our proof of Proposition 1 of Section 7.3).

There are other useful representation theorems. For example, Makkai and Reyes [1977, Theorem 6.3.1] show that any Grothendieck topos can be embedded into the topos of sheaves on a complete Heyting algebra by a functor which is the left adjoint part of a geometric morphism and which preserves all infs as well as $\forall$. They also show [Theorem 6.3.3] that when there are enough points the Heyting algebra can be taken to be the open set lattice of a topological space.

In both Abelian categories and regular categories there is a full embedding theorem, which states that there is a *full*, exact embedding into a standard category. In the case of Abelian categories, the standard categories are the categories of R-modules for rings R, while in the regular case it was $\mathcal{Set}$-valued functor categories. (A $\mathcal{Set}$-valued functor category can be viewed as the category of M-sets where is M is a monoid with many objects, i.e. a category.) The corresponding theorem for toposes would be a full, exact embedding into a functor category. Makkai [unpublished] has given an example of a topos that has no such full embedding. Fortunately, these full embeddings have had very limited usefulness. The existence of an embedding that reflects isomorphisms has allowed all the diagram-chasing arguments that one seems to need.

It should be observed that hypotheses that a topos be small or even countable are not a significant limitation on results used for diagram-chasing. Any diagram involves only a finite number of objects and morphisms and can be taken as being in some countable subtopos. We have already illustrated the technique in Section 7.6.

One of the main thrusts of categorical logic is the exploitation of the insight that each pretopos corresponds to a theory in the sense of model theory (a language, a set of deduction rules and a set of axioms), and vice versa. Under this equivalence, embedding theorems correspond to completeness theorems—theorems to the effect that if a statement made in the language is true in every model of a certain type, then it follows from the axioms. In particular, Deligne's theorem is an easy consequence of Gödel's completeness theorem for finitary first order logic. In fact it is equivalent to that theorem for the case of finitary geometric theories. Barr's theorem can be interpreted as saying that if something follows by classical logic from the axioms, then it follows by intuitionistic logic. See Makkai, Reyes [1977] and Lambek-Scott [to appear] for more details.

Chapter 8.

Cocone Theories

In this chapter, we consider a general type of theory which is left exact and which in addition has various types of cocones included in the structure. In the kinds of theories considered here, the family of cocones which is part of the structure is always induced by the covers of a Grothendieck topology according to a construction which we will now describe.

Let $S = \{A_i \to A\}$ be a sieve indexed by I in a left exact category $\mathcal{C}$. Form a graph $\mathcal{I}$ whose objects are $I + (I \times I)$, with arrows of the form

$$i \xleftarrow{\ l\ } (i,j) \xrightarrow{\ r\ } j$$

We define a diagram $D : \mathcal{I} \to \mathcal{C}$ which takes i to A_i and (i,j) to the fiber product $A_i \times_A A_j$. The **cocone induced by** S is the cocone from D to A whose arrows are the arrows of the sieve S.

The reason for this restriction is not that this is the only conceivable kind of cocone theory, but this the only kind of theory for which there is a generic topos (classifying topos).

8.1. Regular Theories

A **regular sketch** $\mathcal{R} = (\mathcal{G}, U, D, C, E)$ is a sketch $(\mathcal{G}, U, D, C)$ together with a class E of arrows in $\mathcal{G}$. A **model** of $\mathcal{R}$ is a model of the sketch which satisfies the additional condition that every arrow in E is taken to a regular epimorphism.

A **preregular theory** is a left exact theory Th together with a class E of arrows. A model of a preregular theory is a left exact functor which takes every arrow in E to a regular epimorphism. It is clear that if $\mathcal{R}$ is a regular sketch, then the left exact theory generated by $\mathcal{R}$ in the sense of Section 4.4 is a preregular theory, which we will denote $\mathrm{PR}(\mathcal{R})$, with the "same" class E of arrows.

The sieves are, of course, the single arrow sieves and the corresponding cocones consist of diagrams

$$(*) \qquad A \times_B A \rightrightarrows A \xrightarrow{e} B$$

for $e \in E$. If a functor M is left exact, then it is immediate that M takes $(*)$ into a colimit if and only if $M(e)$ is regular epi. Thus models are indeed characterized by the properties of being left exact and taking the corresponding cocones to colimits.

A **regular theory** Th is a regular category, i.e., a left exact category in which regular epis are stable under pullbacks. A model of a regular theory is a regular functor which we will always suppose to take values in a regular category. This is the same as the model of the underlying regular sketch which has all regular epis as its distinguished class of morphisms.

In this section, we show how to begin with a regular sketch $\mathcal{R}$ and construct the regular theory induced by $\mathcal{R}$. In regular categories, it will have the same models as $\mathcal{R}$.

An example of such a theory which arises in real life is the theory of regular rings. (The coincidence of terminology is purely accidental). A **regular ring** is a ring A in which for every element a there is an element b such that $aba = a$. This condition can be rewritten as follows: Define $B = \{(a, b) \mid aba = a\} \subseteq A \times A$. Then A is a regular ring if and only if the composite $B \to A \times A \to A$ (the last map is the first projection) is surjective.

The regular sketch to describe regular rings can be constructed as follows. Begin with the sketch for rings in which A denotes the ring and add an object B together with an arrow $B \to A^2$ and a cone that forces the image of B in any model to be $[(a, b) \mid aba = a]$. Then E consists of the single arrow $B \to A$ that corresponds to the first projection.

Notice that B and the map from B to A can be defined in an arbitrary left exact category (B is an equalizer). We say that a ring object in an arbitrary regular category is a regular ring object if the map $B \to A$ is a regular epi. Of course, other definitions of regular ring object in a category are conceivable; e.g., one could ask that the map be a split epi. However, an attempt to formalize this would almost surely lead to the introduction of a splitting map defined by equations as part of the structure. This would lead to a non-full subcategory of the category of rings. In the case of commutative rings and some classes of non-commutative rings, this equational definition is actually equivalent to the existential one.

Another example of a regular theory is the theory of groups in which every element is an nth power for some fixed $n > 1$ (see Exercise (Div)).

Regular theories from regular sketches

If $\mathcal{R}$ is a regular sketch, then as remarked above, it generates a preregular theory $\mathrm{PR}(\mathcal{R})$ with the same models by following the process of Section 4.4 and taking for the class of arrows the image of the given class in the regular sketch.

This preregular theory $\mathrm{PR}(\mathcal{R})$ generates a site by closing the class E under pullbacks and composition to obtain a topology. The covers in this topology each consist of only one arrow. This site is also a preregular theory, with E consisting of all arrows which are covers in the resulting topology.

Proposition 1. *A model in a regular category of* $\mathrm{PR}(\mathcal{R})$ *is the same as a model for the site generated by* $\mathcal{R}$.

Proof. This follows from the fact that a model preserves pullbacks and composition, and a pullback of a regular epi in a regular category is a regular epi.

Now given a regular sketch $\mathcal{R}$, we define $\mathrm{Reg}(\mathcal{R})$, the regular theory associated to $\mathcal{R}$, to be the full image of the composite

$$\mathcal{A} \xrightarrow{\;Y\;} \mathbf{Set}^{\mathcal{A}^{\mathrm{op}}} \xrightarrow{\;L\;} \mathbf{Sh}(\mathcal{Th})$$

where $\mathcal{A}$ is the site generated by $\mathcal{R}$, Y is the Yoneda embedding and L is sheafification. Observe that $\mathrm{Reg}(\mathcal{A}) = \mathrm{Reg}(\mathcal{R})$.

The induced map $\mathcal{A} \to \mathrm{Reg}(\mathcal{A})$ will be denoted y. Note that y is left exact. If $\mathcal{A}$ is already a regular category, then by Proposition 4 of Section 6.7, $\mathcal{A} = \mathrm{Reg}(\mathcal{A})$.

Proposition 2. *In the notation of the preceding paragraph, the covers in* $\mathcal{A}$ *become regular epis in* $\mathrm{Reg}(\mathcal{A})$.

Proof. Let $f : B \to A$ be a cover. In a left exact category to say that yf is a regular epi is to say that

$$y(B \times_A B) \cong yB \times_{yA} yB \rightrightarrows yB \longrightarrow yA$$

is a coequalizer. This is equivalent to saying that for every sheaf F,

$$\mathrm{Hom}(yA, F) \longrightarrow \mathrm{Hom}(yB, F) \rightrightarrows \mathrm{Hom}(y(B \times_A B), F)$$

is an equalizer. Now $yA = LY(A)$ where L is left adjoint to the inclusion of the sheaf category, so

$$\mathrm{Hom}(y(-), F) \cong \mathrm{Hom}(Y(-), F) \cong F(-).$$

Hence

$$FA \longrightarrow FB \rightrightarrows F(B \times_A B)$$

must be an equalizer, which is exactly the condition that F be a sheaf.

Proposition 3. *Let $\mathcal{A}$ and $\mathcal{B}$ be preregular theories and $f : \mathcal{A} \to \mathcal{B}$ a left exact functor which takes arrows in the distinguished class of $\mathcal{A}$ to arrows in the distinguished class of $\mathcal{B}$. Then there is a unique regular functor* $\mathrm{Reg}(f) : \mathrm{Reg}(\mathcal{A}) \to \mathrm{Reg}(\mathcal{B})$ *for which*

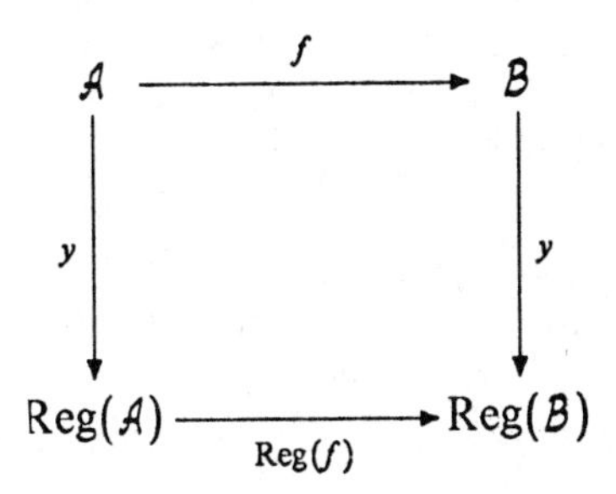

commutes.

Proof. The condition on f is equivalent to the assertion that it is a morphism of the associated sites. The map $f_\#$ constructed in Theorem 2 of Section 7.3 clearly takes yA into yB, and so takes $\mathrm{Reg}(\mathcal{A})$ into $\mathrm{Reg}(\mathcal{B})$.

Corollary 4. *If $\mathcal{B}$ in the diagram above is a regular theory and $\mathcal{A}$ is the site associated to some regular sketch $\mathcal{R}$, then there is a bijection between models of $\mathcal{R}$ in $\mathcal{B}$ and regular morphisms from* $\mathrm{Reg}(\mathcal{R}) = \mathrm{Reg}(\mathcal{A})$ *to* $\mathcal{B}$.

In other words, every regular sketch $\mathcal{R}$ has a model in a universal regular theory $\mathrm{Reg}(\mathcal{R})$ which induces a bijection on models.

Exercise 8.1

(Div). (a) Let $n > 1$ be an integer. Say that a group G is **n-divisible** if for any $a \in G$ there is a $b \in G$ for which $b^n = a$. Show that the category of n-divisible groups (and all group homomorphisms between them) is the category of models for a regular theory.

(b) Show that the group of all 2^n roots of unity is 2-divisible. This group is denoted Z_{2^∞}.

(c) Show that in the category of 2-divisible groups, the equalizer of the zero map and the squaring maps on Z_{2^∞} is $Z/2Z$, which is not divisible. Conclude that there is no left exact theory which has this category as its category of algebras and for which the underlying set functor is represented by one of the types.

8.2. Finite Sum Theories

If we wanted to construct a theory of fields, we could clearly start with the sketch for commutative rings, let us say with an object F representing the ring. Since the inverse is defined only for the subset of nonzero elements, we need to add an object Y to be the nonzero elements and a map $Y \to Y$ together with equations forcing this map to be the inverse map. All this is clear, except how to force Y to be the nonzero elements. We will see in Section 8.4 that this cannot be done in an LE or even a regular theory.

The approach we take is based on the observation that a field is the sum (disjoint union) of Y and a set $Z = \{0\}$. Thus to the LE-sketch of commutative rings we add objects Y and Z and arrows

(1) $$Z \longrightarrow F \longleftarrow Y$$

Besides this, we need an arrow from Y to Y which takes an element to its inverse. and a diagram forcing any element of Z to be zero (remember zero is already given by an arrow from the terminal object to F in the LE-theory of commutative rings). This can be done by techniques of Chapter 4 so we will not give details here.

A field is a model of this theory which takes (1) to a sum diagram. This construction suggests that we vary the concept of regular theory defined in Section 8.1 to allow more general classes of covers. In this section, we develop the idea of a finite-sum theory.

A **finite-sum sketch** or **FS-sketch** $\mathcal{S} = (\mathcal{G}, U, D, C, E)$ is a sketch $(\mathcal{G}, U, D, C)$ with a class E of finite sieves. A model of $\mathcal{S}$ is a LE-model of the sketch $(\mathcal{G}, U, D, C)$ which takes each sieve to a sum diagram. In this book, the models will be in left exact categories with disjoint finite universal sums.

A **pre-FS-theory** is an LE-theory together with a distinguished class of sieves. Again, a model of the theory is an LE model which takes the sieves to sums. An FS-sketch clearly induces a pre-FS-theory by taking the theory to be the LE completion of the sketch and taking as distinguished sieves those corresponding to the distinguished sieves of the original sketch.

An **FS-theory** is a left exact category with finite disjoint universal sums. It will be regarded as a pre-FS-theory by taking all finite sums as distinguished sieves. A model of one FS-theory in another is a left exact functor that preserves finite sums.

An FS-sketch $\mathcal{S}$ induces an FS-theory FS($\mathcal{S}$) in the following way. Begin with the LE-theory and take as covers the images of all the sieves in the original sketch. To this add to the covers, for any sieve $\{A_i \to A\}$ in $\mathcal{S}$, the sieves $\{A_i \to A_i \times_A A_i\}$ (using the diagonal map) and for any $i \neq j$, the empty sieve with vertex $A_i \times_A A_j$. Then add all sieves obtained from the above by pullbacks and composition. Finally, take as FS($\mathcal{S}$) the smallest subcategory closed under finite sums of the category of sheaves for that topology which contains the image of $\mathcal{S}$.

Proposition 1. *Let $\mathcal{S}$ be an FS-sketch. Any model of $\mathcal{S}$ in a left exact category with finite disjoint universal sums extends to a model of the associated FS-theory.*

Proof. It is an easy exercise to show that in a left exact category with disjoint (finite) sums, a cocone $\{A_i \to A\}$ is characterized as a sum by the following assertions:

(i) $\sum A_i \twoheadrightarrow A$
(ii) $A_i \twoheadrightarrow A_i \times_A A_i$
(iii) $0 \twoheadrightarrow A_i \times_A A_j$ for $i \neq j$

Theorem 2. *Any FS-sketch has a model in a universal FS-theory which induces a bijection on models.*

Proof. Essentially the same as Corollary 4 of Section 8.1 (the covers correspond to the regular epis there).

Exercise 8.2

(Toto). Prove that total orderings and non-decreasing maps are models of a finite-sum theory. (Hint: express the order as a strict order and consider trichotomy).

8.3. Geometric Theories

A local ring is charactized as a ring A in which for each $a \in A$ either

a or $1 - a$ has an inverse. Both may be invertible, however, so that $\{a \mid a \text{ is invertible }\}$ and $\{a \mid 1 - a \text{ is invertible }\}$ are not necessarily disjoint. We will describe in this section a more general kind of theory in which such predicates may be stated.

A **geometric sketch** $\mathcal{S} = (\mathcal{G}, U, C, D, E)$ is a sketch $(\mathcal{G}, U, C, D)$ together with a class E of sieves. A model of a geometric sketch is a model of the sketch in a pretopos such that the sieves are sent to regular epimorphic families. (Recall that a pretopos is a regular category with effective equivalence relations and disjoint universal finite sums.) Note that you can still force a particular sieve to go to a sum by adding fiber products forced to be zero as in the preceding section.

A **pre-geometric theory** is a left exact category together with a class of sieves. As above, a model is a left exact functor which takes the given sieves to regular epimorphic families. A geometric sketch induces, in an obvious way, the structure of a sketch on its associated left exact theory.

A **geometric theory** is simply a Grothendieck topos. A model is a functor which is the left adjoint part of a geometric morphism. Using the special adjoint functor theorem, it is not hard to show that a left exact functor between Grothendieck toposes which takes covers to covers has a right adjoint, so that provides an equivalent definition of geometric morphism. A model in $\mathcal{S}et$ is also known as a **point**. We have:

Theorem 1. *Every geometric sketch has a model in a universal geometric theory which induces a bijection on models.*

Proof. It is clear that there is a bijection between the models of the geometric sketch and models of the associated pregeometric theory. The rest is similar to the proof of Corollary 4 of Section 8.1.

An important variation on the notion of geometric theory is that of a coherent sketch. A coherent sketch is a sketch in which all the sieves are finite. A model is required to be a model of the associated LE-theory which takes the sieves to regular epimorphic families. However, the models are permitted to take values in an arbitrary pretopos.

A coherent theory is the same thing as a pretopos. We leave to the reader the task of modifying the constructions given above to find the coherent theory associated to a coherent sketch and of proving the analog of Theorem 1.

The model of the sketch in the universal geometric theory is called the **generic** model. The generic topos is called the **classifying topos** for the theory. Of course, as we have defined things, the classifying topos for the geometric theory *is* the theory. However, there are many kinds of theories besides geometric (FP, LE, Regular, FS, coherent) and they all have classifying toposes.

The classifying topos of a pre-geometric theory is often constructed directly; that is one adds directly all the necessary disjoint sums, quotients of equivalence relations, etc. necessary to have a topos. This bears about the same relation to our construction as does the construction of the free group using words does to the argument using the adjoint functor theorem. In each case, both constructions are useful. The one is useful for getting information about the detailed internal structure of the object while the other is useful for the universal properties. Less obvious is the fact that the syntactic construction provides a convenient locus for the semantic one to take place in. Did you ever wonder what x was in the polynomial ring $k[x]$ (which is the free k-algebra on a singleton set)? Even if x is defined, what is a "formal sum of powers of x"?

For more about classifying toposes, see Johnstone [1977], Makkai and Reyes [1977] and Tierney [1976]. The realizability topos, which is the classifying topos of the theory of recursive functions (properly defined), allows topos theory to encompass recursive funtion theory. Details are in Hyland [1982].

Exercises 8.3

(CTFP). Let $\mathcal{S}$ be a single sorted FP-sketch with generic object (generating sort) B. $\mathcal{S}$ generates a geometric theory $\mathcal{G}$ by taking E to be empty. Let $Th = \mathrm{LE}(\mathcal{S})$.

(a) Show that $\mathit{Set}^{\mathcal{A}}$ is the geometric theory associated to $\mathcal{S}$, where $\mathcal{A} = Th^{\mathrm{op}}$ is the category of finitely presented algebras. (See Theorem 5 of Section 4.1.)
(b) Show that the generic model of $\mathcal{S}$ in $\mathcal{G}$ takes B to the underlying functor in $\mathit{Set}^{\mathcal{A}}$.

(MDU). Let $\mathcal{S}$ be an LE-sketch and $\mathcal{G}$ be the geometric theory generated by $\mathcal{S}$, taking E to be empty.

(a) Show that $\mathcal{G}$ is a full subcategory of $\mathit{Set}^{\mathrm{LE}(\mathcal{S})}$.
(b) Show that the generic model of $\mathcal{S}$ in $\mathcal{G}$ takes each object B of $\mathcal{S}$ to the functor from $\mathrm{LE}(\mathcal{S})$ to Set which evaluates at B, and each arrow of $\mathcal{S}$ to a natural transformation between such functors.
(c) Show that if f is a model of $\mathcal{S}$ in a topos $\mathcal{E}$, then the induced model $f^* : \mathcal{G} \to \mathcal{E}$ is determined uniquely by what it does to the objects and arrows of $\mathcal{S}$. (This is the semantic explication of the generic nature of the geometric theory of $\mathcal{S}$, as opposed to the syntactical one of Theorem 1. Note that a similar exercise can

be done for FS- and geometric sketches, with the functors from $\mathbf{LE}(\mathcal{S})$ replaced by sheaves).

(INF). Show that there is a coherent theory whose models correspond to infinite sets. (Hint: For any set S and any natural number n, let S^n denote, as usual, all functions from n to S. If E is an equivalence relation on n, let $S^E \subseteq S^n$ denote all functions whose kernel pair is exactly E. Then S^n is the disjoint union of all the S^E over all the equivalence relations. Moreover, S is infinite if and only for all n and all equivalence relations E in n, S^E is non-empty, meaning its terminal map is epi.)

(DLO). Construct a geometric theory which classifies the category of dense linear orderings and non-decreasing maps. (See Exercise (TOTO) of Section 8.2.) Show that the topos is Boolean.

8.4. Properties of Model Categories

In this section we raise and partly answer questions of recognizing categories as categories of models of different sorts of theories. The answers we give are in the form of properties that categories of models have, so that any category lacking them is not such a category. For example, the category of categories is not a regular category and hence cannot be the category of algebras of any triple (Corollary 4 of Section 3.4 and Theorem 5 of Section 4.3) or even the category of models of any FP-theory (see Theorem 1 below).

In the following theorem we refer to these properties of a category $\mathcal{C}$ of models in $\mathcal{S}et$ of a sketch $\mathcal{S}$. "Underlying functors" are functors which take a model to the set corresponding to a given object B of the sketch.

L: $\mathcal{C}$ has all limits and the underlying functors preserve them.

FC: $\mathcal{C}$ has all filtered colimits and the underlying functors preserve them.

R: $\mathcal{C}$ is regular.

EE: $\mathcal{C}$ has effective equivalence relations and the underlying functors preserve their coequalizers.

Other properties to which we refer require definitions.

For the purpose of the definitions that follow we must give a more general definition of a regular epimorphic sieve in the case that the ambient category lacks the relevant pullbacks. In Exercise (EPIS) you are invited to show that the two definitions agree in the presence of pullbacks.

A sieve $\{f_i : A_i \to A\}$ is said to be **regular epimorphic** if whenever B is any object and for each i there is given a morphism $g_i : A_i \to B$ with the property that for any object C of the category and pair of maps

$d^0 : C \to A_i, d^1 : C \to A_j$, $f_i \circ d^0 = f_j \circ d^1$ implies $g_i \circ d^0 = g_j \circ d^1$, then there is a unique $h : A \to B$ such that $g_i = h \circ f_i$ for all i. The class $\mathcal{C}$ of objects may be replaced without loss of generality by a generating set.

An object G of a category is a **regular projective generator** if

(i) It is regular projective: for any regular epi $A \to B$, the induced map

$$\mathrm{Hom}(G, A) \to \mathrm{Hom}(G, B)$$

is surjective, and

(ii) It is a regular generator: the singleton $\{G\}$ is a regular generating family (Section 6.8).

More generally, a set G of objects is a **regular projective generating set** if it is a regular generating family in which each object is a regular projective.

An **ultrafilter** on a set I is a filter $\mathfrak{f}$ (under inclusion) of subsets of I with the property that for each set J, either J or its complement belongs to $\mathfrak{f}$. It is an easy exercise to see that if $J \cup K \in \mathfrak{f}$, then at least one of J or K must be. In a category $\mathcal{C}$ with all products, an **ultraproduct** is an object constructed this way: Begin with a family $\{(C_i)\}, i \in I$ of objects and an ultrafilter $\mathfrak{f}$. For each set J in the ultrafilter we define an object $N(J) = \prod_{i \in J}(C_i)$. If $K \subseteq J$, then the universal property of products induces an arrow $N(J) \to N(K)$. This produces a filtered diagram and the colimit of that diagram is the ultraproduct induced by $\prod(C_i)$ and $\mathfrak{f}$. However, the concept of ultraproduct does not itself have any universal mapping property. Note that since an ultraproduct is a filtered colimit of products, L and FC together imply that a category has ultraproducts.

In the theorem below, we list a number of properties of categories of models. We would emphasize that the categories may well and often do permit other limit and colimit constructions. The ones mentioned in the theorem below are limited to those which are preserved by the functors which evaluate the models (which are, after all, functors) at the objects of the sketch. Equivalently, they are the constructions which are carried out "pointwise", meaning in the category of sets where the models take values.

Theorem 1. (a) *If $\mathcal{S}$ is a single sorted FP-sketch, then $\mathcal{C}$ has L, FC, R, EE and a regular projective generator.*

(b) *If $\mathcal{S}$ is an FP-sketch, then $\mathcal{C}$ has L. FC, R, EE and a regular projective generating set.*

(c) *If $\mathcal{S}$ is an LE-sketch, then $\mathcal{C}$ has L and FC.*

(d) *If $\mathcal{S}$ is a regular sketch, then $\mathcal{C}$ has FC and all products.*

(e) *If $\mathcal{S}$ is a coherent theory, then $\mathcal{C}$ has all ultraproducts.*

Note that we give no properties distinguishing FS- and coherent theories.

Proof. The category of models in sets of an LE-sketch has all limits and filtered colimits by Theorem 4 of Section 4.4. It is trivial to see that the same is true of models of FP-sketches.

To show that FP-theories have effective equivalence relations and are regular, we need a lemma.

Lemma 2. *Given the diagram*

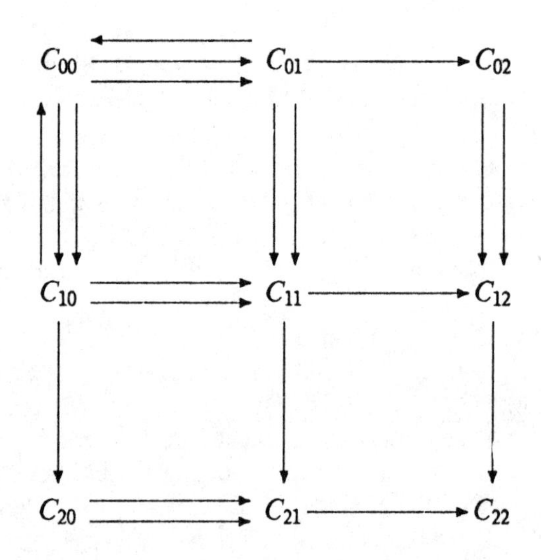

which is serially commutative and in which all three rows and columns are coequalizers and the top row and left column are reflexive, the induced

$$C_{00} \rightrightarrows C_{11} \longrightarrow C_{22}$$

is also a coequalizer.

Proof. Exercise (CoQT).

Let M be a model, E a submodel of M which is an equivalence relation, and C the quotient functor in $\mathcal{S}et^{\mathrm{FP}(\mathcal{S})}$, which we will prove to be a model. This last claim is equivalent to the assertion that C preserves products. Let A and B be objects of $\mathrm{FP}(\mathcal{S})$. We get the diagram

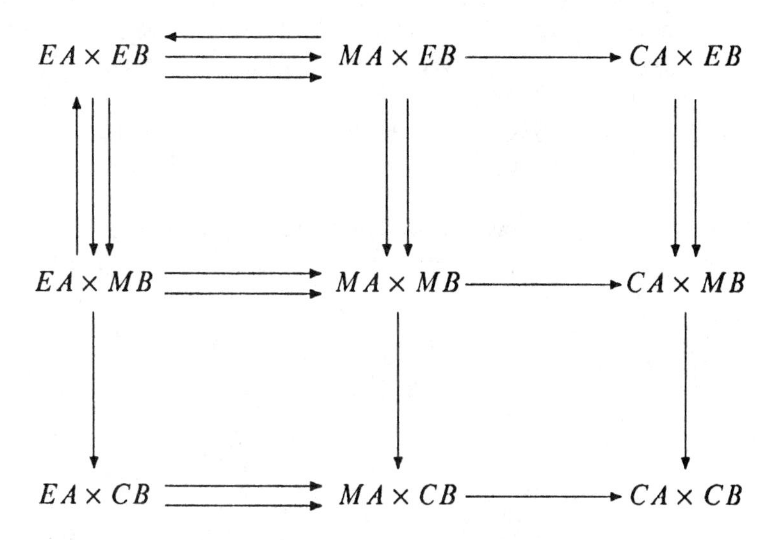

The rows and columns are coequalizers because products in a topos have a right adjoint. The lemma then implies that the diagonal is a coequalizer. But since E and M preserve products, this implies that C does also.

This not only shows that the model category has coequalizers of equivalence relations, but that the evaluation functors preserve them. In view of the Yoneda Lemma, that is just the assertion that the objects of $\mathrm{FP}(\mathcal{S})$ are regular projective. To see that they generate, observe that the maps from representable functors form an epimorphic family in the functor category. But that category is a Grothendieck topos, so those maps form a regular epimorphic family. It is easy to see that they continue to do so in any full subcategory. If the FP-theory is single sorted, that set may be replaced by the generic object to get a single regular projective generator.

Any surjective natural transformation in a $\mathcal{S}et$ -valued functor category is a coequalizer. Since $\mathcal{C}$ is closed under limits, this means that maps between models are regular epis if and only if they are surjective. It follows from this and the fact that the pullback of a surjective map in $\mathcal{S}et$ is surjective that $\mathcal{C}$ is regular. This takes care of (a) and (b). (Note that in fact $\mathcal{C}$ has all colimits, but they are not necessarily preserved by the evaluations).

Let $\mathcal{S}$ be a regular sketch and $\mathcal{R} = \mathrm{Reg}(\mathcal{S})$ be the regular theory associated to it. If $\{M_i\}$ is a family of models then we claim that the pointwise product M is a model. Models must preserve finite limits and take regular epis to regular epis. But a product of finite limits is a finite limit and a product of regular epis is a regular epi (since it is in $\mathcal{S}et$ and we carry out these constructions pointwise). As for filtered colimits, the argument is similar. Filtered colimits commute with finite limits in $\mathcal{S}et$ and in any category with regular epis. This completes the proof of (d).

Let $\mathcal{R}$ denote a coherent theory and $\mathcal{C}$ be the category of models. Let I be an index set, $\{M_i\}$ an I-indexed family of models and $\mathfrak{f}$ an ultrafilter on I. A product of models is not a model, but it is regular. Since this property is preserved by filtered colimits, an ultraproduct of models also preserves it. Thus it suffices to show that the ultraproduct preserves finite sums.

Let A and B be objects of the theory and suppose first that for all $i \in I, M_iA \neq \emptyset$ and $M_iB \neq \emptyset$. Let $M = \prod M_i$. Then if N is the ultraproduct, the canonical morphism $M \to N$ is surjective at A, B and $A + B$. Consider the diagram

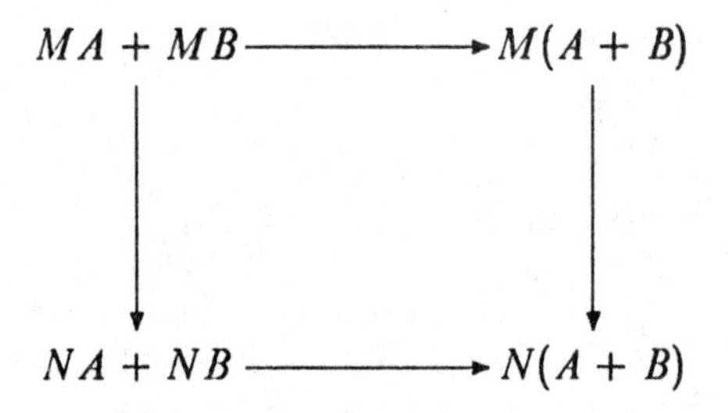

The vertical arrows are quotients and the horizontal arrows the ones induced by the properties of sums. We want to show that the bottom arrow is an isomorphism. Let $x \in N(A + B)$ and choose a $y \in M(A + B)$ lying over it. Since $M(A + B)$ is a colimit of the products over sets in $\mathfrak{f}$, there is a set $J \in \mathfrak{f}$ such that $y = (y_i), i \in J$ is an element of $\prod(M_iA + M_iB)$, the product taken over the $i \in J$. Let $K = \{i \mid y_i \in M_iA\}$ and $L = \{i \mid y_i \in M_iB\}$. Since $K \cup L = J \in \mathfrak{f}$ either K or L belongs to $\mathfrak{f}$. If K does, then $(y_i), i \in K$ is an element of MA whose image in NA goes to $x \in N(A + B)$.

To finish the argument, let $I_A = \{i \in I \mid M_iA \neq \emptyset\}$ and $I_B = \{i \in I \mid M_iB \neq \emptyset\}$. We observe that if $I_A \cap I_B \in \mathfrak{f}$, the proof above may be repeated with $I_A \cap I_B$ replacing I. If neither I_A nor I_B is in $\mathfrak{f}$, then it is clear that $NA = NB = N(A + B) = \emptyset$. Finally, suppose that one of the two, say I_A, is in $\mathfrak{f}$ and the other one isn't. Then we have

$$\{i \mid M_i(A + B) = M_iA + M_iB = M_iA\} \in \mathfrak{f}$$

whence $N(A + B) = NA = NA + NB$.

This theorem demonstrates, for instance, that the category of fields is not the models of any regular theory and similar such negative results. As for positive results, there are few. Beck's tripleableness theorem and its variants give a positive result in terms of a chosen underlying set functor, or equivalently, in terms of chosen representing sketch. It is tempting to speculate that if a category is known to be the category of models of an LE-sketch and is regular then it is the category of models of an FP-sketch, but we know of neither proof nor counterexample.

Exercises 8.4

(CoQT)°. Prove Lemma 2. (Compare Lemma 6 of Section 4.4 for a variation on this assertion.)

(EPIS)°. Show that the definition of regular epimorphic family given in this section is the same as the one given in Section 6.7 in the presence of pullbacks. Show that the class of C used in the definition in this section may be restricted to being in a generating family.

(EEPO). (a) Show that the category of posets and order-preserving maps is a subcategory of the category of graphs which is closed under products and subobjects.

(b) Show that the map which takes the graph below to the graph obtained by identifying A with A' and B with B'

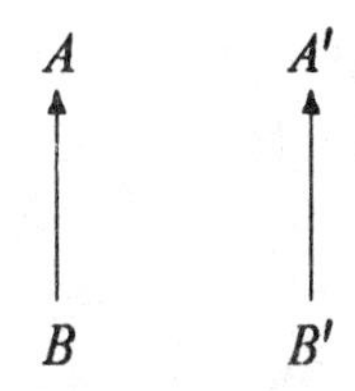

(but *not* identifying the edges) has a kernel pair which is (i) an equivalence relation and (ii) is defined on a poset.

(c) Use (a) and (b) to show that equivalence relations are not effective in the category of posets.

(TORGRP). In this exercise and the next, the categories of groups considered are understood to have all group homomorphisms as arrows. Show that the category of torsion groups is the category of models of a geometric theory. Can it be the category of models of a coherent theory? (Hint: For the first part, consider the theory of groups augmented with types

$$G_n = \{x \in G \mid x^n = 1\}$$

which is an equalizer and hence in the left exact theory. Require that $\{G_n \to G\}$ be a cover.)

(CYCGRP). (a) Show that the category of cyclic groups is not the category of models of any coherent theory. (Hint: Consider a non-principal ultrapower of $\mathbf{Z}$.)

(b) Show that the category of finite cyclic groups is not the category of models of any coherent theory.

(c) (We do not know the answer to this question.) Is the category of cyclic (or of finite cyclic) groups the category of models of a geometric theory?

Chapter 9.

More on Triples

This chapter consists of four independent sections providing additional results about triples. Everything may be read immediately after Chapter 3 except for Lemma 5 of Section 9.3, which depends on a fairly easy result from Chapter 5.

9.1. Duskin's Tripleability Theorem

In this section we state and prove a theorem of Duskin (Theorem 1) giving necessary and sufficient conditions for a functor $U : \mathcal{B} \to \mathcal{C}$ to be tripleable, under certain assumptions on the two categories involved.

If $\mathcal{B}$ is an equationally defined class of algebras (i.e. models of a single sorted FP theory) and $\mathcal{C}$ is $\mathcal{S}et$ then Birkhoff's theorem on varieties says that $\mathcal{B}$ is "closed under" products, subobjects and quotients by equivalence relations in $\mathcal{B}$. The first two closure properties mean little more than that, in our language, U creates limits. The third condition means that U creates coequalizers of parallel pairs which become equivalence relations in $\mathcal{C} = \mathcal{S}et$.

Duskin's Theorem is motivated by the idea that a functor U which satisfy a categorical version of Birkhoff's theorem ought to be tripleable. We begin by studying equivalence relations in a category. Throughout this section we study a functor $U : \mathcal{B} \to \mathcal{C}$ with left adjoint F.

Equivalence pairs and separators

An **equivalence pair** on an object C is a parallel pair $f, g : B \to C$ for which (a) f and g are **jointly monic**, which means that for any elements x and y of B defined on the same object, if $f(x) = f(y)$ and $g(x) = g(y)$ then $x = y$; and (b) the induced pair of arrows from $\mathrm{Hom}(X, B) \to \mathrm{Hom}(X, C)$ is an equivalence relation in $\mathcal{S}et$ for any object X. When B has products, f and g are jointly monic if and only if the arrow (f, g) is monic, and are an equivalence pair if and only if

$(f,g) : B \to C\times C$ is an equivalence relation (see Exercise (ER) of Section 1.7).

Maps $f,g : B \to C$ form a U-**contractible equivalence pair** if Uf and Ug are an equivalence pair which is part of a contractible coequalizer diagram in $\mathcal{C}$.

A **separator** of a parallel pair $f,g : B \to C$ is the limit

$$[(b,b') \mid f(b) = f(b') \text{ and } g(b) = g(b')].$$

Thus it is the intersection of the kernel pairs of f and g. In particular, a category which has separators of all parallel pairs has kernel pairs of all maps.

Duskin's theorem

Theorem 1 (Duskin). *Suppose $\mathcal{B}$ has separators and $\mathcal{C}$ has kernel pairs of split epimorphisms. Then the following two statements are equivalent for a functor $U : \mathcal{B} \to \mathcal{C}$:*

(i) *U is tripleable.*
(ii) *U has an adjoint and reflects isomorphisms, and every U-contractible equivalence pair has a coequalizer that is preserved by U.*

Before we prove this theorem, notice what it says in the case of groups. An equivalence pair $d^0, d^1 : H \to G$ in $\mathcal{G}rp$ forces H to be simultaneously a subgroup of $G \times G$ and a U-contractible equivalence relation on G (every equivalence pair in $\mathcal{S}et$ is part of a contractible coequalizer diagram). It is easy to see that the corresponding quotient set is the set of cosets of a normal subgroup of G, namely the set of elements of G equivalent to 1. The canonical group structure on the quotient set makes the quotient the coequalizer of d^0 and d^1. Note that you can show that the quotient is the coequalizer by showing that d^0 and d^1 are the kernel pair of the quotient map; since the quotient map is a regular epi (it is the coequalizer of the trivial homomorphism and the injection of the kernel of the quotient map into its domain), it follows from Exercise (EQC) of Section 1.8 that the quotient map is the coequalizer of d^0 and d^1. That may not be the method of proof that would have occurred to you, but it is the strategy of the proof that follows.

The argument is even more direct in the case of compact Hausdorff spaces. If R is an equivalence relation in that category on a space X then R is a closed (because compact) subspace of $X \times X$, so corresponds to a compact (because image of compact) Hausdorff (because R is closed) quotient space X/R.

Proof of Theorem 1. Because of Beck's Theorem and Proposition 3 of Section 3.3, the proof that (i) implies (ii) is immediate.

To prove that (ii) implies (i), we first prove three lemmas, all of which assume the hypotheses of Theorem 1 (ii).

Lemma 2. *If $Uf : UA \to UB$ is a split epimorphism, then $f : A \to B$ is a regular epimorphism.*

Proof. Let $h, k : S \to A$ be the kernel pair of f. Then because U preserves limits, (Uh, Uk) is the kernel pair of Uf; hence by Exercise (KPSE) (h, k) is a U-contractible equivalence pair and so by assumption has a coequalizer $A \to C$ in $\mathcal{B}$. Thus there is an induced map from C to B which necessarily becomes an isomorphism under U. Thus because U reflects isomorphisms, f must be the coequalizer of h and k.

Lemma 3. *If $Ud, Ue : UE \to UB$ is an equivalence pair, then so is d, e.*

Proof. For any object B of $\mathcal{B}$, $U\varepsilon B \circ \eta UB = \mathrm{id}_{UB}$ (Exercise (Uco) of Section 1.9), so Lemma 2 implies that εB is a regular epi. Then by Corollary 7 of Section 3.3,

$$FUFUB \underset{\varepsilon FUB}{\overset{FU\varepsilon B}{\rightrightarrows}} FUB \xrightarrow{\varepsilon B} B$$

is a coequalizer.

For any object C of $\mathcal{C}$, the induced maps

$$\mathrm{Hom}(C, UE) \rightrightarrows \mathrm{Hom}(C, UB)$$

are an equivalence relation in $\mathcal{S}et$, and by adjointness so is

$$\mathrm{Hom}(FC, E) \rightrightarrows \mathrm{Hom}(FC, B)$$

Putting the facts in the preceding two paragraphs together, we have that the rows in

(1)
$$\begin{array}{ccccc}
\mathrm{Hom}(B', E) & \longrightarrow & \mathrm{Hom}(FUB', E) & \rightrightarrows & \mathrm{Hom}(FUFUB', E) \\
\downarrow\downarrow & & \downarrow\downarrow & & \downarrow\downarrow \\
\mathrm{Hom}(B', B) & \longrightarrow & \mathrm{Hom}(FUB', B) & \rightrightarrows & \mathrm{Hom}(FUFUB', B)
\end{array}$$

are equalizers and the right hand and middle columns are both equivalence relations. It follows from an easy diagram chase that the left hand column is an equivalence relation, too.

Lemma 4. *In the notation of Lemma 3, if $f : B \to Y$ is a map for which Ud and Ue form the kernel pair of Uf, then d and e form the kernel pair of f.*

Proof. The proof follows the same outline as that of Lemma 3. One has that for any object C, the parallel pair in

$$\mathrm{Hom}(FC, E) \rightrightarrows \mathrm{Hom}(FC, B) \longrightarrow \mathrm{Hom}(FC, B'')$$

is the kernel pair of the right arrow, so that in the diagram below
(2)

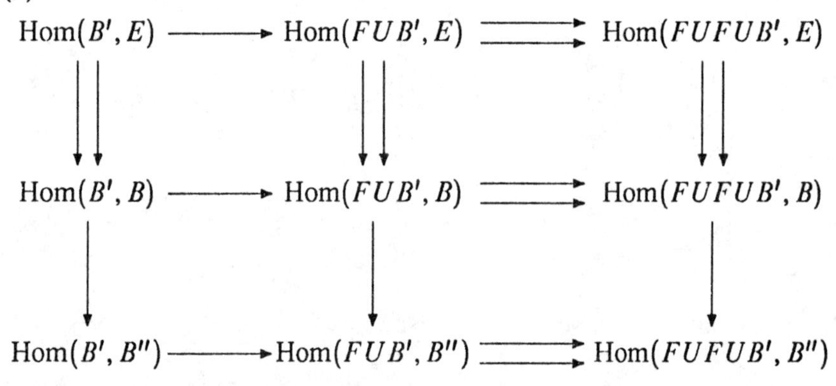

the middle and right hand vertical parallel pairs are kernel pairs of the corresponding arrows and the horizontal sides are all equalizers. Then a diagram chase shows that the left hand column is a kernel pair diagram too.

Now, to prove that (ii) of Theorem 1 implies (i). By Beck's Theorem, we must prove that if

$$A \underset{d^1}{\overset{d^0}{\rightrightarrows}} B \tag{3}$$

is a reflexive U-split coequalizer pair, then it has a coequalizer which is preserved by U.

In $\mathcal{B}$, we construct the following diagram, in which (p^0, p^1) is the kernel pair of d^1 and S is the separator of $d^0 \circ p^0$ and $d^0 \circ p^1$. The object E and the arrows into and out of it will be constructed later.

(4)

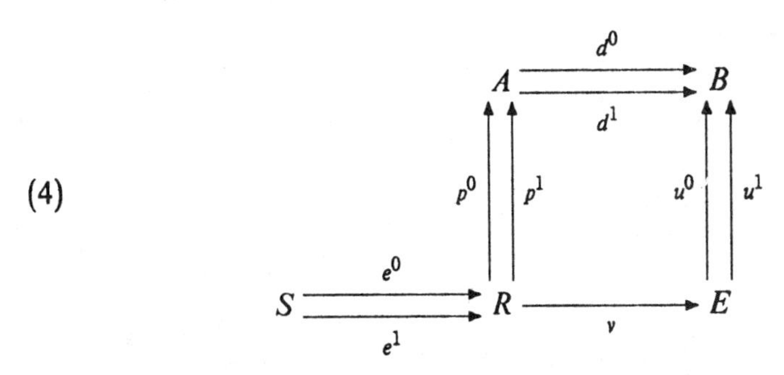

In $\mathcal{C}$, we have diagram (5), in which c is the coequalizer of Ud^1 and Ud^2 with contracting maps s and t and (q^0, q^1) is the kernel pair of of the split epi c.

(5)

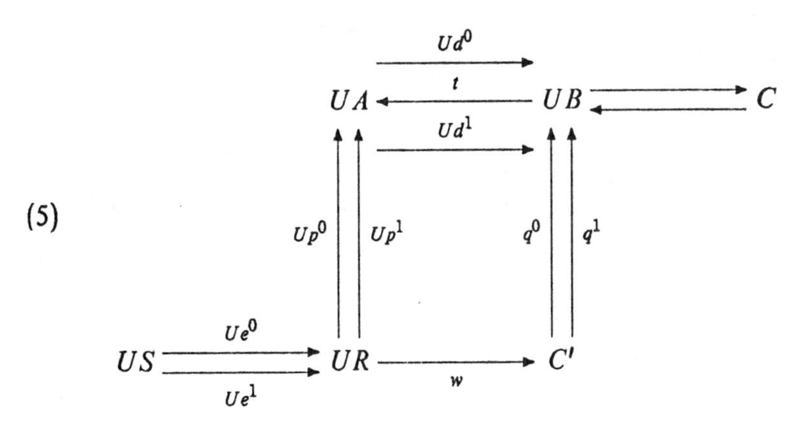

This is how the proof will proceed: (a) We will construct w and (b) show that it is a split epi with (c) kernel pair (Ue^0, Ue^1). It then follows from Lemma 3 that (e^0, e^1) is a U-contractible equivalence relation which must by hypothesis have a coequalizer $v : R \to E$ from which we conclude that up to isomorphism, UE can be taken to be C'. (d) We then construct u^0, u^1 for which $u^0 \circ v = d^0 \circ p^0$ and $u^1 \circ v = d^0 \circ p^1$ and also $U(u^0) = q^0$ and $U(u^1) = q^1$. Now c is the coequalizer of (q^0, q^1) (Exercise (EQC) of Section 1.8: if a regular epi has a kernel pair then it is the coequalizer of its kernel pair), so (u^0, u^1) is a U-split equivalence pair, hence has a coequalizer x. By changing c to an isomorphic arrow if necessary, we may assume that $U(x) = c$. Finally, (e) we show that x is also the coequalizer of (d^0, d^1). It follows easily from the fact that U reflects isomorphisms that U takes this coequalizer to an arrow isomorphic to c, as required.

(a) It follows from the identities for a contractible coequalizer that c coequalizes $Ud^0 \circ Up^0$ and $Ud^0 \circ Up^1$. Let w be the unique map (by virtue of (q^0, q^1) being the kernel pair of c) for which $q^0 \circ w = Ud^0 \circ Up^0$ and $q^1 \circ w = Ud^0 \circ Up^1$.

(b) To see that w is a split epi (surjective on elements), suppose that $(b^0, b^1) \in_{UB \times UB} C'$; then $s \circ c \circ b^0 = s \circ c \circ b^1$, so $Ud^1 \circ t \circ b^0 = Ud^1 \circ t \circ b^1$, which means that $(t \circ b^0, t \circ b^1) \in UR$ (Up^0 and Up^1 are the kernel pair of Ud^1). We claim that $w(t \circ b^0, t \circ b^1) = (b^0, b^1)$. The first coordinate is

$$q^0 \circ w(t \circ b^0, t \circ b^1) = Ud^0 \circ Up^0(t \circ b^0, t \circ b^1) = Ud^0 \circ t \circ b^0 = b^0,$$

and similarly the second coordinate is b^1, as required. Hence w is a split epi.

(c) Since e^0 and e^1 form the separator of $d^0 \circ p^0$ and $d^0 \circ p^1$ and U preserves limits, Ue^0 and Ue^1 are the separator of $Ud^0 \circ Up^0$ and $Ud^0 \circ Up^1$. It is straightforward to calculate, using the fact that q^0 and q^1 are jointly monic, that the kernel pair of w is the intersection of the kernel pairs of $Ud^0 \circ Up^0$ and $Ud^0 \circ Up^1$. Hence Ue^0 and Ue^1 are the kernel pair of w.

(d) Because e^0 and e^1 form the kernel pair of $d^0 \circ p^0$ and $d^0 \circ p^1$, they coequalize e^0 and e^1, so for $i = 0, 1$ there are induced maps u^i for which $u^i \circ v = d^0 \circ p^i$. Then

$$q^i \circ w = U(d^0 \circ p^i) = U(u^i) \circ U(v) = U(u^i) \circ w,$$

so because w is epi, $U(u^i) = q^i$.

To complete the proof, we must show (e) that x, which by definition is the coequalizer of u^0 and u^1 is also the coequalizer of d^0 and d^1. Now $c = Ux$ is the coequalizer of Ud^0 and Ud^1, so by Exercise (EQC) of Section 1.8, Ud^0 and Ud^1 is the kernel pair of c. Hence by Lemma 4, d^0 and d^1 form the kernel pair of x. Since x is a regular epi, it is the coequalizer of d^0 and d^1 as required.

Variation on Duskin's theorem

The following version of Duskin's theorem is the form in which it is most often used.

Proposition 5. *If*

(i) $\mathcal{B}$ *has separators,*
(ii) $\mathcal{C}$ *has kernel pairs of split epis.*
(iii) $U : \mathcal{B} \to \mathcal{C}$ *has an adjoint* F,

(iv) *U reflects isomorphisms and preserves regular epis, and*

(v) *U-contractible equivalence pairs in $\mathcal{B}$ are effective and have coequalizers,*

then U is tripleable.

Proof. All that is necessary is to show that under these hypotheses, U preserves coequalizers of U-contractible equivalence pairs. If such an equivalence pair (d^0, d^1) has coequalizer x and is the kernel pair of y, then by Exercise (EQC) of Section 1.8, x has kernel pair (d^0, d^1). Since U preserves kernel pairs, Ux is a regular epi which has (Ud^0, Ud^1) as kernel pair. Again using Exercise (EQC), Ux is therefore the coequalizer of (d^0, d^1).

Tripleability over $\mathcal{S}et$

If X is a set and G is an object in a cocomplete category, we write $X \cdot G$ for the coproduct of X copies of G. In particular, if A is another object, there is an obvious induced map $\mathrm{Hom}(G, A) \cdot G \to A$ (the arrow from the copy of G corresponding to f is f).

An object P of a category is a **regular projective** if whenever $f : A \to B$ is a regular epi then the induced map $\mathrm{Hom}(P, A) \to \mathrm{Hom}(P, B)$ is surjective. (If this is true for all epis then P is a **projective**.) An object G is a **generator** if for each object A, the induced map $\mathrm{Hom}(G, A) \cdot G \to A$ is an epi. If the induced map is a regular epi, then G is a **regular generator.** P is a **regular projective generator** if it is both a regular projective and a regular generator. If it is both a generator and a projective then it is a **projective generator.** The dual of "projective" is "injective".

Theorem 5. *A category $\mathcal{C}$ is tripleable over $\mathcal{S}et$ if and only if it is regular, has effective equivalence relations and has a regular projective generator P for which $X \cdot P$ exists for all sets X.*

Proof. If $U : \mathcal{C} \to \mathcal{S}et$ is tripleable, then $F(1)$ is a regular projective generator (Exercise (RPF)). Then for any set X, $F(X) = X \cdot F(1)$ because $X = X \cdot 1$ in $\mathcal{S}et$ and F preserves colimits. (Compare Theorem 1(a), Section 8.4.)

For the converse, the functor which is tripleable is $U = \mathrm{Hom}(P, -)$, which has a left adjoint taking X to $X \cdot P$ (Exercise (RPA)).

U reflects isomorphisms: Suppose $f : A \to B$ and Uf is an isomorphism. The top arrow in this diagram

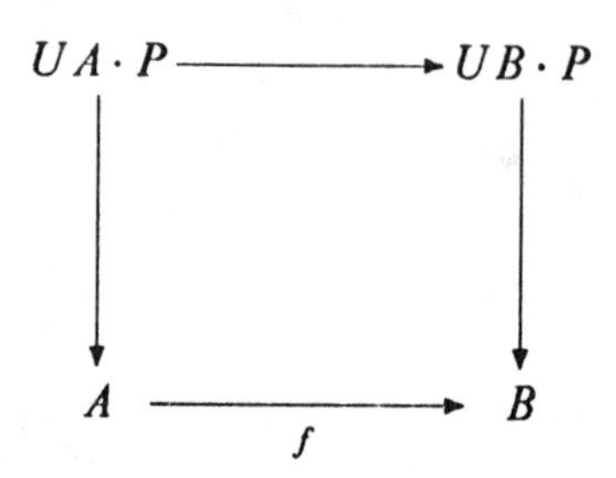

is an isomorphism and the vertical arrows are regular epis, so f is a regular epi. On the other hand, in this diagram

$$E \xrightarrow{\ e\ } X \underset{d^1}{\overset{d^0}{\rightrightarrows}} A \xrightarrow{\ f\ } B$$

let (d^0, d^1) be the kernel pair of f and e the equalizer of d^0 and d^1. Since Uf is an isomorphism, $Ud^0 = Ud^1$, so that Ue is an isomorphism. Hence by the first part of this argument, e is epi, whence $d^0 = d^1$ and so f is an isomorphism.

Now suppose $e^0, e^1 : A \to B$ is a U-contractible equivalence relation. Since equivalence relations are effective, there is a coequalizer/kernel pair diagram

$$A \underset{e^1}{\overset{e^0}{\rightrightarrows}} B \xrightarrow{\ c\ } C$$

Ue^0 and Ue^1 become the kernel pair of Uc because U preserves limits. Uc is epi because P is a regular projective generator, so because we are in $\mathcal{Set}$, Uc is the coequalizer of Ue^0 and Ue^1, as required.

As an application of Theorem 5, observe that the Tietze Extension Theorem says that the unit interval I is an injective cogenerator in $\mathcal{CptHaus}$. Thus it is a projective generator in the opposite category; in fact it is a regular projective generator since every monomorphism in $\mathcal{CptHaus}$ is regular (take two copies of the codomain and amalgamate the subspace).

Corollary 6. $\mathrm{Hom}(-, I) : \mathcal{CptHaus}^{\mathrm{op}} \to \mathcal{Set}$ *is tripleable.*

Proof. The only complicated thing to prove is that $\mathcal{CptHaus}$ has effective coequivalence relations. We leave the rest of the verifications to the reader.

Suppose

$$X \underset{d^1}{\overset{d^0}{\rightrightarrows}} Y$$

is a coequivalence relation. Then there is a map $r : Y \to X$ for which $r \circ d^0 = r \circ d^1$. (In fact, that is the only property of coequivalence relations we use. See Exercise (MAL2).) Let $d : Z \to X$ be the equalizer of d^0 and d^1. We must show that d^0 and d^1 form the cokernel pair of d. We do this by showing that the map $X +_Z X \to Y$ is bijective. It is clearly surjective because $[d^0, d^1] : X + X \to Y$ is. For $j = 0, 1$, d^j is injective because it has a left inverse r. It follows that if (for $j = 0$ or 1) $[d^0, d^1](i_j x) = [d^0, d^1](i_j x')$, then $x = x'$. Clearly if $[d^0, d^1](i_0 x) = [d^0, d^1](i_1 x')$, then $x = x'$, hence $d^0 x = d^1 x$ so that the map from $X +_Z X$ to Y is injective as well as surjective, and so is an isomorphism.

Exercises 9.1

(SEPKP)$^\diamond$. Show that if $f : A \to B$ is a map, then the kernel pair of f is the the separator of f and f. Show that if $f, g : A \to B$, then the separator of f and g is $\mathrm{kerp}(f) \cap \mathrm{kerp}(g)$.

(KPSE)$^\diamond$. Show that if g is a split epi and (h, k) is its kernel pair, then h, k and g form a contractible coequalizer diagram.

(RPA)$^\diamond$. Show that if $\mathcal{C}$ is a category with a regular projective generator P, then $\mathrm{Hom}(P, -)$ has a left adjoint taking a set X to the sum $X \cdot P$ (assuming that sum exists).

(RPF)$^\diamond$. Show that if $U : \mathcal{C} \to \mathcal{S}et$ is tripleable, then $F(1)$ is a regular projective generator in $\mathcal{C}$.

(MAL2). A single-sorted equational theory is called a Mal'cev theory if there is in the theory a ternary operation μ satisfying the equations $\mu(a, a, c) = c$ and $\mu(a, b, b) = a$. A theory which includes a group operation is automatically a Mal'cev theory for one can define $\mu(a, b, c) = ab^{-1}c$. Thus the *dual* of the category of sets is the category of algebras for a Mal'cev theory, since $\mathcal{S}et^{\mathrm{op}}$ is equivalent to the category of complete atomic Boolean algebras which, like any rings, include the group operation of addition. See Exercise (MAL) of Section 5.6 for an example that doesn't arise from a group (or any other binary) operation. Show that in the category of algebras for a Mal'cev theory every reflexive relation is an equivalence relation.

The converse is also true. The construction of μ in such a category can be sketched as follows. On the free algebra on two generators which we will denote by 0 and 1 consider the relation generated by $(0,0)$, $(1,1)$ and $(0,1)$ which is the image of a map from the free algebra on three generators (the map taking the three generators to the three above mentioned elements). This image is a reflexive relation which is hence an equivalence relation. In particular it is symmetric which means there is an element μ in the free algebra on three generators that maps to $(1,0)$. Given the correspondence between n-ary operations and elements of the free algebra on n elements developed in Section 4.3 (see Exercise (OPS)), this element corresponds to a ternary operation which (you will have no doubt already guessed) is the required Mal'cev operation. Remarkably, one need only assume that every reflexive relation is symmetric to derive the conclusion.

These operations were first studied (you will have no doubt also guessed) by Mal'cev [1954] who derived most of their interesting properties, *not* including what was probably their most interesting property: every simplicial object in a category of algebras for a Mal'cev theory satisfies Kan's condition.

9.2. Distributive Laws

It is not ordinarily the case that when $\mathbf{T}_1 = (T_1, \eta_1, \mu_1)$ and $\mathbf{T}_2 = (T_2, \eta_2, \mu_2)$ are triples on the same category, there is a natural triple structure on $T_2 \circ T_1$. But that does happen; for example, the free ring triple is the composite of the free monoid triple followed by the free Abelian group triple. The concept of distributive law for triples was formulated by Jon Beck to formalize the fact that the distribution of multiplication over addition explains this fact about the free ring triple.

Another way of seeing the distributive law in a ring is that it says that multiplication is a homomorphism of the Abelian group structure. This underlies the concept of a lifting of a triple, which we will see is equivalent to having a distributive law.

Distributive laws

Using the notation above, a **distributive law** of $\mathbf{T}_1$ over $\mathbf{T}_2$ is a natural transformation $\lambda : T_1 \circ T_2 \to T_2 \circ T_1$ for which the following diagrams commute. We omit the "∘" to simplify notation.

(D1)

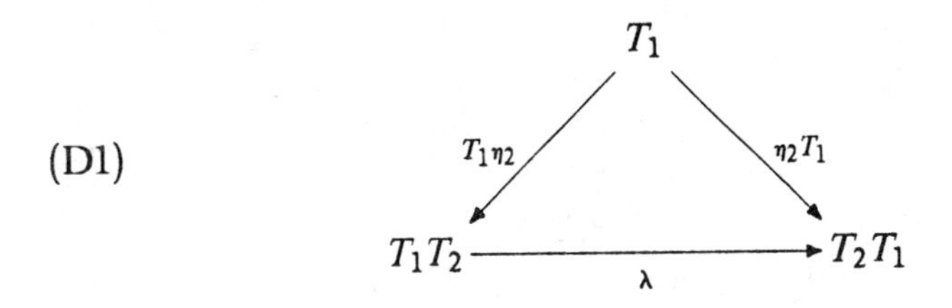

(D2)

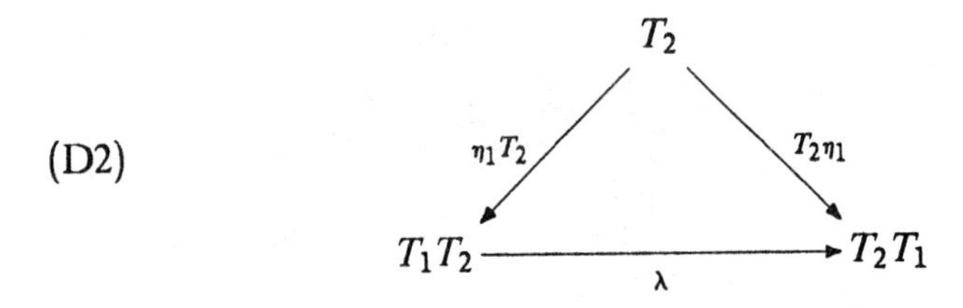

(D3)

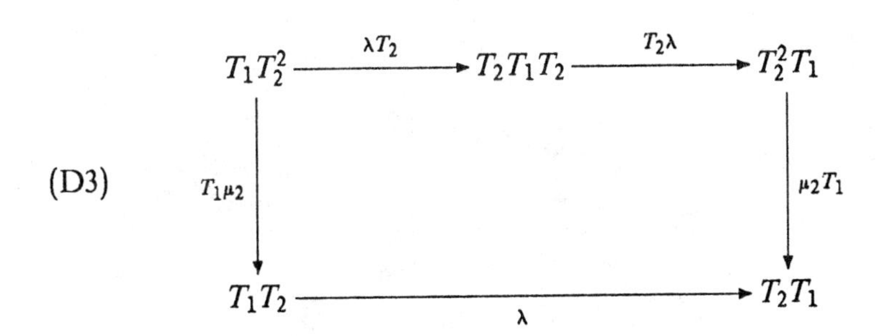

(D4)

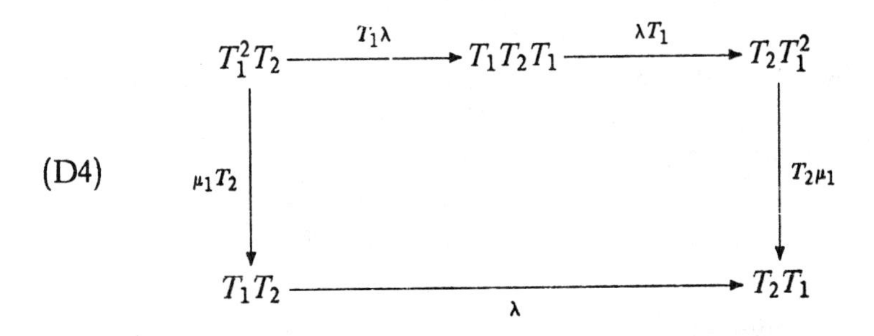

For example, when $\mathbf{T}_1$ is the free monoid triple on $\mathcal{Set}$ and $\mathbf{T}_2$ is the free Abelian group triple, we have a distributive law $\lambda : T_1 \circ T_2 \to T_2 \circ T_1$ which takes an element of T_1T_2X of the form

$$(\sum \alpha_1(x) \cdot x)(\sum \alpha_2(x) \cdot x) \cdots (\sum \alpha_m(x) \cdot x)$$

(each sum only having the integer $\alpha_i(x)$ nonzero for a finite number of terms) to

$$\sum \alpha_1(x_1)\alpha_2(x_2)\cdots\alpha_m(x_m)\cdot(x_1x_2\cdots x_m)$$

the sum over all strings of length m in elements of X.

Another example in $\mathcal{S}et$ has T_1 the free semigroup triple and T_2 defined by $T_2(X) = 1 + X$ (the disjoint union—see Example 3 of Section 3.1) Then $T_2 \circ T_1$ is the free monoid triple. The distributive law λ takes

$$1 + X + (1 + X)^2 + (1 + X)^3 + \cdots = 1 + X + 1 + X + X + X^2 \\ + 1 + X + X + X + X^2 + X^2 + X^2 + X^3 + 1 + \cdots$$

to $1 + X + X^2 + X^3 + \cdots$ by the map which takes each summand on the left to the same thing on the right.

Lifting

If T_1 and T_2 are triples on a category $\mathcal{C}$, a **lifting** of T_2 to $\mathcal{C}^{\mathsf{T}_1}$ is a triple $\mathsf{T}_2^* = (T_2^*, \eta_2^*, \mu_2^*)$ on $\mathcal{C}^{\mathsf{T}_1}$ for which

(L1) $U^{T_1} \circ T_2^* = T_2 \circ U^{T_1}$,

(L2) $U^{T_1}\eta_2^* = \eta_2 U^{T_1} : U^{T_1} \to U^{T_1} \circ T_2^*$ and

(L3) $U^{T_1}\mu_2^* = \mu_2 U^{T_1} : U^{T_1} \circ (T_2^*)^2 \to U^{T_1} \circ T_2^*$.

In understanding L2 and L3, observe that by L1, $U^{T_1} \circ T_2^* = T_2 \circ U^{T_1}$ and $U^{T_1} \circ (T_2^*)^2 = T_2 \circ U^{T_1} \circ T_2^* = T_2^2 \circ U^{T_1}$.

Note that given a functor T_2^* satisfying (L1), rules (L2) and (L3) determine what η_2^* and μ_2^* must be, and the resulting $\mathsf{T}_2^* = (T_2^*, \eta_2^*, \mu_2^*)$ is trivially a triple.

A lifting of T_2 is equivalent to having a natural way of viewing $T_2 A$ as a T_1 algebra whenever A is a T_1 algebra, in such a way that η_2 and μ_2 are algebra morphisms.

Given a distributive law $\lambda : T_1 \circ T_2 \to T_2 \circ T_1$, define $T_2^*(A, a) = (T_2 A, T_2 a \circ \lambda A)$, and define T_2^* to be the same as T_2 on algebra morphisms.

Proposition 1. *T_2^* is the functor part of a lifting of* T_2 *to* $\mathcal{C}^{\mathsf{T}_1}$.

Proof. (L1) is clear, so (L2) and (L3) determine η_2^* and μ_2^*.

These diagrams show that $T_2^*(A, a)$ is a T_1 algebra.

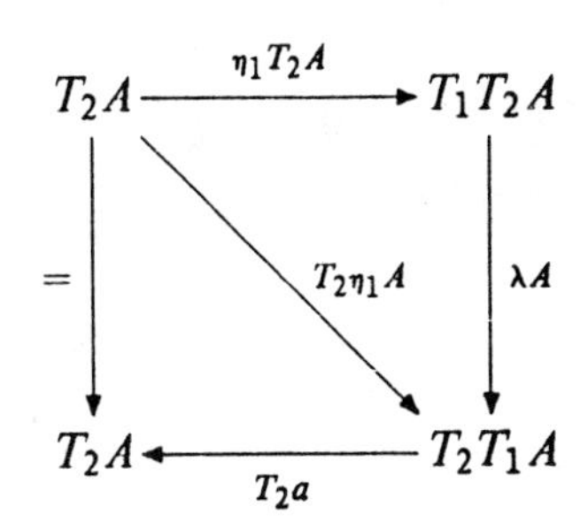

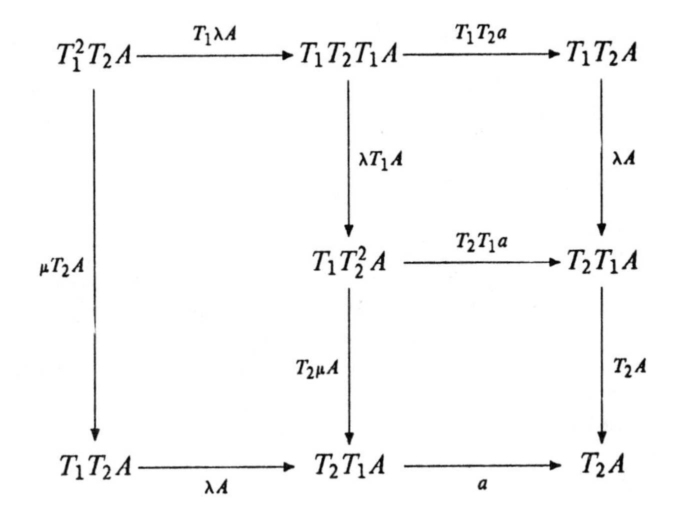

Furthermore,

$$\eta_2 A \circ a = T_2^* a \circ \lambda A \circ T_1 \eta_2 A$$

by (D1) and naturality, so η_2^* is an algebra morphism, and

$$\mu_2 A \circ T_2^2 a \circ T_2 \lambda A \circ \lambda T_2 A = T_2 a \circ \lambda A \circ T_1 \mu_2 A : TT_2^2 A \to T_2 A$$

by D3 and naturality, so μ_2^* is an algebra morphism.

Compatibility

The following definition captures the idea that a triple with functor $T_2 \circ T_1$ is in a natural way the composite of triples with functors T_2 and T_1.

The triple $\mathsf{T} = (T_2 \circ T_1, \eta, \mu)$ is **compatible** with triples $\mathsf{T}_1 = (T_1, \eta_1, \mu_1)$ and $\mathsf{T}_2 = (T_2, \eta_2, \mu_2)$ if

(C1) $\eta = T\eta_2 \circ \eta_1 = \eta_1 T \circ \eta_2 : \mathrm{id} \to T_1 \circ T_2$.

(C2) $\mu \circ T_2 T_1 \eta_2 T_1 = T_2 \mu_1 : T_2 \circ T_1^2 \to T_2 \circ T_1$.
(C3) $\mu \circ T_2 \eta_1 T_2 T_1 = \mu_2 T_1 : T_2^2 \circ T_1 \to T_2 \circ T_1$.
(C4) $\mu_2 T_1 \circ T_2 \mu = \mu \circ \mu_2 T_1 T_2 T_1 : T_2^2 \circ T_1 \circ T_2 \circ T_1 \to T_2 \circ T_1$, and
(C5) $T_2 \mu_1 \circ \mu T_1 = \mu \circ T_2 T_1 T_2 \mu_1 : T_2 \circ T_1 \circ T_2 \circ T_1^2 \to T_2 \circ T_1$.

Given triples T_1 and T_2 on a category $\mathcal{C}$ and a lifting T_2^* on $\mathcal{C}^{\mathsf{T}_1}$, we get

$$(\mathcal{C}^{\mathsf{T}_1})^{\mathsf{T}_2^*} \underset{U_2^*}{\overset{F_2^*}{\rightleftarrows}} \mathcal{C}^{\mathsf{T}_1} \underset{U_1}{\overset{F_1}{\rightleftarrows}} \mathcal{C}$$

whence $F_2^* \circ F_1$ is left adjoint to $U_1 \circ U_2^*$, so

$$T_2 \circ T_1 = T_2 \circ U_1 \circ F_1 = U_1 \circ T_2^* \circ F_1 = U_1 \circ U_2^* \circ F_2^* \circ F_1$$

produces a triple $\mathsf{T} = (T_2 \circ T_1, \eta, \mu)$ on $\mathcal{C}$.

Proposition 2. *With the notation of the preceding paragraph, the triple* T *is compatible with* T_1 *and* T_2.

Proof. We will check one of the hardest compatibility conditions, namely C5, and leave the rest to you. In the diagram, we have carried out the replacement of $U_2 F_2 U_1$ by $U_1 U_2^* F_2^*$ and replaced μ by its definition so that the diagram that has to be shown to commute is

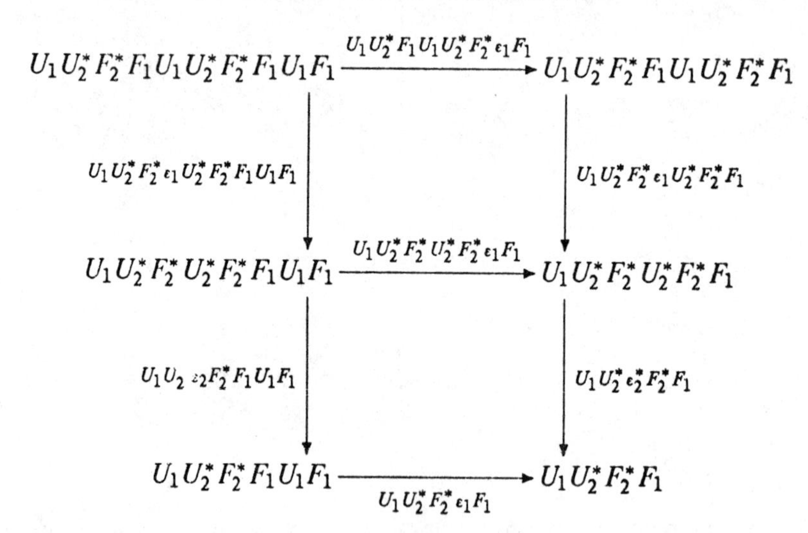

Now suppose that the triple T is compatible with T_1 and T_2. Define $\lambda : T_1 \circ T_2 \to T_2 \circ T_1$ as the composite

$$T_1 T_2 \xrightarrow{\eta_2 T_1 T_2 \eta_1} T_2 T_1 T_2 T_1 \xrightarrow{\mu} T_2 T_1$$

Proposition 3. λ *is a distributive law.*

Proof. D1 and D2 follow from this diagram (note that $T\eta = T_2T_1\eta_2\eta_1$).

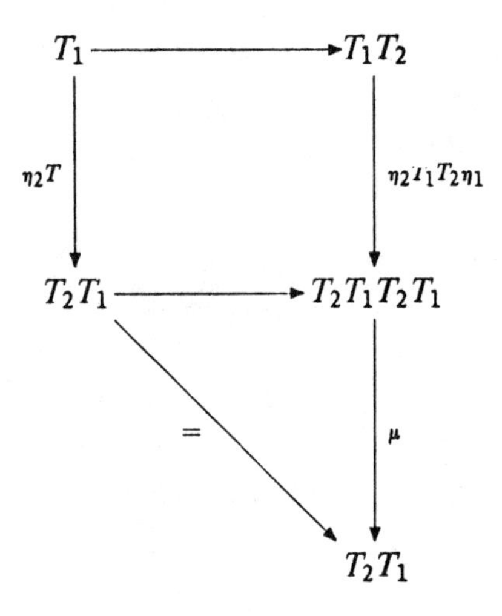

To get D4, consider this diagram.

$$
\begin{array}{ccccc}
T_1^2T_2 & \xrightarrow{T_1\eta_2T_1T_2\eta_1} & T_1T_2T_1T_2T_1 & \xrightarrow{T_1\mu} & T_1T_2T_1 \\
\downarrow{\scriptstyle T_1^2T_2\eta_1} & & \downarrow{\scriptstyle \eta_2T_1T_2T_1T_2T_1} & & \downarrow{\scriptstyle \eta_2T_1T_2T_1} \\
T_1^2T_2T_1 & \xrightarrow{\eta_2T_1\eta_2T_1T_2T_1} & T_2T_1T_2T_1T_2T_1 & \xrightarrow{T_2T_1\mu} & T_2T_1T_2T_1 \\
\downarrow{\scriptstyle \mu_1T_2T_1} & & \downarrow{\scriptstyle \mu T_2T_1} & & \downarrow{\scriptstyle \mu} \\
T_1T_2T_1 & \xrightarrow[\eta_2T_1T_2T_1]{} & T_2T_1T_2T_1 & \xrightarrow[\mu]{} & T_2T_1
\end{array}
\tag{1}
$$

where the top left square commutes by definition, the top right square by naturality, and the bottom right square by a triple identity. The bottom left square is the following square applied to $T_2 \circ T_1$:

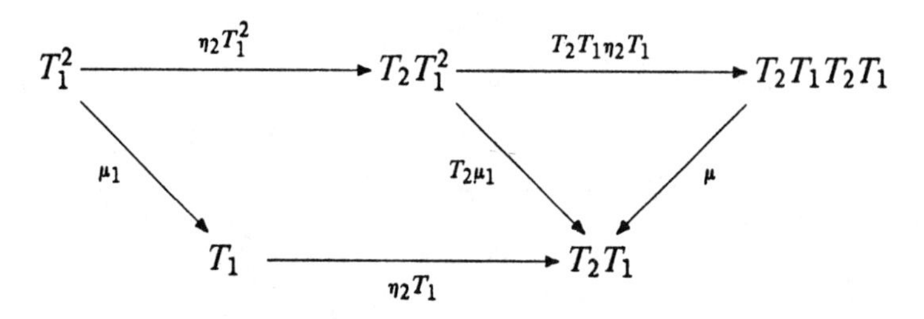

Now the left and bottom route around square (1) is $\lambda \circ \mu_1 T_2$ by definition of λ and the fact that

$$u_1 T_2 T_1 \circ T_1^2 T_2 \eta_1 = T_1 T_2 \eta_1 \circ \mu_1 T_2$$

by naturality, and the top and right route is $T_2 \mu_1 \circ \lambda T_1 \circ T_1 \lambda$ because

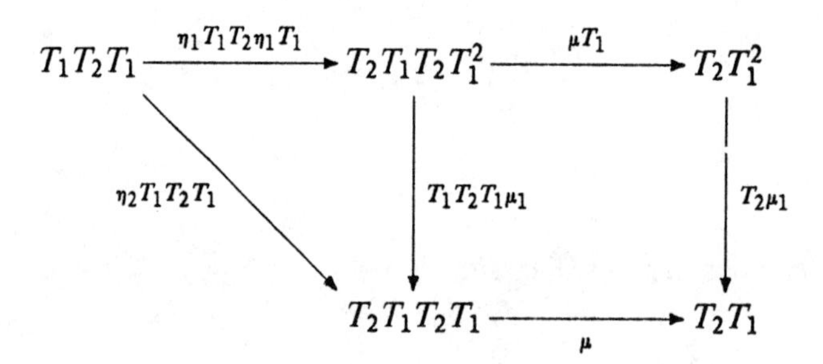

The left triangle is a triple identity and the square is C5.
We leave the rest to you.

Exercise 9.2

(DL). Prove that for the constructive processes described in the text, all composites in the following triangle are the identity.

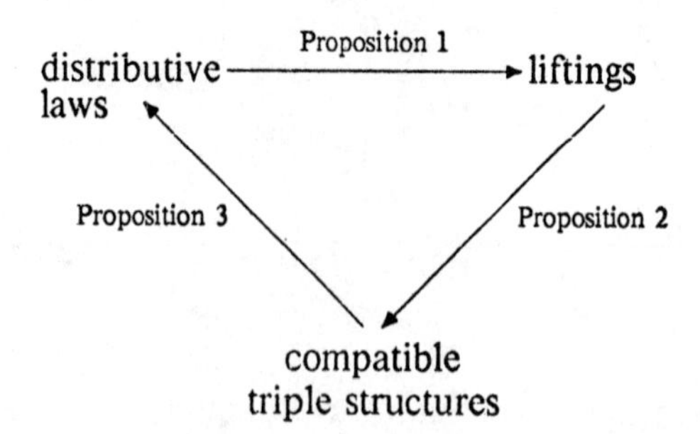

9.3. Colimits of Triple Algebras

Theorem 1 of Section 3.4 gives what is perhaps the best possible result on the completeness of categories of triple algebras. In this section, we investigate cocompleteness, with results which are informative but less satisfactory. First we need a lemma.

Lemma 1. *Let $\mathcal{B}$ be a category and $\mathcal{I}$ a small diagram. Then the diagonal functor $\Delta : \mathcal{B} \to \mathcal{B}^{\mathcal{I}}$ has a left adjoint L, if and only if every functor $F : \mathcal{I} \to \mathcal{B}$ has a colimit in $\mathcal{B}$, and in that case LF is that colimit.*

Proof. The isomorphism $\mathrm{Hom}_{\mathcal{B}}(LF, B) \cong \mathrm{Hom}_{\mathcal{B}^{\mathcal{I}}}(F, K_B)$, where K_B is the constant functor, applied to the identity on LF yields a cocone from F to LF which is universal by definition.

The converse is true by pointwise construction of adjoints.

Dually, a right adjoint, if it exists, takes a functor to its limit.

Theorem 2 (Linton). *Let $U = \mathcal{B} \to \mathcal{C}$ be of descent type and $\mathcal{I}$ a small category. In order that every $D : \mathcal{I} \to \mathcal{B}$ have a colimit, it suffices that every $UD : \mathcal{I} \to \mathcal{C}$ have colimits and that $\mathcal{B}$ have coequalizers.*

Note that there is no claim that any colimits are created by U, only that they exist.

Proof. We give a slick proof under the additional assumption that $\mathcal{C}$ is cocomplete. See Linton [1969c] for a proof without this hypothesis. Apply Theorem 3(b) of Section 3.7 and Lemma 1 to the diagram

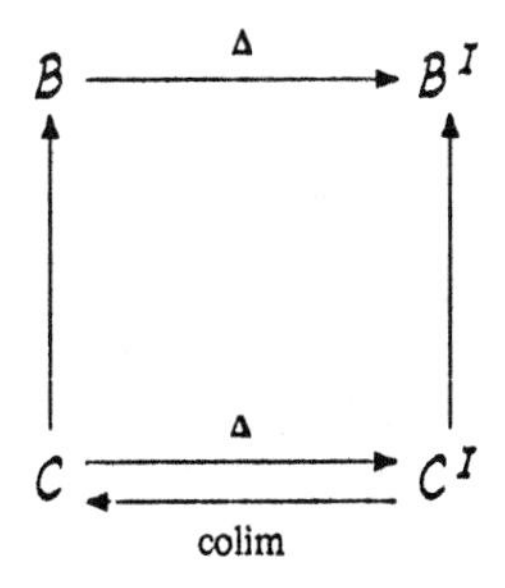

where the colimit functor on the bottom exists by Exercise (FCL) of Section 1.7.

Corollary 3. *Let $\mathcal{B} \to \mathcal{C}$ be of descent type and $\mathcal{C}$ have finite colimits. Then $\mathcal{B}$ has finite colimits if and only if it has coequalizers.*

A similar statement can be made for countable colimits, all colimits, etc.

Proposition 4 (Linton). *Let* T *be a triple on* $\mathcal{S}et$. *Then* $\mathcal{S}et^{\mathsf{T}}$ *is cocomplete.*

Proof. Given

$$(1) \qquad (A,a) \underset{d^1}{\overset{d^0}{\rightrightarrows}} (B,b)$$

we must construct its coequalizer. Since the underlying functor from $\mathcal{S}et^{\mathsf{T}}$ creates limits, $B \times B$ underlies a unique T-algebra. Let $E \subseteq B \times B$ be the intersection of all subsets of $B \times B$ which

(a) are T-subalgebras,
(b) are equivalence relations, and
(c) contain the image of $R = (d^0, d^1) : A \to B$.

E is clearly a subset satisfying all three conditions. Thus in $\mathcal{S}et$,

$$E \rightrightarrows C$$

is a contractible pair because any equivalence relation in $\mathcal{S}et$ is. Thus U, being tripleable, must lift the coequalizer of this contractible pair to a coequalizer in $\mathcal{S}et^{\mathsf{T}}$ which is also the coequalizer of d^0 and d^1. Hence $\mathcal{S}et^{\mathsf{T}}$ has all colimits by Corollary 3.

When do triple algebras have coequalizers?

Given a triple T in a category $\mathcal{C}$, to force $\mathcal{C}^{\mathsf{T}}$ to have coequalizers seems to require some sort of preservation properties for T. For example, if T preserves coequalizers and $\mathcal{C}$ has them, then $\mathcal{C}^{\mathsf{T}}$ will have them too (Exercise (CCTA)). However, that happens only rarely.

Theorem 7 below is an example of what can be proved along these lines. We need another result first; it says that the quotient of an algebra map is an algebra map, if by quotient we mean the image of a regular epimorphism. We use the definition of epi-mono factorization system of Exercises (FAC1) and (FAC2) of Section 5.5. Recall from Section 7.1 that a category is regular if it has finite limits and coequalizers and if pullbacks of regular epis are regular epis.

Lemma 5. *If $\mathcal{C}$ is regular, then the classes E of regular epis and M of all monos form a factorization system.*

Proof. This has the same proof as Theorem 2 of Section 5.5. (The assumption of regularity makes r in diagram (1) of Section 5.5 an epi).

Given an arrow $F : C \to B$ in such a category, its image via the factorization into a regular epi followed by a mono is called its **regular image.**

One defines equivalence classes of epis the same way as for monos—if $C \to A$ and $C \to B$ are epi, they are equivalent if there are (necessarily unique) maps for which

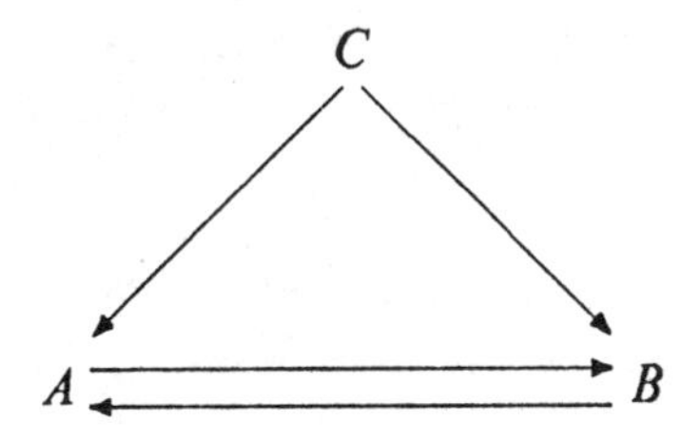

commutes both ways.

A category $\mathcal{C}$ is **regularly co-well-powered** if for each object C there is a set R consisting of regular epimorphisms of $\mathcal{C}$ with the property that every regular epimorphism of $\mathcal{C}$ is equivalent to one in R.

Proposition 6. *Let $\mathcal{C}$ be a regular, regularly co-well-powered category with coequalizers, and T a triple which preserves regular epis. Then the regular image of an algebra morphism is a subalgebra of its codomain.*

Notice that the condition that T preserve regular epis is automatically satisfied in categories such as $\mathcal{S}et$ and categories of vector spaces over fields in which every epi is split.

Proof. Let an algebra map $(C, c) \to (B, b)$ factor as in the bottom line of the following diagram. The vertical arrows make coequalizers as in diagram (13), Section 3.3. The arrow a, and hence Ta, exists because of the diagonal fill-in property of the factorization system.

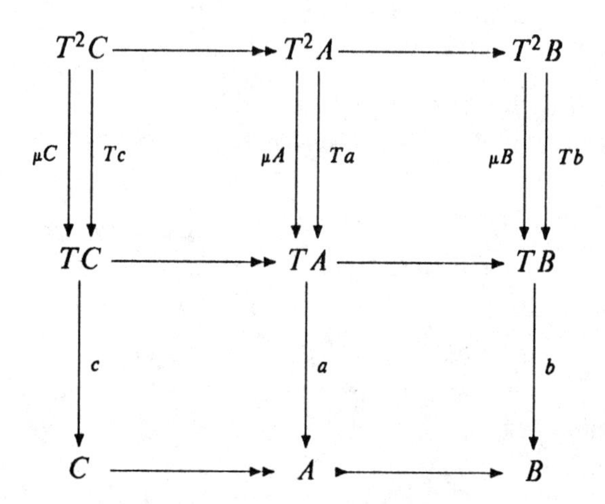

The associative law for algebra follows because it works when preceded by the epimorphism $T^2C \to T^2A$. A similar diagram with η's on top gives the unitary law.

Proposition 7. *Let* $\mathcal{C}$ *and* $\mathbf{T}$ *satisfy the hypotheses of Proposition 6, and suppose in addition that* $\mathcal{C}$ *is complete. Then* $\mathcal{C}^{\mathbf{T}}$ *has finite colimits.*

Proof. We construct coequalizers in $\mathcal{B} = \mathcal{C}^{\mathbf{T}}$ and use Corollary 3.
Given a parallel pair

$$(B, b) \underset{d^1}{\overset{d^0}{\rightrightarrows}} (C.c) \tag{2}$$

let $C \to C_i$ run over all regular quotients of C which are algebras and which coequalize d^0 and d^1. Then form the image $d : C \to C_0$ which is a subalgebra of $\prod C_i$. Clearly d coequalizes d^0 and d^1. If $f : (C, c) \to (C', c')$ coequalizes d^0 and d^1, the image of f is among the C_i, say C_j. Then

$$C_0 \longrightarrow \prod C_i \xrightarrow{\ p_j\ } C_j \rightarrowtail^{\text{inclusion}} C'$$

is the required arrow. It is unique because if there were two arrows, their equalizer would be a smaller subobject of C_0 through which $C \to \prod C_i$ factors.

The following theorem provides another approach to the problem.

Theorem 8. *Suppose $\mathcal{C}$ has finite colimits and equalizers of arbitrary sets of maps (with the same source and target). Let $\mathbb{T}$ be a triple in $\mathcal{C}$ which preserves colimits along countable chains. Then $\mathcal{B} = \mathcal{C}^{\mathbb{T}}$ has coequalizers.*

Proof. Again we use Corollary 3. Let (2) be given, and let $e : C \to C_0$ be the coequalizer in $\mathcal{C}$ of d^0 and d^1.

For $i > 0$, define each C_i in the diagram below to be the colimit of everything that maps to it.

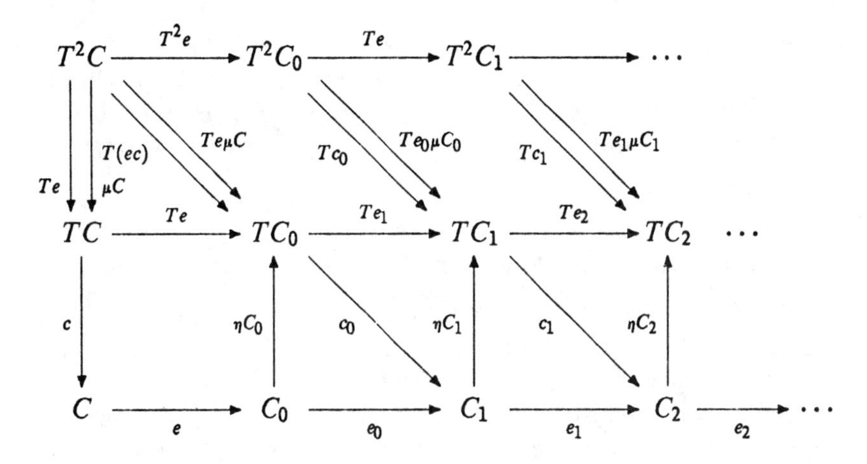

It follows that $c_{i+1} \circ Tc_i = c_{i+1} \circ Te_i \mu C_i$.

If $f : (C, c) \to (B, b)$ coequalizes d^0 and d^1, then there is a unique $g : C_0 \to B$ for which $g \circ e = f$. Then

$$g \circ e \circ c = f \circ c = b \circ Tf = b \circ Tg \circ Te.$$

Also,

$$\begin{aligned} b \circ TgT(e \circ c) &= b \circ Tf \circ Tc = f \circ c \circ Tc \\ &= f \circ c \circ \mu C = b \circ Tf \circ \mu C = b \circ Tg \circ Te \circ \mu C, \end{aligned}$$

and by a similar calculation,

$$b \circ Tg \circ \eta C_0 \circ e \circ c = b \circ Tg \circ Te.$$

It follows that there is a unique $g_1 : C_1 \to B$ for which $g_1 \circ e_0 = g$ and $g_1 \circ c_0 = b \circ Tg$.

Now assume $g_i : C_i \to B$ has the property that $g_i \circ e_{i-1} = g_{i-1}$ and $b \circ Tg_{i-1} = g_i \circ c_{i-1}$. Then,

$$\begin{aligned} b \circ Tg_i \circ Tc_{i-1} &= b \circ T(b \circ Tg_{i-1}) = b \circ Tb \circ T^2 g_{i-1} \\ &= b \circ \mu B \circ T^2 g_{i-1} = b \circ Tg_{i-1} \circ \mu C_{i-1} \\ &= b \circ T(g_i \circ e_{i-1}) \circ \mu C_{i-1} = b \circ Tg_i \circ Te_{i-1} \circ \mu C_{i-1}, \end{aligned}$$

and similar but easier calculations show that

$$b \circ Tg_i \circ Te_{i-1} = g_i \circ e_{i-1}$$

and

$$b \circ Tg_i \circ \eta C_i \circ c_{i-1} = b \circ Tg_i \circ Te_{i-1}.$$

This means that there is a unique $g_{i+1} : C_{i+1} \to B$ for which $g_{i+1} \circ e_i = g_i$ and $b \circ Tg_i = g_{i+1} \circ c_i$.

Now we go to the colimit of the chain. Let $C'' = \operatorname{colim} C_i$, so $TC'' = \operatorname{colim} TC_{i-1}$ and $T^2C'' = \operatorname{colim} T^2C_{i-2}$. We get $g' : C'' \to B$ whose "restriction" to C_i is g_i. The maps c_i induce a map $c'' : TC'' \to C''$. This is an algebra structure map making g an algebra morphism:

(a) $g' \circ c'' = b \circ Tg'$ because $g_i \circ c_i = b \circ Tg_{i-1}$.
(b) $c'' \circ Tc'' = c'' \circ \mu C''$ because $c_i \circ Tc_{i-1} = c_i \circ Te_{i-1} \circ \mu C_{i-1}$ and the e_i commute with the transition maps in the diagram.

However, we are not done. This map g' need not be unique. To make it unique, we pull a trick similar to the construction in the proof of the Adjoint Functor Theorem. We have a map $e' : (C, c) \to (C'', c'')$ induced by the e_i which is an algebra morphism. The equalizer of all the endomorphisms m of (C'', c'') for which $me' = e'$ is the coequalizer. This equalizer exists because it exists in $\mathcal{C}$ and tripleable functors create limits. By copying the argument of the Adjoint Functor Theorem, one gets a map $e'' : C \to E$ which is the required coequalizer.

Exercises 9.3

(CCTA). Show that if T is a triple in $\mathcal{C}$, $\mathcal{C}$ has coequalizers and T preserves them, then $\mathcal{C}^{\mathsf{T}}$ has coequalizers.

(TRANS). Show that Theorem 8 may be generalized to show that if $\mathcal{C}$ has finite colimits and equalizers of arbitrary sets of maps and if T is a triple in $\mathcal{C}$ which preserves colimits along chains indexed by some cardinal α, then $\mathcal{B} = \mathcal{C}^{\mathsf{T}}$ has coequalizers.

9.4. Free Triples

Given an endofunctor $R : \mathcal{C} \to \mathcal{C}$ on a category $\mathcal{C}$, the **free triple** generated by R is a triple $\mathsf{T} = (T, \eta, \mu)$ together with a natural transformation $\alpha : R \to T$ with the property that if $\mathsf{T}' = (T', \eta', \mu')$ is a triple and

$\beta : R \to T'$ is a natural transformation, then there is a triple morphism $\mathsf{T} \to \mathsf{T}'$ for which

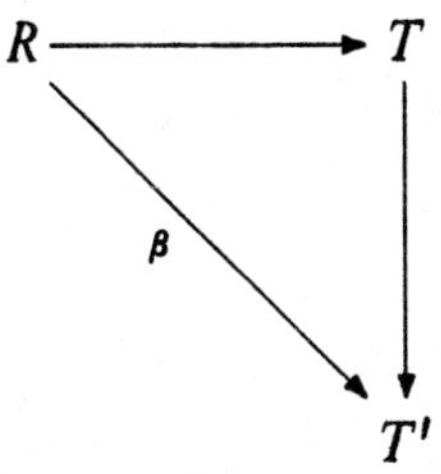

commutes.

The concept of free triple on R is clearly analogous to the concept of free monoid on a set, with composition of functors playing the role of Cartesian product of sets. In one special situation, we can construct the free triple on R in very much the same way as for the free monoid. If $\mathcal{C}$ has countable sums and R preserves them, let $T = \mathrm{id} + R + R \circ R + R \circ R \circ R + \cdots$. Then $T \circ T \cong T$, and with the identity map and the obvious map onto the first summand serving as μ and η respectively, one obtains a triple which is easily seen to be the free triple generated by R. (See Exercise (FTS)).

To get a more general construction, we form the category which will be $\mathcal{C}^{\mathsf{T}}$. Let $(R : \mathcal{C})$ denote the category whose objects are pairs (C, f) where C is an object of $\mathcal{C}$ and $f : RC \to C$. This is a full subcategory of the comma category $(R, \mathcal{C})$ whose objects are 3-tuples (C, f, C') with $f : RC \to C')$. Thus a morphism $f : (A, a) \to (A', a')$ in $(R : \mathcal{C})$ is an arrow $f : A \to A'$ for which

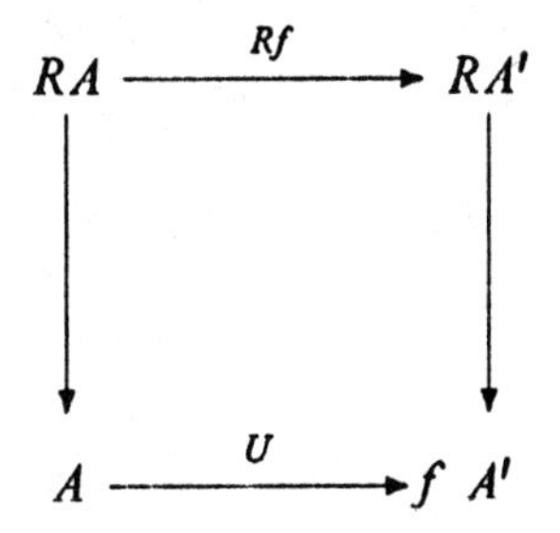

commutes. The **canonical underlying functor** $U : (R : \mathcal{C}) \to \mathcal{C}$ takes (A, a) to A and f to f. Note that if R should happen to be the functor part of a triple, then the category of algebras for the triple is a full subcategory of $(R : \mathcal{C})$.

Proposition 1. *The underlying functor $U : (R : \mathcal{C}) \to \mathcal{C}$ satisfies the condition of the PTT except possibly for the existence of a left adjoint. Thus if it has a left adjoint, it is tripleable.*

Proof. It is clear that U reflects isomorphisms. If

$$(A,a) \underset{d^1}{\overset{d^0}{\rightrightarrows}} (B,b)$$

lies over a contractible coequalizer

$$A \rightrightarrows B \xrightarrow{d} C \quad (s : B \to A,\ t : C \to B)$$

then $c = d \circ b \circ Rt$ is a structure on C for which d is a morphism in $(R : \mathcal{C})$ which is the required coequalizer.

Proposition 2. *Let $R : \mathcal{C} \to \mathcal{C}$ be a functor and $\mathsf{T} = (T, \eta, \mu)$ a triple on $\mathcal{C}$. Then there is a one to one correspondence between natural transformations from R to T and functors $\Phi : \mathcal{C}^{\mathsf{T}} \to (R : \mathcal{C})$ which commute with the underlying functors.*

Proof. Given a natural transformation $\alpha : R \to T$, define Φ to take an algebra $(A, a : TA \to A)$ to $(A, a \circ \alpha A)$, and a morphism to itself. It is easy to see that this gives a functor.

Going the other way, let Φ be given. Since $(TA, \mu A)$ is a T-algebra, $\Phi(TA, \mu A) = (TA, \varphi A)$ for some arrow $\varphi A : RTA \to TA$. Now given $f : A \to B$, $Tf : (TA, \mu A) \to (TB, \mu B)$ is a morphism between (free) algebras, so $\Phi Tf = Tf : (TA, \varphi A) \to (TB, \varphi B)$ must be a morphism in $(R : \mathcal{C})$. It is immediate from the definition that this is the same as saying that $\varphi : RT \to T$ is a natural transformation, whence so is $\alpha = \varphi \circ R\eta : R \to T$.

We must show that this construction is inverse to the one in the preceding paragraph. For this we need

Lemma 3. *Any functor $\Phi : \mathcal{C}^{\mathsf{T}} \to (R : \mathcal{C})$ which commutes with the underlying functors preserves the coequalizers of U^{T}-contractible coequalizer pairs.*

Proof. Such a functor clearly takes a U^{T}-contractible coequalizer pair to a U-contractible coequalizer pair, where $U : (R : \mathcal{C}) \to \mathcal{C}$ is the canonical underlying functor. Then by Proposition 1 and the fact that U^{T} is tripleable, Φ must preserve the coequalizer.

Now suppose that Φ is given, α is constructed as above, and Φ' is constructed from α. We must show that Φ and Φ' agree on T-algebras;

to do this, we use the standard technique of showing they agree on free algebras, so that by Lemma 3 they must agree on coequalizers of U-contractible diagrams of free algebras; but those are all the algebras by Proposition 4 of Section 3.3.

A free algebra has the form $(TA, \mu A)$, and $\mu : (T^2A, \mu TA) \rightarrow (TA, \mu A)$ is a T-algebra morphism. Thus $\mu : (T^2A, \varphi TA) \rightarrow (TA, \varphi A)$ is a morphism in $(R : C)$, whence $\varphi \circ R\mu = \mu \circ \varphi T$. Then

$$\begin{aligned}\Phi'(TA, \mu A) &= (TA, \mu A \circ \alpha TA) = (TA, \mu A \circ \varphi TA \circ R\eta TA) \\ &= (TA, \varphi A \circ R\mu A \circ R\eta TA) = (TA, \varphi A) = \Phi(TA, \mu A).\end{aligned}$$

Thus Φ and Φ' agree on free algebras, and so since both U^{T} and U preserve coequalizers, Φ and Φ' must agree on all algebras. (Since both U^{T} and U create coequalizers of U-contractible coequalizer pairs, Φ and Φ' do not merely take an algebra to isomorphic objects of $(R : C)$, but are actually the same functor.)

Conversely, suppose we start with a natural transformation α, construct a functor Φ, and then from Φ construct a natural transformation α'. As before, let $\Phi(TA, \mu A) = (TA, \varphi A)$. Since by definition $\Phi(TA, \mu A) = (TA, \mu A \circ \alpha TA)$, we have $\varphi = \mu \circ \alpha T)$. Then

$$\alpha' = \varphi \circ R\eta = \mu \circ \alpha T \circ R\eta = \mu \circ T\eta \circ \alpha = \alpha.$$

Theorem 4. *If $U : (R : C) \rightarrow C$ has a left adjoint F, then the resulting triple is the free triple generated by R.*

Proof. The comparison functor $(R : C) \rightarrow C^{\mathsf{T}}$ is an equivalence, so its inverse corresponds via Proposition 2 to a morphism $\eta : R \rightarrow T$. If $\lambda : R \rightarrow T'$ is a natural transformation, the composite of the corresponding functor with the comparison functor yields a functor from $C^{\mathsf{T}'}$ to C^{T} which by Theorem 3 of Section 3.6 corresponds to natural transformation $\alpha : T \rightarrow T'$. Since $\alpha \circ \eta$ corresponds to the same functor from C^{T} to $(R : C)$ as λ, they must be equal.

The converse of Theorem 4 is true when C is complete:

Proposition 5. *Let C be a complete category and R an endofunctor on C which generates a free triple T. Then $U : (R : C) \rightarrow C$ has a left adjoint and the Eilenberg-Moore category C^{T} is equivalent to $(R : C)$.*

Proof. To construct the left adjoint we need a lemma.

Lemma 6. *If C is complete, then $(R : C)$ is complete.*

Proof. The proof is the same as that of Theorem 1 of Section 3.4, which does not require the existence of an adjoint.

Now let B be any *set* of objects of $(R : \mathcal{C})$. Let $\mathcal{B}^{\#}$ be the full subcategory of $(R : \mathcal{C})$ consisting of all subobjects of products of objects in B. Then the composite $\mathcal{B}^{\#} \to (R : \mathcal{C}) \to \mathcal{C}$ has an adjoint F by the Special Adjoint Functor Theorem, the objects of B being the solution set.

Now form $\bar{\mathcal{B}}$ from $\mathcal{B}^{\#}$ by adding any object (B, b) for which there is a morphism $(B_0, b_0) \to (B, b)$ for which $B_0 \to B$ is a split epi. If a map $C \to B$ in $\mathcal{C}$ is given, it lifts via the splitting to a unique map $C \to B_0$ for which

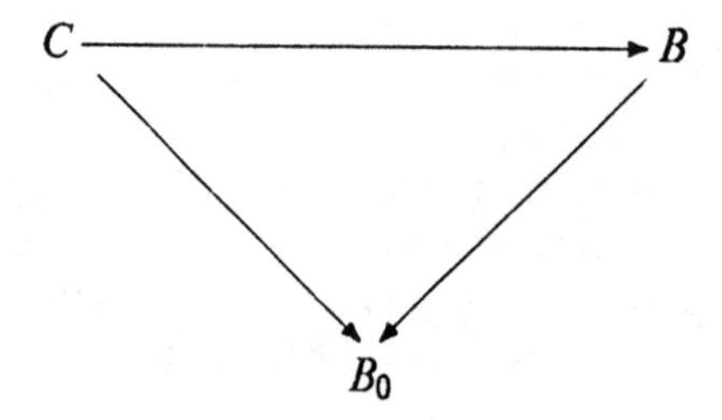

commutes. This lifts to the unique map $FC \to (B_0, b_0)$ given by the definition of left adjoint. Composition with $B_0 \to B$ gives a map $FC \to (B, b)$, and this sequence of constructions gives an injection from $\mathrm{Hom}_{\mathcal{C}}(C, B)$ to $\mathrm{Hom}_{\bar{\mathcal{B}}}(FC, (B, b))$. The fact that $B_0 \to B$ is split makes it surjective, so that we have shown that the underlying functor $\bar{U} : \bar{\mathcal{B}} \to \mathcal{C}$ has a left adjoint.

If

$$(B', b') \rightrightarrows (B, b)$$

is a U-contractible coequalizer diagram with both algebras in $\bar{\mathcal{B}}$, then it has a coequalizer (B'', b'') in $(R : \mathcal{C})$ which by definition belongs to $\bar{\mathcal{B}}$. Thus since $U : (R : \mathcal{C}) \to \mathcal{C}$ satisfies the requirements of PTT except for having a left adjoint, and $\bar{U}$ is a restriction of U to a subcategory which has the requisite coequalizers, $\bar{U}$ must be tripleable. Let $\mathsf{T}' = (T', \eta', \mu')$ be the resulting triple.

The inclusion $\mathcal{C}^{\mathsf{T}'} \to (R : \mathcal{C})$ corresponds to a natural transformation $R \to T'$ which by the definition of free triple gives a morphism $\mathsf{T} \to \mathsf{T}'$

of triples which makes

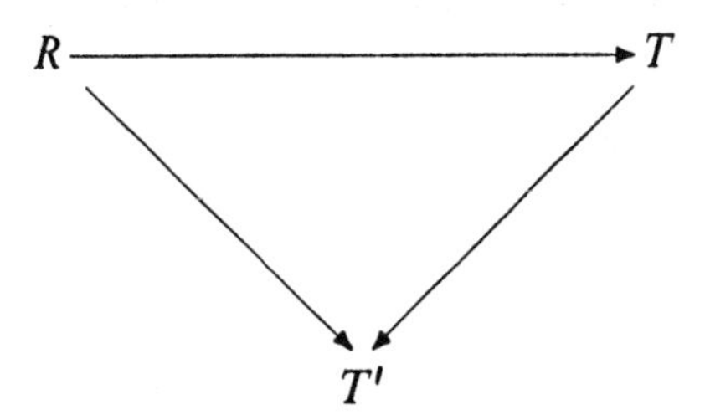

commute, the top transformation being the one given by the definition of free triple. By the correspondence between morphisms of triples and morphisms of triple algebras (Theorem 3 of Section 3.6), this yields the commutative diagram

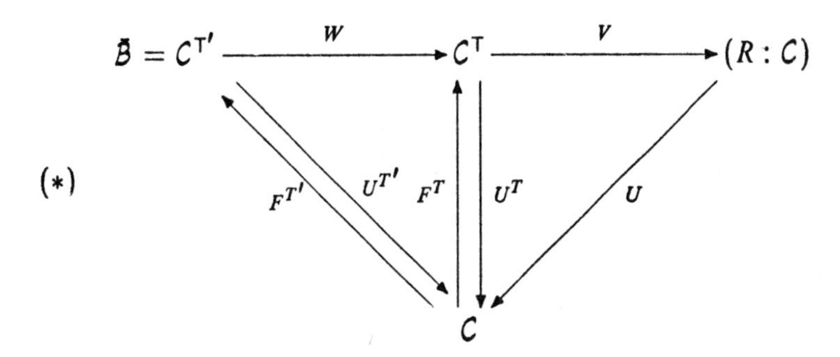

in which the top row is the inclusion.

Thus every object and every map of $\mathscr{B}$ is in the image of $\mathscr{C}^{\mathsf{T}}$. Since we began with any set of objects, it follows that V is surjective on objects and maps.

Now take an object B of $(R : \mathscr{C})$ and a $\mathscr{B}$ contained in $(R : \mathscr{C})$ as constructed above which contains B. We have, in the notation of diagram (*),

$$\begin{aligned}\mathrm{Hom}(VF^TC, B) &\cong \mathrm{Hom}(VF^TC, VWB) \cong \mathrm{Hom}(F^TC, WB)\\ &\cong \mathrm{Hom}(C, U^{T'}B) \cong \mathrm{Hom}(C, UB),\end{aligned}$$

which proves that U has a left adjoint.

The last statement in the proposition then follows from Proposition 1 and Theorem 4.

Proposition 7. *Let $\mathscr{C}$ be complete and have finite colimits and colimits of countable chains. Let R be an endofunctor of $\mathscr{C}$ which commutes with colimits of countable chains. Then R generates a free triple.*

Proof. By Theorem 4, we need only construct a left adjoint to $U : (R : \mathcal{C}) \to \mathcal{C}$. Let C be an object of $\mathcal{C}$. Form the sequence

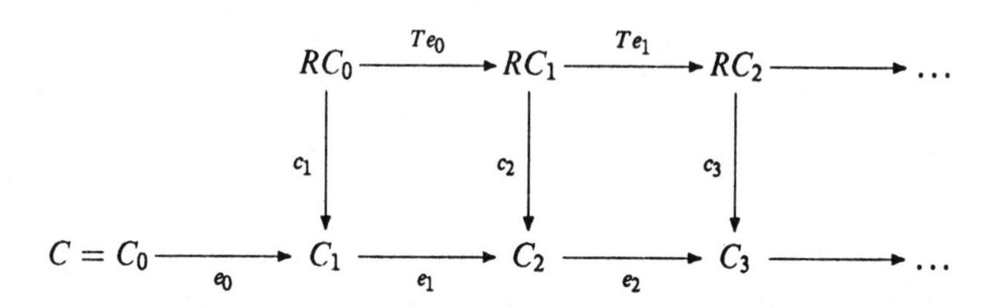

in which each square is a pushout and $C_1 = C_0 + RC_0$. Let $C' = \operatorname{colim} C_n$. Then $RC' = \operatorname{colim} RC_n$ and there are induced maps $e : C \to C'$ and $c' : RC' \to C'$.

Lemma 8. *For any diagram*

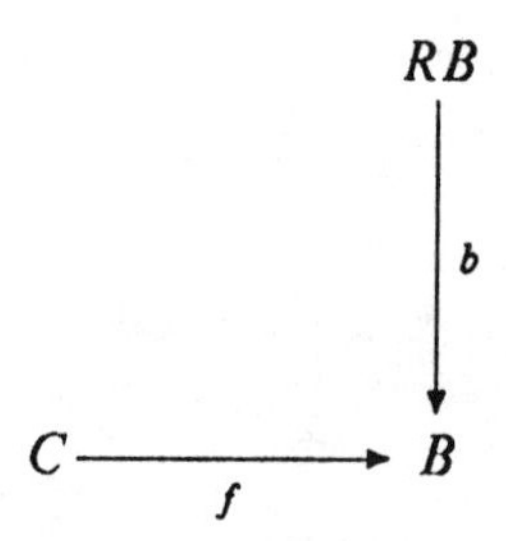

there is an arrow $f' : C' \to B$ for which

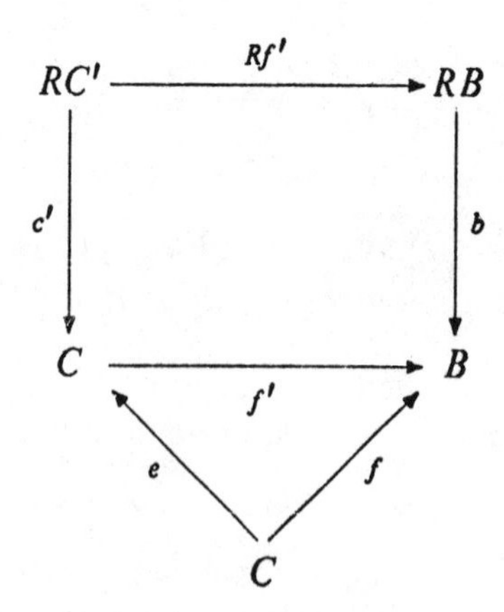

commutes.

Proof. Begin with $f_0 = f$. Let $f_1 : C_1 \to B$ be defined using the defining property of a sum by $f_1 \circ e_0 = f$ and $f_1 \circ c_1 = b \circ Rf$. We will define f_n for $n \geq 1$ inductively; notice that after f_1 we are using a pushout rather than a sum, so our induction hypothesis will have to carry with it a commutativity condition.

So assume that $f_0, \cdots, f_n$ have been defined in such a way that

(a) $f_i : C_i \to B$;
(b) $f_i \circ c_i = b \circ Rf_{i-1}$; and
(c) $f_i \circ e_{i-1} = f_{i-1}$.

Then $b \circ Rf_i \circ Re_{i-l} = b \circ Rf_{i-1}$, so we can legitimately define f_{i+1} by requiring that $f_{i+1} \circ c_{i+1} = b \circ Rf_i$ and $f_{i+1} \circ e_i = f_i$. The induced map f' then clearly satisfies the required identities.

By Lemma 8, the objects $(C', c' : RC' \to C')$ form a solution set for the underlying functor $U : (R : \mathcal{C}) \longrightarrow \mathcal{C}$, which therefore has a left adjoint as required.

Exercises 9.4

(FTS). Prove that if $\mathcal{C}$ is a category with countable sums and R an endofunctor which preserves countable sums, then R generates a free triple.

(FTIN). (Lambek)

(i) Let R be an endofunctor of a category $\mathcal{C}$. Show that if $a : RA \to A$ is an initial object in $(R : \mathcal{C})$, then a is an isomorphism.
(ii) Prove that an endofunctor which has no fixed points does not generate a free triple.
(iii) Prove that the covariant power set functor which takes a map to its direct image does not generate a free triple.

(TRANSF). Formulate and prove a transfinite generalization of Theorem 7 analogous to the way in which Exercise (TRANS) of Section 9.3 generalizes Theorem 8 of that section.

Bibliography

H. Applegate, *Categories with models.* Dissertation, Columbia University (1965).

H. Applegate, M. Tierney, *Categories with models.* Springer Lecture Notes in Mathematics 80 (1969) 156–244.

M. Barr, *Exact categories.* In Exact Categories and Categories of Sheaves, Springer Lecture Notes in Mathematics 236 (1971), 1–120.

M. Barr, *Toposes without points.* J. Pure and Applied Algebra 5 (1974) 265–280.

M. Barr, *Coequalizers and free triples.* Math. Z. 116 (1970), 307–322.

M. Barr, J. Beck, *Homology and standard constructions.* In Seminar on Triples and Categorical Homology Theory. Springer Lecture Notes in Mathematics 80 (1969), 245–335.

M. Barr, R. Diaconescu, *Atomic Toposes.* J. Pure Applied Algebra, 17 (1980), 1–24.

M. Barr, R. Paré, *Molecular Toposes.* J. Pure Applied Algebra, 17 (1980), 127–152.

A. Bastiani, C. Ehresmann, *Categories of sketched structures.* Cahiers de Topologie et Géometrie Différentielle 13 (1973), 1–105.

J. M. Beck, *Triples, Algebras, and Cohomology.* Ph. D. Thesis, Columbia University (1967).

J. Bénabou, *Structures algébriques dans les catégories.* Cahiers de Topologie et Géometrie Différentielle 10 (1968), 1–126.

J. Bénabou, *Structures algébriques dans les catégories.* Cahiers de Topologie et Géometrie Différentielle 13 (1972), 103–214.

G. Birkhoff, *On the structure of abstract algebras.* Proc. Cambridge Phil. Soc. 31 (1935), 433–454.

A. Blass, *The interaction between category theory and set theory.* In Mathematical Applications of Category Theory, J. Gray, ed. A.M.S. Series in Contemporary Mathematics 30 (1984), 5-29.

A. Blass, A. Ščedrov, *Boolean classifying topoi.* To appear in J. Pure and Applied Algebra.

A. Boileau, A. Joyal, *La logique des topos.* J. Symbolic Logic 46 (1981), 6–16.

M. C. Bunge, *Toposes in logic and logic in toposes.* (To appear in Topoi.)

C. C. Chang, H. J. Keisler, Model Theory. North-Holland, (1973).

P. M. Cohn, Universal Algebra. D. Reidel Publishing Company, (1982).

A. Day, *Filter monads, continuous lattices and closure systems.* Canadian J. Math. 27 (1975), 50–59.

B. Day, *Note on the Stone representation theorem.* (Undated mimeographed notes.)

P. Deligne, *La conjecture de Weil I.* I.H.E.S. Publ. Math. 43 (1974), 273–307.

R. Diaconescu, *Axiom of choice and complementation.* Proc. Amer. Math. Soc. 51 (1975), 176–178.

R. Diaconescu, *Change of base for toposes with generators.* J. Pure Applied Algebra 6 (1975), 191–218.

E. Dubuc, G. M. Kelly, *A presentation of topoi as algebraic relative to categories or graphs.* J. Algebra, 81 (1983), 420–433.

J. Duskin, *Variations on Beck's tripleability criterion.* Springer Lecture Notes in Mathematics 106 (1969) 74–129.

B. Eckmann, P. Hilton, *Group-like structures in general categories.* Math. Ann. I. 145 (1962), 227–255; II. 151 (1963), 165–187; III. 150 (1963), 535–596.

S. Eilenberg, S. Mac Lane, *General theory of natural equivalences.* Trans. Amer. Math. Soc. 58 (1945), 231–244.

S. Eilenberg, J. C. Moore, *Adjoint functors and triples.* Illinois J. Math. 9 (1965) 381–398.

G. D. Findlay, *Reflexive homomorphic relations.* Can. Math. Bull. 3 (1960), 131–132.

M. Fourman, *The logic of topoi.* In Handbook of Mathematical Logic, J. Barwise, ed. North-Holland (1977).

P. Freyd, Abelian Categories: An introduction to the theory of functors. Harper and Row (1964).

P. Freyd, *Several new concepts: lucid and concordant functors, pre-limits, pre-completeness, the continuous and concordant completions of categories.*

In Category theory, homology theory and their applications III: Proceedings of the Batelle Institute Conference, Seattle, 1968. Springer Lecture Notes in Mathematics 99 (1969), 196–241.

P. Freyd, *Aspects of topoi*. Bull. Austral. Math. Soc. 7 (1972), 1–72 and 467–480.

P. Freyd, G. M. Kelly, *Categories of continuous functors I*. J. Pure Applied Algebra 2 (1972), 169-191.

P. Gabriel, F. Ulmer, *Lokal präsentierbare Kategorien*. Springer Lecture Notes in Mathematics 221. Springer-Verlag, (1971).

R. Godement, Théorie des faisceaux. Hermann (1958).

J. Goguen, J. Meseguer, *An initiality primer*. Preprint, SRI International, Computer Science Laboratory, 333 Ravenswood Ave., Menlo Park, CA, USA, 94025 (1983a).

J. Goguen, J. Meseguer, *Completeness of many-sorted equational logic*. To appear in Houston J. Math. (1983b).

J. W. Gray, Formal Category Theory I: Adjointness for 2-Categories. Springer Lecture Notes in Mathematics 391 (1974).

J. W. Gray ed., Mathematical Applications of Category Theory. A.M.S. Series in Contemporary Mathematics 30 (1984).

A. Grothendieck, *Sur quelques points d'algèbre homologique*. Tohôku Math. Journal 2 (1957), 199–221.

A. Grothendieck, Cohomologie étale des schémas. Seminaire de Géometrie Algébrique de L'I.H.E.S. (1964).

A. Grothendieck, J. L. Verdier, *Théorie des topos*. In SGA 4 Exposés I–VI (1963–4), reprinted in Springer Lecture Notes in Mathematics (1972) 269–270.

P. Huber, *Homotopy theory in general categories*. Math. Ann. 144 (1961) 361–385.

J. M. E. Hyland, *The effective topos*. In The L. E. J. Brouwer Symposium, A. S. Troelstra and D. van Dalen, eds., North-Holland (1982), 165–216.

J. R. Isbell, *General functorial semantics*. Amer. J. Math. 94 (1972), 535–596.

P. T. Johnstone, Topos Theory. Academic Press (1977).

P. T. Johnstone, G. Wraith, *Algebraic theories in toposes*. In Indexed Categories and their Applications, Springer Lecture Notes in Mathematics 661 (1978).

D. M. Kan, *Adjoint functors*. Trans. Amer. Math. Soc. 87 (1958), 294–329.

G. M. Kelly, *On the essentially-algebraic theory generated by a sketch.* Bulletin Australian Mathematical Society 26 (1982), 45–56.

G. M. Kelly, *A unified treatment of transfinite constructions for free algebras, free monoids, colimits, associated sheaves, and so on.* Bulletin Australian Mathematical Society 22 (1980), 1–83.

G. M. Kelly, *Structures defined by finite limits in the enriched context I.* Cahiers de Topologie et Géometrie Différentielle 23 (1982), 3–42.

J. F. Kennison, *On limit-preserving functors.* Illinois Jour. Math. 12 (1968) 616–619.

J. F. Kennison, D. Gildenhuys, *Equational completion, model induced triples and pro-objects.* J. Pure Applied Algebra 1 (1971), 317–346.

H. Kleisli, *Comparaison de la résolution simpliciale à la bar-résolution.* In Catégories Non-Abéliennes, L'Université de Montréal (1964), 85–99.

H. Kleisli, *Every standard construction is induced by a pair of adjoint functors.* Proc. Amer. Math. Soc. 16 (1965) 544–546.

A. Kock, G. E. Reyes, *Doctrines in categorical logic.* In Handbook of Mathematical Logic, J. Barwise, editor. North-Holland (1977).

A. Kock, G. Wraith, Elementary toposes. Aarhus Lecture Notes 30 (1971).

J. Lambek, P. Scott, Cartesian closed categories and λ-calculus. To appear.

J. Lambek, P. Scott, *Aspects of higher order categorical logic.* In Mathematical Applications of Category Theory, J. Gray, ed. A.M.S. Series in Contemporary Mathematics 30 (1984), 145-174.

F. W. Lawvere, *Functorial semantics of algebraic theories.* Dissertation, Columbia University (1963). Announcement in Proc. Nat. Acad. Sci. 50 (1963), 869–873.

F. W. Lawvere, An elementary theory of the category of sets. Preprint (1965).

F. W. Lawvere, *The category of categories as a foundation for mathematics.* Proceedings of the Conference on Categorical Algebra at La Jolla. Springer-Verlag (1966).

F. W. Lawvere, *Quantifiers and sheaves.* Actes du Congrès International des Mathématiciens, Nice (1970), 329–334.

F. W. Lawvere, *Continuously variable sets: algebraic geometry = geometric logic.* Proc. A.S.L. Logic Colloquium, Bristol, 1973. North-Holland (1975), 135–156.

F. W. Lawvere, *Introduction.* Toposes, Algebraic Geometry and Logic. Springer Lecture Notes in Mathematics 445 (1975), 3–14.

F. W. Lawvere, Variable sets, étendus, and variable structures in topoi (Notes by S. E. Landsburg). Lecture Notes, University of Chicago (1976).

F. W. Lawvere, *Variable quantities and variable structures in topoi.* In Algebra, Topology and Category Theory: a collection of papers in honor of Samuel Eilenberg. Academic Press (1976), 101–131.

F. E. J. Linton, *Some aspects of equational categories.* Proceedings of the Conference on Categorical Algebra at La Jolla. Springer-Verlag (1966).

F. E. J. Linton, *An outline of functorial semantics.* Springer Lecture Notes in Mathematics 80 (1969a) 7–52.

F. E. J. Linton, *Applied functorial semantics.* Springer Lecture Notes in Mathematics 80 (1969b) 53–74.

F. E. J. Linton, *Coequalizers in categories of algebras.* Springer Lecture Notes in Mathematics 80 (1969c) 75–90.

S. Mac Lane, *Duality for groups.* Bull. Amer. Math. Soc. 56, (1950), 485–516.

S. Mac Lane, Categories for the Working Mathematician. Springer-Verlag (1971).

A. I. Mal'cev, *On the general theory of algebraic systems.* Math. Sbornik, N. S. 35 (1954), 3–20.

M. Makkai, G. Reyes, First Order Categorical Logic. Lecture Notes in Mathematics 611. Springer-Verlag (1977).

C. J. Mikkelson, Lattice Theoretic and Logical Aspects of Elementary Topoi. Aarhus University Various Publications Series 25 (1976).

B. Mitchell, *The full embedding theorem.* Amer. J. Math. 86, (1964), 619–637.

B. Mitchell, Theory of Categories. Academic Press (1965).

R. Paré, *Connected components and colimits.* J. Pure Applied Algebra, 3 (1973), 21–42.

R. Paré, D. Schumacher, *Abstract families and the adjoint functor theorem.* In Indexed Categories and their Applications, Springer Lecture Notes in Mathematics 661 (1978).

J. Peake, G. R. Peters, *Extensions of algebraic theories.* Proc. Amer. Math. Soc. 32 (1972), 356–362.

J. A. Power, *Butler's Theorems.* Doctoral dissertation, McGill University (1984).

H. Rasiowa, R. Sikorski, The Mathematics of Metamathematics. Polish Scientific Publishers, Warsaw (1963).

G. Reyes, *From sheaves to logic.* In Studies in Algebraic Logic, A. Daigneault, ed. Mathematical Association of America Studies in Math, 9 (1974).

R. Rosebrugh, *On algebras defined by operations and equations in a topos.* J. Pure and Applied Algebra 17 (1980), 203–221.

A. Ščedrov, Forcing and classifying topoi. Memoirs Amer. Math. Soc. 295 (1984).

J. Shonfield, Mathematical Logic. Addison-Wesley (1967).

R. Street, *Notions of topos.* Bull. Australian Math. Soc. 23 (1981), 199–208.

R. Street, *The family approach to total cocompleteness and toposes.* Reprint, School of Mathematics and Physics, McQuarrie University, North Ryde, NSW 2113, Australia (1983).

R. Street, R. Walters, *The comprehensive factorization of a functor.* Bull. Amer. Math. Soc. 79 (1973), 936–941.

B. R. Tennison, Sheaf Theory. London Math. Soc. Lecture Notes 20. Cambridge University Press (1975).

M. Tierney, *Forcing topologies and classifying topoi.* In Algebra, Topology and Category Theory, A. Heller and M. Tierney, eds., Academic Press (1976).

A. S. Troelstra, *Aspects of Constructive Mathematics.* In Handbook of Mathematical Logic, J. Barwise, ed., North-Holland (1977).

C. Wells, *A triple in Cat*. Proc. Edinburgh Math. Soc. 23 (1980), 261–268.

G. C. Wraith, *Algebras over theories.* Colloq. Math. 23 (1971), 180–190.

O. Wyler, *Algebraic theories of continuous lattices.* In Continuous lattices, Springer Lecture Notes in Mathematics 871 (1981), 390–413.

Index to Exercises

Index

Grundlehren der mathematischen Wissenschaften

Continued from page ii